SHIELDED METAL ARC WELDING

SHIELDED METAL ARC WELDING

LEONARD KOELLHOFFER
Union County Technical Institute and Vocational Center

JOHN WILEY & SONS
New York   Chichester
Brisbane   Toronto   Singapore

Copyright © 1983, by John Wiley & Sons, Inc.

All rights reserved. Published simultaneously in Canada.

Reproduction or translation of any part of this work beyond that permitted by Sections 107 and 108 of the 1976 United States Copyright Act without the permission of the copyright owner is unlawful. Requests for permission or further information should be addressed to the Permissions Department, John Wiley & Sons.

Library of Congress Cataloging in Publication Data:

Koellhoffer, Leonard.
   Shielded metal arc welding.

   Includes indexes.
   1. Shielded metal arc welding.   I. Title.
TK4660.K564   1983   671.5'212   82-8629
ISBN 0-471-05048-2   AACR2

Printed in the United States of America

10 9 8 7 6 5 4 3 2 1

I dedicate this book to my wife Doris, who made it possible.

PREFACE

This book is primarily intended for use in technical and vocational schools, proprietary trade schools, and industry. It is also suitable for self-teaching. Designed for the student who wishes to learn welding for the purpose of earning a living, the book offers a comprehensive treatment of the shielded metal arc process.

The text includes all of the theory the welder would need to know about the shielded metal arc process. The main strength of the book, however, is the practical lessons in welding. These lessons contain all of the information required to complete a particular joint. Also included is flame cutting, air carbon arc cutting, and welding of plate, sheetmetal, and pipe in many joint configurations in all positions. The primary electrodes used for these lessons are the E–6010, E–6011, E–6012, E–6013, E–7014, E–7024, and the E–7018 low hydrogen electrode. There is also a chapter on welding stainless steel.

Although gas tungsten arc and gas metal arc welding have made tremendous strides in the industry, a significant amount of the welding performed today is accomplished with the shielded metal arc process. This process has historically required the most extensive training, and industry will continue to need the services of intelligent, skillful shielded metal arc welders far into the future.

Training welders requires a great deal of individual instruction, which places a strain on the instructor because of limited class time. Additionally, many welding instructors are faced with very large class enrollments. This text has been designed to provide the student with a ready reference and practical guide to facilitate self-learning. It is written to give answers to many of the questions that students would normally ask of an instructor, to help students who have fallen behind, and to enable students to go ahead on their own.

Each lesson is complete within itself. All the information and figures pertaining to a particular lesson are contained in that lesson. This eliminates the need for any cross-referencing. Only the information a welder should know to succeed in industry is included. "Nice to know" material has not been included.

This book reflects my thirty-four years as a welder and welding instructor. I have used the material in this book in my welding classes over a period of several years.

The lessons are designed in the mode of operation, or job sheets. The text follows AWS and ASME procedures, since the student who learns to weld to those criteria will be able to perform well in any setting.

Chapters 15 and 16 contain more than 20 lessons on welding in the vertical and overhead positions.

Chapters 22 through 26 deal with pipe welding, starting with the preparation of the pipe, fitting, tacking, and welding. The lessons cover the 1–G, 2–G, 5–G, and 6–G test positions.

Chapter 27 is concerned with welding branch connections.

The lessons are arranged systematically, but the instructor may rearrange them to suit individual class needs. For example, the instructor may wish to eliminate those lessons that do not fit the kinds of industry in a particular geographical area.

Lessons are organized as prescribed by the vocational education departments of most states. They include lesson title, measurable objectives, list of materials, and general instructions and procedures. Safety is stressed throughout.

A word about methodology. The electrode angles and manipulation suggested in this book will do the job correctly. But each welding instructor has developed an individual style, and the instructor's preferences must certainly supersede those in the book. In the final analysis, the ultimate reference and authority must be the instructor.

Leonard Koellhoffer

Welder using the shielded metal arc welding process to repair a leak on an oil transmission line.

ACKNOWLEDGMENTS

I thank the following members of the Board of Education of the Vocational and Technical Schools of Union County, New Jersey for their valuable assistance.

Mathew C. Bistus
James J. Clancy
Fredrick E. Hahn

Charles S. Mancuso
James J. Scanlon
James V. Spagnoli/Esquire

The generous help of the following companies and educational institutions is gratefully acknowledged.

Airco Welding Products
American Chain and Cable Co., Inc.
Arcair Corp.
American Society of Mechanical Engineers
Aronson Machine Co.
American Welding Society
Elliot Glove Co., Inc.
Harris Calorfic
Hobart Brothers
Jackson Products
Lane Community College
Lenco Inc.
The Lincoln Electric Co.

Phoenix Products Co. Inc.
Prentice-Hall, Inc.
Robvon Backing Ring Co.
Sellstrom Manufacturing Co.
Solen Industries
Tempil Division Big Three Industries
Tinus Olsen Testing Machine Co., Inc.
Tweco Co.
United Air Specialists Inc.
United States Steel
Victor Equipment Co.
Wilson Instrument Division of Acco.

I would also like to thank the following former students for their assistance.

Gregory Ruane
Donald Barry

David Myers
Donald Gilford

L. K.

CONTENTS

CHAPTER 1
INTRODUCTION TO THE SHIELDED METAL ARC WELDING (SMAW) PROCESS — 1
- LESSON 1A FUNDAMENTALS OF THE SHIELDED METAL ARC WELDING (SMAW) PROCESS — 2
- LESSON 1B PROCESS CAPABILITIES AND USAGE — 4

CHAPTER 2
SAFETY AND SAFE PRACTICES IN WELDING AND CUTTING — 7
- LESSON 2A PERSONAL SAFETY IN SHIELDED METAL ARC WELDING — 7
- LESSON 2B PROTECTIVE CLOTHING AND EQUIPMENT — 8
- LESSON 2C POWER SOURCE AND EQUIPMENT SAFETY — 12
- LESSON 2D SAFE PRACTICES IN WELDING AND CUTTING — 15

CHAPTER 3
SHIELDED METAL ARC WELDING POWER SOURCES AND EQUIPMENT — 19
- LESSON 3A THE TRANSFORMER WELDER — 19
- LESSON 3B ALTERNATING CURRENT (AC) AND DIRECT CURRENT (DC) POWER SOURCES — 22
- LESSON 3C EQUIPMENT REQUIRED FOR SHIELDED METAL ARC WELDING — 23

CHAPTER 4
THE WELDING CIRCUIT, POLARITY, AND ARC BLOW — 28
- LESSON 4A THE WELDING CIRCUIT — 28
- LESSON 4B ARC BLOW: ITS CAUSES AND HOW TO REDUCE ITS EFFECTS — 29

CHAPTER 5
METALS AND WELDING — 32
- LESSON 5A THE MECHANICAL AND PHYSICAL PROPERTIES OF METAL — 32
- LESSON 5B CLASSIFICATION OF IRON AND STEEL — 36
- LESSON 5C IDENTIFICATION OF METALS — 38
- LESSON 5D STRUCTURAL SHAPES — 43
- LESSON 5E DISTORTION: ITS CAUSES AND CURES — 45
- LESSON 5F PREHEATING AND POSTWELD HEAT TREATMENT — 49
- LESSON 5G WELDMENT DEFECTS — 54

CHAPTER 6
SHIELDED METAL ARC WELDING ELECTRODES — 58
- LESSON 6A ELECTRODE COATING FUNDAMENTALS — 58
- LESSON 6B ELECTRODE SELECTION — 60
- LESSON 6C ELECTRODE CLASSIFICATION — 63
- LESSON 6D STAINLESS STEEL ELECTRODES — 65

CHAPTER 7
PROCEDURES AND PREPARATION FOR PLATE WELDING — 67
- LESSON 7A TYPES OF JOINT CONFIGURATIONS — 67
- LESSON 7B WELD POSITIONS FOR TEST PURPOSES — 70
- LESSON 7C PRACTICE AND TEST PLATE PROCEDURES — 71
- LESSON 7D PLATE PREPARATION FOR TEST PURPOSES — 72
- LESSON 7E PROCEDURES FOR TACK WELDING PRACTICE AND TEST PIECES — 74
- LESSON 7F THE USE OF BACK-UP BARS AND BACKING RINGS — 75
- LESSON 7G PROCEDURE AND WELDER QUALIFICATION TESTS — 78

CHAPTER 8
WELDING SYMBOLS — 84
- LESSON 8A CORRECT USAGE OF WELDING SYMBOLS — 84

CHAPTER 9
WELDING INSPECTION AND QUALITY CONTROL — 94
- LESSON 9A THE IMPORTANCE OF WELDING INSPECTION — 94
- LESSON 9B VISUAL INSPECTION OF WELDS — 95
- LESSON 9C DESTRUCTIVE TESTING METHODS — 98
- LESSON 9D ROOT, FACE, AND SIDE BENDS — 102
- LESSON 9E NONDESTRUCTIVE TESTING METHODS — 105

CHAPTER 10
OXY-FUEL FLAME CUTTING (OC) — 112
- LESSON 10A SAFE USAGE OF THE EQUIPMENT — 112
- LESSON 10B MAKING STRAIGHT CUTS ON STEEL PLATE — 116
- LESSON 10C BEVELING PLATE — 121
- LESSON 10D FLAME CUTTING AND BEVELING PIPE — 123

CHAPTER 11
THE AIR CARBON ARC CUTTING PROCESS (AAC) — 126

LESSON 11A EQUIPMENT AND FUNDAMENTALS OF THE PROCESS — 126

LESSON 11B PRECAUTIONS, TROUBLESHOOTING, AND MAINTENANCE — 128

CHAPTER 12
AIR CARBON ARC PROCEDURES — 131

LESSON 12A ASSEMBLING THE EQUIPMENT — 131

LESSON 12B GOUGING PLATE — 132

LESSON 12C CUTTING LIGHT PLATE — 133

LESSON 12D CUTTING HEAVY PLATE — 134

CHAPTER 13
FLAT OR DOWNHAND WELDING — 136

LESSON 13A STRIKING AND MAINTAINING AN ARC — 138

LESSON 13B WELDING STRINGER BEADS — 140

LESSON 13C THE ESSENTIALS OF PROPER WELDING PROCEDURES — 141

LESSON 13D WELDING A PAD OF STRINGERS (E–6010 OR E–6011 ELECTRODES) — 144

LESSON 13E T-JOINT FILLET, 1–F POSITION, STRINGER BEADS (E–6010 OR E–6011 ELECTRODES) — 145

LESSON 13F T-JOINT FILLET, 1–F POSITION, STRINGER BEADS (E–7014 AND E–7024 ELECTRODES) — 147

LESSON 13G FILLING CRATERS AT THE END OF WELDS — 148

LESSON 13H SQUARE GROOVE BUTT JOINT WITH BACK-UP BAR, STRINGER BEADS (E–7024 IRON POWDER ELECTRODES) — 150

LESSON 13I OUTSIDE CORNER JOINT, 1–G POSITION, STRINGER AND WEAVE BEADS (E–6010 OR E–6011 ELECTRODES) — 152

CHAPTER 14
HORIZONTAL WELDING OF PLATE — 154

LESSON 14A PAD OF STRINGERS (E–6010 OR E–6011 ELECTRODES) — 154

LESSON 14B PAD OF STRINGERS (E–7014 AND E–7018 LOW HYDROGEN ELECTRODES) — 156

LESSON 14C LAP JOINT FILLET (2–F) POSITION, MULTIPASS (E–6010 OR E–6011 ELECTRODES) — 157

LESSON 14D LAP JOINT FILLET (2–F) POSITION, MULTIPASS (E–7014 AND E–7024) ELECTRODES — 159

LESSON 14E T-JOINT FILLET (2–F) POSITION, MULTIPASS (E–6010 AND E–6011) ELECTRODES — 160

LESSON 14F T-JOINT FILLET (2–F) POSITION, SINGLE PASS (E–7024) ELECTRODES — 162

LESSON 14G SINGLE BEVEL BUTT JOINT, WITH BACK-UP BAR (E–7018 ELECTRODES) — 163

LESSON 14H V-GROOVE BUTT JOINT, OPEN ROOT (2–G) POSITION (E–6010 OR E–6011 ELECTRODES) — 165

CHAPTER 15
VERTICAL WELDING OF PLATE — 167

LESSON 15A T-JOINT FILLET, VERTICAL, 3–F POSITION, WEAVE BEADS (UPHILL), E–6010 OR E–6011 ELECTRODES — 167

LESSON 15B T-JOINT FILLET, VERTICAL, 3–F POSITION, WEAVE BEADS (UPHILL), E–7018 ELECTRODE — 169

LESSON 15C LAP JOINT FILLET, VERTICAL, 3–F POSITION, STRINGER AND WEAVE BEADS (UPHILL), E–6010 AND E–6011 ELECTRODES — 171

LESSON 15D LAP JOINT FILLET, VERTICAL, 3–F POSITION, WEAVE BEADS (UPHILL), E–7018 ELECTRODES — 172

LESSON 15E V-GROOVE BUTT JOINT VERTICAL, 3–G POSITION, WITH BACK-UP BAR WEAVE BEADS (UPHILL), E–7018 ELECTRODE — 173

LESSON 15F OPEN CORNER JOINT, VERTICAL, 3–G POSITION, STRINGER AND WEAVE BEADS (UPHILL), E–6010 OR E–6011 ELECTRODES — 175

LESSON 15G OPEN CORNER JOINT, VERTICAL, 3–G POSITION, WEAVE BEADS (DOWNHILL), E–6010 OR E–6011 ELECTRODES — 177

LESSON 15H V-GROOVE BUTT JOINT, VERTICAL, 3–G POSITION, STRINGER AND WEAVE BEADS (UPHILL), E–6010 OR E–6011 ELECTRODES — 178

LESSON 15I SQUARE GROOVE BUTT JOINT, VERTICAL, 3–G POSITION, WITH BACK-UP BAR (UPHILL), 3–G POSITION, E–7018 ELECTRODES, AND (DOWNHAND) 1–G POSITION, E–7024 ELECTRODES — 180

LESSON 15J V-GROOVE BUTT JOINT, VERTICAL, 3–G POSITION, WEAVE BEADS, E–6010 OR E–6011 ELECTRODE. FIRST PASS DRAGGED (DOWNHILL) SECOND AND OUTER PASSES WEAVED (UPHILL) — 181

LESSON 15K V-GROOVE BUTT JOINT, VERTICAL, 3–G POSITION, AWS STRUCTURAL WELDING CODE D1.1 WELDER QUALIFICATION TEST FOR LIMITED PLATE THICKNESS (E–6010 AND E–7018 ELECTRODES) — 183

CHAPTER 16
OVERHEAD WELDING OF PLATE — 185

LESSON 16A T-JOINT FILLET, OVERHEAD, 4–F POSITION, STRINGER BEADS (E–6010 OR E–6011 ELECTRODES) — 185

CONTENTS

LESSON 16B T-JOINT FILLET, OVERHEAD, 4–F POSITION, STRINGER BEADS (E–7018 ELECTRODES)   187

LESSON 16C LAP JOINT FILLET, OVERHEAD, 4–F POSITION, STRINGER BEADS (E–6010 AND E–7018 ELECTRODES)   188

LESSON 16D OPEN CORNER JOINT, OVERHEAD, 4–G POSITION, STRINGERS AND WEAVE BEADS (E–6010 OR E–6011 ELECTRODES)   190

LESSON 16E V-GROOVE BUTT JOINT, OVERHEAD, 4–G POSITION, OPEN ROOT, FIRST PASS DRAGGED, SECOND AND OUTER PASSES STRINGERS (E–6010 OR E–6011 ELECTRODES)   191

LESSON 16F V-GROOVE BUTT JOINT, OVERHEAD, 4–G POSITION, OPEN ROOT, IN ACCORDANCE WITH SECTION IX ASME BOILER AND PRESSURE VESSEL CODE (E–6010 AND E–7018 ELECTRODES)   193

CHAPTER 17
WELDING SHEET METAL IN THE FLAT POSITION   195

LESSON 17A WELDING SHEET METAL, FLAT (DOWNHAND) POSITION, OPEN CORNER JOINT (E–6013 ELECTRODES)   195

CHAPTER 18
WELDING SHEET METAL IN THE HORIZONTAL POSITION   197

LESSON 18A WELDING SHEET METAL, T-JOINT FILLET, 2–F HORIZONTAL POSITION (E–6012 ELECTRODES)   197

LESSON 18B WELDING SHEET METAL, LAP JOINT FILLET, 2–F HORIZONTAL POSITION (E–6013 ELECTRODES)   198

CHAPTER 19
WELDING SHEET METAL IN THE VERTICAL POSITION   200

LESSON 19A WELDING SHEET METAL, OPEN CORNER JOINT, 3–G VERTICAL POSITION (DOWNHILL) (E–6011 ELECTRODE)   200

LESSON 19B WELDING SHEET METAL, OPEN CORNER JOINT, 3–G VERTICAL POSITION (DOWNHILL) (E–6013 ELECTRODES)   201

LESSON 19C WELDING SHEET METAL, LAP JOINT FILLET, 3–F VERTICAL POSITION (DOWNHILL) (E–6013 ELECTRODES)   202

LESSON 19D WELDING SHEET METAL, LAP JOINT FILLET, DISSIMILAR METAL THICKNESS, 3–F VERTICAL POSITION (DOWNHILL) (E–6013 ELECTRODES)   203

CHAPTER 20
WELDING SHEET METAL IN THE OVERHEAD POSITION   205

LESSON 20A WELDING SHEET METAL, LAP JOINT FILLET, 4–F OVERHEAD POSITION (E–6013 ELECTRODES)   205

LESSON 20B WELDING SHEET METAL, OPEN CORNER JOINT, 4–G OVERHEAD POSITION (E–6012 ELECTRODE)   206

LESSON 20C WELDING SHEET METAL, WELDING A SIX-SIDED BOX IN THE 2–G, 3–G, AND 4–G POSITIONS (E–6011 AND E–6013 ELECTRODES)   207

CHAPTER 21
WELDING WITH STAINLESS STEEL ELECTRODES   210

LESSON 21A PADDING A PLATE, FLAT OR DOWNHAND, AND HORIZONTAL POSITION, SERIES 300 CHROME-NICKEL ELECTRODES   211

LESSON 21B LAP JOINT FILLET, 4–F OVERHEAD POSITION, SERIES 300 CHROME-NICKEL ELECTRODES   213

LESSON 21C V-GROOVE BUTT, 3–G VERTICAL POSITION, SERIES 300 CHROME-NICKEL ELECTRODES   214

CHAPTER 22
PREPARATIONS FOR WELDING PIPE   216

LESSON 22A PREPARATION OF PIPE FOR TEST PURPOSES   216

LESSON 22B ALIGNMENT, FIT UP, AND TACKING PROCEDURES   218

CHAPTER 23
WELDING PIPE IN THE 1–G POSITION (AXIS OF THE PIPE IN THE HORIZONTAL POSITION)   220

LESSON 23A WELDING PIPE, AXIS OF THE PIPE HORIZONTAL, PIPE ROTATED ONE-QUARTER TURN AT A TIME (E–6010 OR E–6011 ELECTRODES)   221

CHAPTER 24
WELDING PIPE IN THE 2–G TEST POSITION (AXIS OF THE PIPE IN THE VERTICAL POSITION)   223

LESSON 24A WELDING PIPE, 2–G TEST POSITION (AXIS OF PIPE VERTICAL) FIRST PASS WHIPPED OR DRAGGED (E–6010 OR E–6011 ELECTRODES)   223

LESSON 24B WELDING PIPE, 2–G TEST POSITION (AXIS OF PIPE VERTICAL) FIRST PASS

DRAGGED (E-6010 OR E-6011 ELECTRODES, SECOND AND OUT E-7018 ELECTRODES) **225**

CHAPTER 25
WELDING PIPE IN THE 5–G TEST POSITION (AXIS OF PIPE IN THE HORIZONTAL POSITION) **227**

LESSON 25A WELDING PIPE, 5–G TEST POSITION (AXIS OF PIPE HORIZONTAL) FIRST PASS WHIPPED, SECOND PASS AND OUT WEAVED (ALL UPHILL) (E–6010 OR E–6011 ELECTRODES) **227**

LESSON 25B WELDING PIPE, FIRST 5–G TEST POSITION (AXIS OF PIPE HORIZONTAL) FIRST PASS DRAGGED (DOWNHILL) (E–6010 OR E–6011 ELECTRODES) SECOND AND OUT (UPHILL) (E–7018 ELECTRODES) **229**

LESSON 25C WELDING PIPE, 5–G TEST POSITION (AXIS OF PIPE HORIZONTAL) FIRST PASS DRAGGED, SECOND AND OUT WEAVED (ALL DOWNHILL) (E–6010 OR E–6011 ELECTRODES) **230**

CHAPTER 26
WELDING PIPE IN THE 6–G TEST POSITION **232**

LESSON 26A WELDING PIPE, 6–G TEST POSITION (AXIS OF PIPE AT A 45° ANGLE), FIRST PASS WHIPPED, SECOND AND OUT STRINGERS (ALL UPHILL) (E–6010 OR E–6011 ELECTRODES) **232**

LESSON 26B WELDING PIPE, 6–G TEST POSITION (AXIS OF PIPE AT A 45° ANGLE) FIRST PASS DRAGGED, SECOND AND OUT STRINGERS (ALL DOWNHILL) (E–6010 OR E–6011 ELECTRODES) **234**

LESSON 26C WELDING PIPE, 6–G TEST POSITION (AXIS OF PIPE AT A 45° ANGLE) FIRST PASS DRAGGED (DOWNHILL) (E–6010 OR E–6011 ELECTRODES) SECOND AND OUT STRINGERS (UPHILL) (E–7018 ELECTRODES) **236**

CHAPTER 27
WELDING BRANCH CONNECTIONS ON PIPE **238**

LESSON 27A PREPARATION, FITTING AND TACKING OF BRANCH CONNECTIONS **238**

LESSON 27B SADDLE TYPE BRANCH CONNECTION, AXIS OF THE MAIN HORIZONTAL. AXIS OF THE BRANCH VERTICAL AT A 90° ANGLE ABOVE THE MAIN (E–6010, E–6011, OR E–7018 ELECTRODES) **240**

LESSON 27C SADDLE-TYPE BRANCH CONNECTION, BRANCH BEVELED, AXIS OF THE MAIN VERTICAL, AXIS OF THE BRANCH HORIZONTAL AT A 90° ANGLE TO THE SIDE OF THE MAIN (E–6010 OR E–6011 ELECTRODES) **242**

GLOSSARY **245**

INDEX **267**

SHIELDED METAL ARC WELDING

CHAPTER 1
INTRODUCTION TO THE SHIELDED METAL ARC WELDING (SMAW) PROCESS

The Shielded Metal Arc Welding process is commonly known as *Stick Welding* or covered electrode welding. It is one of the most popular methods of welding and has the capability to weld many types and thicknesses of metal. It can do this in all positions, with a minimum investment in equipment.

Because of its widespread use and its low cost, the Shielded Metal Arc process is taught in most schools. In addition, the Shielded Metal Arc process requires a high degree of welding skill. The welder who learns the skill of Stick Welding will have little difficulty in mastering the other types of arc welding.

A large percentage of welding in construction, fabrication, maintenance, job shops, refineries, and chemical, pharmaceutical, and power plants is accomplished using the Shielded Metal Arc process. Figure 1-1 shows a fine example of an intricate weldment on which the Shielded Metal Arc Welding process was used to great advantage.

WELDERS IN INDUSTRY

Trained **welders** can usually find jobs, especially if they are willing to travel. Welding is used in almost every industry. Even when a company does not employ welders it may use welded products. Almost everything we use is connected in some way to welding. Welders are used by city, county, state, and federal governments, and the armed forces. Welders are employed by the transportation industry, public utilities, metal fabricators, automobile manufacturers, shipyards, railroads, steel mills and metal forming plants, electrical equipment manufacturers, tool builders, the petrochemical industry, and the construction trades.

In some trades the **welder** is not called a welder because welding is used as a "tool of the trade." For instance, in the construction trades the person actually doing the welding could be called a boilermaker, pipefitter, steamfitter, ironworker, plumber, sheetmetal mechanic, or millwright.

In the construction trades a person might work for one employer for his or her entire career or only as long as it takes to complete a job or contract. People in the construction trades normally are hired out of a union hall. They return there for their next assignment when their present job is over.

EMPLOYMENT OPPORTUNITIES

The job opportunities for a welder are limitless. The welder can work just about anywhere in the country. Even during a recession welders seem to be able to find employment. A welder can branch out into other phases of the industry. There are always opportunities for welding technicians, inspectors, and supervisors. Some may even start their own business. If you are willing to work hard, the financial rewards are there.

ENVIRONMENT

Welding is performed under a great variety of conditions. Some welders may work in a laboratory, whereas others

Figure 1-1 The shielded metal arc welding process was chosen as the joining method used in the fabrication of the stern section and drive shaft housing of this giant Super Tanker. (Photo by Paul Abelson.) SMAW was chosen so that distortion could be held to the minimum.

may work in shops, out in the open, inside tanks and vessels, on buildings, on bridges, and on ships. Some jobs are clean and some are not. If you are adventurous, there are jobs that will take you to the Middle East, Africa, South America, the Gulf of Mexico, the North Sea, or wherever a construction site or a drilling rig might be. With special training you might even work beneath the sea repairing oil transmission lines. However, most people tend to stay closer to home. You are sure to find a number of possible employers in the area in which you live.

APPLYING FOR EMPLOYMENT

You cannot talk your way into a welding position. Most employers check your qualifications by giving you a test. A few may give an oral or written test, but almost all give a welder qualification test. Usually, the test will be on the type of metal and joint to be welded on the job. Scoring well on this test of your welding skill can qualify you to start at a much higher rate of pay than normal.

For those of us who are not comfortable during job interviews, the **qualification test** is a lifesaver. Many employers, whose first impression of an applicant was poor, have changed their mind when the good results of the tests were reported.

Good welders are always in demand. The ability to pass a difficult welding test will improve your employment opportunities.

LESSON 1A
FUNDAMENTALS OF THE SHIELDED METAL ARC WELDING (SMAW) PROCESS

OBJECTIVE

Upon completion of this lesson you should be able to:

1. Describe the Shielded Metal Arc Welding Process.
2. Explain "what takes place in the arc."
3. List functions of the **electrode coating**.

DEFINITION

The American Welding Society defines shielded metal arc welding (SMAW) as a process in which coalescence of metals is produced by heat from an electric arc maintained between the tip of a **covered electrode** and the surface of the base metal to be welded.

Coalescence means the melting together of metals. In the shielded metal process it means a melting together of the **base metal**, or **workpieces**, and the filler metal core of the covered electrode. This melting together creates a weld which, when performed correctly, can be stronger than the metal being joined.

HOW AN ARC IS ESTABLISHED

A shielded metal arc electrode consists of a metal rod covered by a flux coating. This coating burns off during the welding operation. It releases gases and materials that help stabilize the arc and protect the molten metal from contamination.

The bare end of the **electrode** is clamped into an **electrode holder** that is connected to a **power source** (see Figure 1A-1). Current begins to flow when the bare tip of the electrode is touched lightly to the workpiece. The arc is started when the contact is broken. One good way to break the contact is to pull back a distance approximately the diameter of the metal core wire (Figure 1A-2).

The instant the bare metal tip of the electrode breaks contact with the metal workpiece an arc takes place. When done properly, the arc seems to start at the first touch. Actually it starts on the first bounce. The term *striking an arc* comes from the similarity to striking a match. In order to maintain or keep the arc going, the welder must hold the electrode steady enough to keep the **arc gap**, between the tip of the electrode and the work, equal to the electrode diameter.

When the gap is too large, the arc

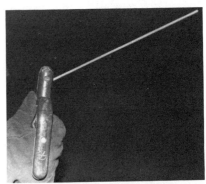

Figure 1A-1 Electrode secured between the jaws of the holder.

FUNDAMENTALS OF THE SHIELDED METAL ARC WELDING (SMAW) PROCESS

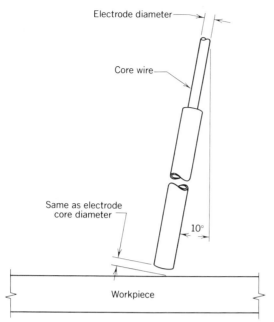

Figure 1A-2 The electrode should be held so the tip is approximately the same distance from the workpiece as the diameter of the core wire.

will go out. Normal power supplies cannot keep long arcs running. Should the arc gap be too short, the electrode may stub into the workpiece and put the arc out. Also, the electrode may stick to the base metal when the molten metal freezes.

When ignition takes place, there will be a brilliant arc between the core wire and the workpiece. Never look at any arc with the naked eye. The light is very intense and can cause injury to the eye. Always wear proper eye protection. Arcs give off strong ultraviolet and infrared rays.

WHAT TAKES PLACE IN THE ARC?

Once the arc is established tiny drops of molten metal are transferred forcibly across the arc gap. They will vary in size. You will be able to see some of the large drops as they join the **molten weld puddle.** The path followed by these globules is called the **arc stream** (Figure 1A-3). These molten metal drops become part of the weld bead.

This drop transfer, is caused by a combination of forces such as those from gravity, expansion of gases, electromagnetism, and surface tension. In your out of position welding practice lessons you will find that surface tension helps you keep the molten metal in place.

The arc also melts the flux coating on the electrode. The molten flux performs a variety of functions that help make a sound weld. For example, some of the flux coating decomposes and gives off gases that surround the arc zone. The gases shield the molten weld metal from contamination by air. (See Figure 1A-3.) This is how Shielded Metal Arc got its name.

There are gases in the air that can cause porosity if they come into contact with the molten weld metal. For example, oxygen in the air has an affinity for molten metal. Oxygen can combine with the molten metal and form harmful oxides. The coating also helps to stabilize the arc and keeps it from going out. In addition, the coating of the electrode, due to some higher melting point ingredients, melts more slowly than the core wire. Because of this, the coating extends slightly beyond the core wire and tends to channel the arc stream. The coating acts in much the same way that the barrel of a shotgun controls the path of the shot.

The coating helps to direct the arc force, and the arc is steadier. Some of the flux ingredients enter the molten pool and pick up impurities. It floats the impurities to the surface where they collect in the protective slag blanket that covers the weld.

The type of flux covering on the electrode influences the welding characteristics of the electrode, as well as the **mechanical properties** of the welds produced with the electrode.

Figure 1A-3 What takes place in the area of the welding arc. (Courtesy of The Lincoln Electric Company)

The welding characteristics of an electrode include the welding current range, welding position, weld puddle fluidity, penetration, and the overall quality of the weld deposit.

The mechanical properties will be discussed under those properties dealing with the strength, ductility, hardness, and metallurgical factors that make the electrode desirable for use under specific conditions. The mechanical properties are controlled by the addition of alloying elements to the electrode coating. These mix with the base metal and the filler metal, to form a weld of the required mechanical properties. Sometimes **iron powder** is mixed with the other materials of the electrode coating before it is extruded and baked onto the electrode. The amount of iron powder added will vary, depending on the type of electrode. The powder is melted during the welding operation. It mixes into the weld puddle and becomes part of the weld.

Iron powder, used in this manner, increases the amount of filler metal added to a joint. Added iron powder requires that more arc current be used. However, the increase in current is not as much as it would have been if the iron powder was in the form of solid wire. Due to the fine iron powder particles less current is required, per unit of weight, to melt the metal. Iron powder electrodes can be used to increase the rate of metal deposit by as much as 50 percent.

Before continuing, look again at Figure 1A-3. It will give you a good idea of what happens in the arc zone. Note the arc not only melts the surface of the plate, but it also penetrates below the surface. The amount of penetration depends upon the type of electrode, type of welding current, amount of current, rate of travel, and the angle of the electrode. You must learn to deal with all these things. Also, note that the molten pool is pushed away from the direction of travel. The molten metal flows into the weld along with the slag. The slag floats to the top of the weld metal and covers it. The slag protects the molten metal from the atmosphere.

The finished weld may be higher than the surface of the plate. The metal that forms the weld is a mixture of melted base metal, the core wire of the electrode, and any additional **filler metal** that may have been added. In addition to the slag, the weld metal is protected by the shielding gases provided by the coating.

LESSON 1B
PROCESS CAPABILITIES AND USAGE

OBJECTIVE
Upon completion of this lesson you should be able to:
1. List the type of welding that can be accomplished with the shielded metal arc process.
2. List the types of metal that are readily welded with the SMAW process.
3. List the limitations of the SMAW process.
4. List the equipment required to weld using the shielded metal arc process.

WHERE USED
The shielded metal arc welding process is probably the most popular and the most widely used of the arc welding processes. It is both versatile and flexible. The welder can work at great distances from the power source and the process does not require the use of compressed gases for shielding.

The process is excellent for repetitive work, short runs, maintenance, repair, fabrication, and field construction. A great deal of routine arc welding is done with the shielded metal arc process.

Almost any metal thickness and joint configuration can be readily welded with this process. Electrodes are available for use with carbon and low alloy steels, stainless steel, high alloy steels, corrosion resistant steels, quenched or tempered steels, cast iron, and malleable iron. Though not used to a great extent, electrodes are also available for use in the welding of copper and nickel and other alloys. Some welding of heavy aluminum is also done, but in very small quantities.

Most small shops use the shielded metal arc welding process. It uses relatively low-cost equipment and it is versatile. The welder can easily change from one type of weldment to another.

LIMITATIONS
The shielded metal arc welding process does not lend itself to use with semiautomatic or automatic equipment. Basically it is done manually. The electrodes come in relatively short lengths: 9 to 28 inches (230 to 700 mm). It takes only a few minutes to completely deposit one electrode.

Because of the short time needed to consume an electrode, the welder must stop to change electrodes every few minutes, and also clean the starting point before beginning the next electrode. Usually less than half of the time is used as actual arc time. But even with this downtime, an accomplished welder can be very productive.

EQUIPMENT REQUIRED

First of all, to weld with the shielded metal arc welding process, you need a **power source**. There are many excellent power sources on the market, and they are well constructed and easy to use. A welder (power supply) that is maintained properly has a long life expectancy. It is not unusual to see machines that are 20 years old or more.

There are different types of power sources available. They range from simple AC transformers to AC/DC rectifiers, DC motor-generators, or DC engine driven generators, as shown in Figures 1B-1 to 1B-4.

There is a welder for every purpose. The process requirements dictate which one is selected.

Figure 1B-1 AC transformer. (Courtesy of The Lincoln Electric Company)

Figure 1B-2 AC/DC type power supply. (Courtesy of The Lincoln Electric Company)

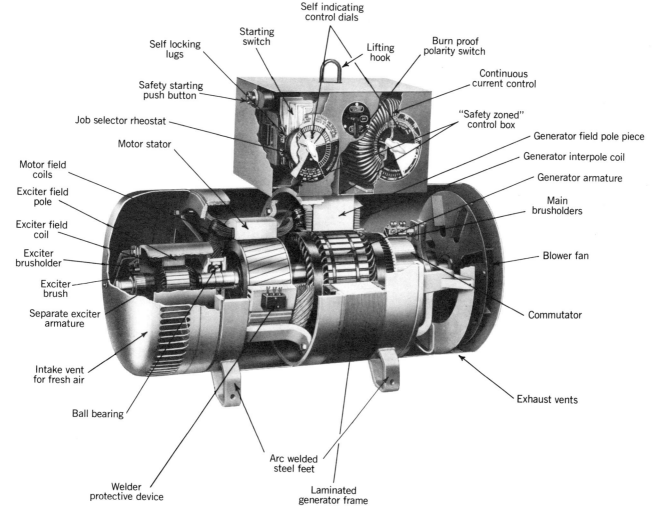

Figure 1B-3 DC motor generator power supply, exploded view. (Courtesy of The Lincoln Electric Company)

Figure 1B-4 DC engine-driven power supply. (Courtesy of The Lincoln Electric Company)

The welding power supply is the major expenditure; with the **workpiece lead, electrode lead, electrode holder,** and **workpiece clamp** rounding out the equipment needed to complete the electrical circuit.

In addition to the electrical equipment the welder needs personal welding equipment such as a face shield, safety shoes, and a cap. Some small tools such as a chipping hammer and a wire brush should also be on hand (Figure 1B-5).

ELECTRODES

Electrodes should be selected with care. It is important that the composition of the electrode is suitable for the composition of the metal to be welded. If the electrode and weld metal are not suited to each other, the weld will probably fail. Welds made with the wrong electrode cannot be expected to withstand the load placed upon them. Use of the wrong electrode can cause porosity, poor corrosion resistance, weak welds, brittle welds, and other types of defects.

New electrodes are constantly being developed. Electrodes used during the First World War are completely different from electrodes we have available now. It is the improved composition of the electrodes and their coatings that is responsible for the quality weld deposits possible today.

Some electrodes can be used successfully with either alternating current or direct current. Some coatings have been developed to increase the amount of filler metal that can be deposited in a given time. Other electrode coatings contain additives that increase strength and improve weld performance.

Although most coatings make the electrodes easy to use, some coatings require greater skill. When you can use any electrode well, you can be called a welder.

Figure 1B-5 Welder properly outfitted.

CHAPTER 2
SAFETY AND SAFE PRACTICES IN WELDING AND CUTTING

Safety is one of the most important practices for anyone to learn. This chapter will review the possible causes of common accidents, the types of clothing and protective equipment a welder should wear and use, the hazards found in the work area, and precautions that can be taken to reduce the probability of an accident taking place.

Students should not be allowed to participate in any of the practical applications unless they have demonstrated a good working knowledge of the material contained in this chapter. This chapter does not tell you all there is to know about safety. Be sure you read and understand the safety information provided by the manufacturer of the equipment you use. Request safety training if it is not already provided by your employer. Do not begin a job if you do not have a clear understanding of the safe practices you should follow.

LESSON 2A
PERSONAL SAFETY IN SHIELDED METAL ARC WELDING

OBJECTIVE
Upon completion of this lesson you should be able to:
1. List the possible causes of accidents.
2. Describe the precautionary steps that will prevent accidents.

THE IMPORTANCE OF WORKING SAFELY

Shielded metal arc welding is not particularly dangerous, but as in any job unsafe work habits can cause injuries. You should never work alone. It is your responsibility to see that you do not injure your fellow students or instructor, or your future fellow workers.

Safety is an important factor for both the employee and employer. Large sums of money are spent each year on safety programs and safety equipment to keep injuries to a minimum.

Accidents are expensive and troublesome for both the employee and the employer. Safety is everyone's job. Both the employer and the employee need to develop a consciousness of its importance.

Everyone involved loses if you are injured severely enough so that you are unable to work. The employer loses productivity due to the loss of your services. You, on the other hand, in addition to the pain and suffering of the injury, may lose all or a portion of your income.

To make matters worse, your whole family is affected. The household becomes disrupted, and plans have to be revised. With loss of income, the problems are even greater.

The mathematics of an accident are simple, an accident:

($\div$) Divides your income

($-$) Subtracts from your pleasure

($+$) Adds to your miseries and

($\times$) Multiplies your worries.

HOW ACCIDENTS ARE CAUSED

Consider the word *accident*. An accident is something that happens unexpectedly; we do not plan for it. It always comes as an unwelcome and unpleasant surprise. After an accident we hear phrases like: "I just took my eyes off it for a second"; "I wasn't looking"; "I didn't see it"; "It just happened"; "It happened so fast"; "I forgot"; "I was just going to put them on"; and the one most often heard: "I was trying to hurry."

When we are involved in an accident, the first feelings we have besides the pain, are those of anger and foolishness. We are usually angry at ourselves because we did something foolish. It's not that we are stupid; it's just that we did not think for the moment, or did not use the proper safety equipment or procedure.

Accidents do happen unexpectedly and they do follow a pattern. Most happen when we are not concentrating on the task. We allow our thoughts to wander or maybe we take *the easy way*.

The easy way usually becomes the hard way. Short cuts can cause accidents. "I'll only be a second", "I don't have to wear my safety glasses", or "I don't have to tie the ladder, I'll only be up on it for a few seconds," are examples of taking the easy way.

Forget the easy way. Take the time to learn the safety rules. Do the job safely. The safe way may take a little longer, but the job will get done without an accident.

Safe practices can be taught and the best safety equipment can be purchased, but you are the most important item. You must know and understand the rules to perform your job in a safe manner. Further, ask for the equipment that you require to perform the job safely, then use it when it's provided.

In 1970, the Congress of the United States passed the Occupational Safety and Health Act (**OSHA**) which requires your employer to provide you with safe work as well as a safe place to work. You are required to comply with the safety and health standards set up by the act.

This act provides for the publication of standards as well as penalties for employers who fail to comply with the standards. As a result working conditions have improved, but there is still a long way to go. However, it still remains up to you to do the job safely.

If you see others working in a careless manner, remind them of the safe way. You may save them from serious injury. Some people don't like to be told what to do. Possibly the person you are trying to help will get annoyed. Help them anyhow. They will know you are trying to protect them from having an accident.

Remember laws and rules are fine. They have their place, but there is no substitute for good judgment. You should have the equipment to do your job safely, but you are responsible for the safety of your own work station and of those working nearby.

LESSON 2B
PROTECTIVE CLOTHING AND EQUIPMENT

OBJECTIVE
Upon completion of this lesson you should be able to:
1. List the safety equipment and clothing a welder should use.
2. List the reasons for using each piece of equipment and clothing.

EYE SAFETY
Protect your eyes from the infrared and ultraviolet rays given off by the welding arc. Also, protect the eyes of nearby workers. There are a few simple rules that can help you. Rays emitted by the welding arc are harmful to the naked eye. It is important to wear proper eye protection at all times. Wear your safety glasses any time you are in a welding shop or near an arc. Welders should wear **flash goggles** such as shown in Figure 2B-1. Side shields of leather, plastic, or metal mesh protect the eyes from arc radiation and spatter that might come from the side. Flash goggle lenses are tinted to filter the arc rays.

Clear **safety glasses** do not protect you from the direct glare of the arc, but they can give you a little protection from indirect rays that strike the lens at an angle. The rays of the arc are very intense; the arc is hotter than the sun. When you are close to the arc, it is possible to burn your eyes if you do not wear the correct eye protection.

FLASH OR ARC EYE
In a few seconds the arc can cause a flash burn to unprotected eyes. A **flash burn** or **arc eye** is a burn on the surface of the eye. The burn is similar to sunburn and is very painful. It takes quite a bit of exposure to the sun before you get a sunburn, but you can get a flash burn from an arc in a fraction of a second. Unprotected skin can also be sunburned by an arc. Be sure to cover all bare skin to protect yourself from arc burn.

Flash burns of the eye are painful and cause quite a bit of discomfort. When treated properly, there will not be any permanent injury from flash burns. You can eliminate burns of the eye by always using the correct eye protection.

You do not always feel the flash burn effects immediately. It may take several hours before you feel the symptoms. At first your eyes may itch or burn. Later they may feel as though there is hot sand in them. In the case of a severe burn you may not be able to open your eyes.

When properly treated, flash burns rarely last more than 24 to 48 hours.

FIRST AID TREATMENT
There are a number of things that you can do to relieve the pain of eye burns. Always consult a doctor as soon as possible.

A simple emergency treatment for an arc burn is to cover the eyes with

Figure 2B-1 Various types of safety glasses. (Courtesy of Sellstrom Manufacturing Company

PROTECTIVE CLOTHING AND EQUIPMENT

clean cold water compresses. Place some water in a clean bowl or other receptacle along with ice cubes, to keep the water as cold as possible. Soak clean cotton pads in the water. Keep your eyes closed and apply the cold compresses.

As with other burns, the cold compresses tend to numb the area and lessen the pain to some extent. Special eye drops prescribed by a doctor are also helpful. Also, there are prescription ointments that guard against infection in the eye.

Always, as with any accident, report it to your instructor or class supervisor as soon as possible.

EYE PROTECTION WHEN WELDING

Your welding shield also provides eye protection. (See Figure 2B-2.) The welding shield must be worn at all times when you are welding or watching someone else weld. This includes watching demonstrations by your instructor.

Do not remove your flash goggles when you put on your welding shield. Safety glasses or flash goggles must be worn at all times.

Prescription safety glasses should be used if you wear eyeglasses. Never use contact lenses while welding. Severe eye injuries have resulted when the contact lenses melted from the intense rays of the arc.

Before you use a welding shield, check it carefully. Make sure there are no cracks where arc rays might enter. Also, check the filter lens. Be sure it is in good condition and of the proper shade. Figure 2B-3 shows the recommended filter shades for various electrodes and processes. Consult your instructor for the shade you should use in your school welding shop.

Curtains or screens should be used to keep the arc rays from reaching other workers (Figures 2B-4 and 2B-5).

GLOVES

Wear heat-resistant gloves while handling or preparing metal, stringing cable, and welding. Welding gloves should be of good quality material. They should have gauntlets to protect the wrists from sparks, especially during welding in the vertical and overhead positions. There are many varieties of gloves that you can purchase. Always use good quality gloves, because poor quality gloves will not protect you from burns of the fingers and hand.

Never pick up hot metal with your gloves; use pliers or tongs. Hot metal will burn the gloves and cause them to become hard and brittle. Misuse will ruin the gloves. They will become uncomfortable to use and you will have to replace them.

Heat-resistant sleeves, capes, and bibs as well as full jackets should be worn to protect you from sparks while you are welding, especially when welding out of position (Figure 2B-6).

PROTECTIVE CLOTHING

Wear fire-retardant clothing. Many synthetic materials, though easy to keep neat, should never be used. Some can burst easily into flames, and some fabrics can melt and stick to your skin, causing painful burns. Cotton clothing that has been treated with flame retardant chemicals is available.

Trousers should be of heavy cotton, wool or canvas and should have no cuffs or frayed edges. Cuffs have a tendency to catch the sparks and burn, and frayed edges catch fire easily. Trousers should be long enough to cover the top of the shoes when you are sitting down.

Welders should wear long-sleeve cotton or wool shirts of dark mate-

Curved shell

Narrow shell

Straight shell

Figure 2B-2 Welding shields are available in various shapes. A few of the more popular styles are shown here. (Courtesy of Jackson Products)

Figure 2B-3 Lens shade selector. Shade numbers are given as a guide only and may be varied to suit individual needs. (American Welding Society. Reprinted by permission)

Operation[1]	Electrode Size, Metal Thickness, or Current		Shade Number
	in.	mm	
Shielded Metal-Arc Welding	under 5/32	under 4.0	10
	5/32 to 1/4	4.0 to 6.4	12
	over 1/4	over 6.4	14
Gas Metal-Arc Welding			
Nonferrous	—	—	11
Ferrous	—	—	12
Gas Tungsten-Arc Welding	—	—	12
Atomic Hydrogen Welding	—	—	12
Carbon Arc Welding	—	—	14
Gas Welding			
Light	under 1/8	under 3.2	4 or 5
Medium	1/8 to 1/2	3.2 to 12.7	5 or 6
Heavy	over 1/2	over 12.7	6 or 8
Oxygen Cutting			
Light	under 1	under 25	3 or 4
Medium	1 to 6	25 to 150	4 or 5
Heavy	over 6	over 150	5 or 6
Air Carbon-Arc Gouging and Cutting[2]			
Light	—	—	12
Heavy	pad washing with large electrodes		14
Plasma-Arc Cutting[3]			
Light	under 300 A	—	9
Medium	300 to 400 A	—	12
Heavy	over 400 A	—	14
Torch Brazing	—	—	3 or 4
Torch Soldering	—	—	2
Thermal Spraying[4]			
Wire Flame Spraying (Except Molybdenum)			2 to 4
Molybdenum Wire Spraying			3 to 6
Flame Spraying of Metal Powder			3 to 6
Flame Spraying of Exothermics, Ceramic Powder, or Rod			4 to 8
Plasma- and Electric-Arc Spraying			9 to 12
Plasma- and Electric-Arc Spraying (when equipment is provided with its own shield)			3 to 6
Electric Bonding			2 to 4
Fusing Operations			4 to 6

1. Except as noted, all data are from ANSI Z49.1-1973 Safety in Welding and Cutting, published by the American Welding Society, which provides further explanation on eye protection and other aspects of welding safety.
2. Recommended Practices for Air Carbon-Arc Gouging and Cutting AWS C5.3.
3. Recommended Safe Practices for Plasma Arc Cutting AWS A6.3-69.
4. Recommended Safe Practices for Thermal Spraying AWS C2.1-73.

PROTECTIVE CLOTHING AND EQUIPMENT

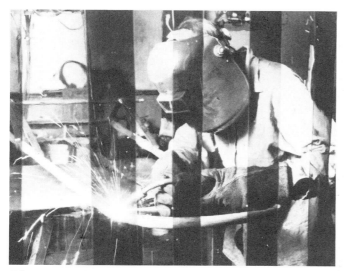

Figure 2B-4 One type of protective curtain is comprised of heavy strips of fire retardant, transparent material hung vertically so that one can pass through by parting them. (Courtesy of Solen Industries)

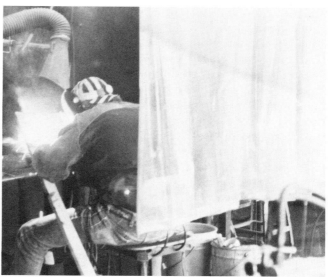

Figure 2B-5 Another type of curtain, pictured above, is made from fire-resistant plastic. It too is transparent and can be opened by sliding along the rod from which it hangs.

Figure 2B-6 Left is a representative group of the various types of protective clothing worn by welders. (Courtesy of Lenco, Inc.)

rial. The rays of the welding arc can be reflected by light-colored fabrics.

Buttons should be intact and the top button should remain closed while welding to keep the sparks from entering. Pocket flaps must be kept buttoned for the same reason, and all materials removed from them. Do not carry cigarettes and books of matches in shirt pockets. They may catch fire and burn you. Pocket combs have been known to catch fire.

High top, steel toe work shoes with a smooth surface over the front of the shoe should be worn. Wear long trousers outside of the high top shoes. Low-cut shoes or short trousers can let sparks enter your shoes. Spatter burns are painful and heal slowly.

Cotton or wool socks help protect your feet from burns. Acetate or other synthetic fibers melt on contact with sparks and also cause burns.

Welders should wear a cap or head cover. Some welders like caps with short peaks. Construction and pipeline workers prefer caps with a long peak, worn backward. This protects the back of their necks from welding spatter. Some types of welder's caps are shown in Figure 2B-7.

Figure 2B-7 Some types of welders caps.

Figure 2B-8 Two types of ear protection: The head set covers the entire ear while the plugs are inserted in the ear. (Courtesy of Sellstrom Manufacturing Company)

EAR PROTECTION

Two types of ear protection are occasionally needed by welders. One protects you from flying sparks and spatter, the other from loud noise.

Loud sounds can upset you as well as cause temporary hearing loss. Long-term exposure to loud noise can cause permanent hearing damage. Typical protective devices are shown in Figure 2B-8.

LESSON 2C
POWER SOURCE AND EQUIPMENT SAFETY

OBJECTIVE
Upon completion of this lesson you should be able to:
1. Describe typical hazards found in the work area.
2. Describe the precautions to be taken with respect to input current or line voltage.
3. Explain the safe handling of welding current.

GENERAL PRECAUTIONS IN THE WORK AREA
Usually you are responsible for the housekeeping in your work area. Keep the weld area free of combustibles and unnecessary materials. The possibility of an accident or fire is increased by cluttered work areas.

Keep aisles and exits clear to make escape easy in an emergency.

Do not throw electrode stubs on the floor. Put them into some sort of a receptacle to prevent slipping or tripping.

When working off the ground, on a structure or a scaffold, care must be taken that electrode stubs are not thrown or dropped. They might hit and injure someone below.

Do not try to become a "marks-

POWER SOURCE AND EQUIPMENT SAFETY

Figure 2C-1 Workers covering a sewer in a refinery with tar paper and clay prior to welding or burning.

Figure 2C-2 A commercial plastic pool being utilized to cover a sewer prior to performing "hot work."

man" and flip the stubs at a target or another worker. Horseplay and safety do not mix.

HAZARDS IN THE AREA

There are many fire hazards that may be found in a work area. For example, discarded lunch bags or newspapers will fuel a fire that can spread rapidly.

Before you start to weld, or perform any "hot work" such as flame cutting or heating, check the area thoroughly. Remove anything that can start a fire. Assume all liquids are flammable, unless you are sure they are water. Many a welder has ignited a pail of paint thinner or other flammable liquid mistakenly assumed to be water.

Remove all empty drums that may have contained flammable liquids. "EMPTY" drums have been known to explode. Any empty container of combustible liquids, unless properly cleaned, is a potential bomb. The fumes or gases can ignite with explosive force.

Protect your oxy-fuel hoses. Place them where sparks, falling plates, or other hot material will not endanger them. Gas cylinders should be secured in an upright position. Move them away if sparks or spatter might be blown toward them. If you cannot move them, cover them with a fire blanket or other flameproof material.

If you work near a sewer opening, cover it with tar paper and damp clay or other protective cover. This provides a seal so hot sparks cannot enter the sewer and ignite the sewer gases. (See Figure 2C-1.) You can also use special seals that are made for this purpose. (See Figure 2C-2.)

Finally, before starting to weld or cut, check the direction of the wind. The wind might blow sparks toward a distant fire hazard. The wind might even blow fumes into your breathing zone.

You are responsible for the safety in your area. Don't take this responsibility lightly; remember the people around you. They may not be aware of the potential hazards of welding or cutting.

Because of your training and knowledge you have the ability to see that the work is performed safely.

ELECTRICAL EQUIPMENT SAFETY

Some welding machines are generators driven by engines. You may have seen one mounted in the back of a pickup truck. Engine-driven generators are used mainly for fieldwork, where there is no available electric power.

The term *primary input* is used for electrical power entering the welding machine. Secondary output is used to describe the welding power delivered to the arc. The primary input can vary from the single phase, 115 volts AC, 60 cycle per second power to three phase, 460 volts AC, 60 cycle power.

These input voltages can kill you if you accidentally touch a bare wire or terminal.

Only welding machines that meet the standards of the National Electrical Manufacturers Association (NEMA), the Underwriters Laboratories Inc., or other recognized agency should be used.

In addition, welding machines must be installed in accordance with the National Electric Code as well as all local codes. All electrical equipment must be properly grounded.

Most welding machines use electrical power from a public utility company. They can be either transformer/rectifiers or electric motor driven. These welders convert the power from the utility power lines into welding power.

WELDING CURRENT SAFETY

The current that passes through the welding leads to the **electrode holder** and the *workpiece clamp* is called the secondary current. When used properly, this current presents few problems. The welding voltage is normally less than ordinary household voltage (115 VAC). With reasonable care you won't receive a shock. Do not touch live electrical parts. Do not wear wet gloves or shoes. Any shock you feel has the possibility to be harmful. Even a shock too weak to kill you, may startle you and cause a fall or injury.

It is good practice to cover all exposed connections. See the example shown in Figure 2C-3.

Cable connections should be checked periodically to be sure they are tight. Loose connections not only can cause sparking, but due to high resistance can reduce the welding current and make it difficult to obtain a good weld.

The welder should always make sure that the workpiece cable and clamp are secured firmly to the workpiece.

The electrode holder jaws should also be checked before each use. If the grooves are not clean, they will not make good contact with the electrode or hold it firmly.

The insulation should be intact and in good repair. Inspect welding cables frequently. Avoid shocks: Never wrap a cable around any part of your body.

Always remove electrodes from the holder when you stop welding; this will prevent accidental arcs between the electrode and any metal touching the workpiece. Be sure to turn off the arc power supply when you finish welding. Do not hang electrode holders on gas cylinders. Accidental arcs on cylinders can lead to explosive cylinder rupture.

Welding power is relatively safe because of its relatively low voltage, but the welder should keep dry. Never work in water or in damp areas without protective gear that insulates you properly.

MAINTENANCE OF EQUIPMENT

Unless you are properly trained do not repair welding equipment. This is an unsafe practice and is not recommended. Installation and repair of welding machines and equipment should be done by qualified technicians. You should become acquainted with **ANSI standard Z49.1**. This standard, known as "Safety in Welding and Cutting," is published by the American Welding Society. Also, read and understand the manuals and other information provided by the manufacturer before beginning any task.

Figure 2C-3 A type of insulator used to cover the welder terminal. These offer complete insulation against electric shock. (There are other types that are designed for the various terminals.) (Courtesy of Lenco, Inc.)

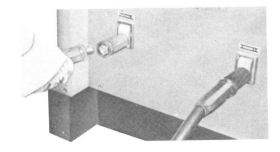

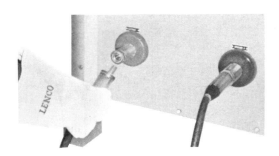

LESSON 2D
SAFE PRACTICES IN WELDING AND CUTTING

OBJECTIVE
Upon completion of this lesson you should be able to pass a test on all aspects of safety as it pertains to a welder who is engaged in welding or cutting operation in either shop or field conditions.

THE WELDER'S RESPONSIBILITIES
As a welder you should assume the responsibility for both your safety and the safety of those working in the immediate area. The reason for this is that welders are more concerned with hot work and its dangers than the average worker. As a welder you are closest to the point where heat or flame is applied. You should know, better than anyone else, the hazards involved. You should also know the precautions to take to minimize the hazards.

JOB PREPARATION
Welding can take place in the shop or the field. Shop work does not necessarily have to take place in a shop. Generally, it means working in the same place and under the same daily conditions.

Fieldwork, on the other hand, is work that is usually performed at the assembly location rather than in the shop or at your home base. Fieldwork is usually carried out under conditions that differ greatly from those encountered in shop work. These conditions are never stable; they change from day to day.

You must make sure that your personal equipment and tools are in good order. All helmets, head shields, and goggles should be checked and cleaned before use. Dirty or pitted cover glass should be replaced whenever necessary. Clean glass improves visibility and reduces eye strain. Good housekeeping is important. A cluttered work area can cause accidents.

Also, before starting work you should check the area for proper ventilation. Air handling or exhaust systems should be in operation. If used, the suction hose of portable exhaust units should be placed as close to the welding arc as possible without disturbing the arc itself. (See Figure 2D-1.)

Hot work in the field presents other safety problems that require your attention. Wherever you work, know the person responsible for safety in the area. This individual might be from the local fire department or someone hired for this purpose. This individual should be knowledgeable in the field of safety. His or her recommendations should be followed.

This does not mean that you should follow directions blindly. You should attempt to work in partnership with this person. You have the right to make suggestions you feel may make

Figure 2D-1 Welder utilizing a portable fume exhauster to remove welding fumes from a confined area.

Figure 2D-2 A worker using a gas testing device to determine the presence of harmful gases in the area prior to starting "hot work."

the job safer. In addition, tests must be taken to determine if the area has any flammable or combustible liquids or gases present.

All activities, required to make the area safe, must be completed before any hot work can begin. (See Figure 2D-2.)

Many companies use a hot work permit similar to that shown in Figure 2D-3. The permit system is used to make sure everything is in order before you start to weld. Even after the permit has been approved, also make sure you are satisfied. Check the permit carefully. Discrepancies or errors should be corrected. If you are not satisfied with the safety precautions, do not start work. Don't hesitate to ask questions.

If the permit is acceptable and you feel that everything is in good order, the work can begin. From time to time you should check to see if conditions have changed. Never take anything for granted. If something is unsafe, stop work until it is put right again.

Precautionary measures take time and effort. They can slow the progress of the job, but they are needed.

SAFETY AND SAFE PRACTICES IN WELDING AND CUTTING

Figure 2D-3 A hot work permit used in industry today.

Figure 2D-4 A welder with safety harness and line prior to entering a vessel.

radio, arrange a set of signals beforehand. Don't leave anything to chance.

If you stop work, remove any gas hoses and torches. Place them outside the vessel until you are ready to go back in. Gases from a leaking hose or valve could accumulate while you were outside the confined area. The leaking gas might explode when an arc is struck, or it might asphyxiate you.

Fuel gas equipment is not indestructible; even new hose can be cut by sharp metal edges or be burned through by hot metal.

CYLINDERS

Cylinders should always be stored and fastened in an upright position so they cannot be knocked over. Prevent arc burns on cylinders. Arc burns can weaken the cylinder and cause it to explode. Place cylinders where they cannot become part of an electrical circuit. Protect them from the sparks from the welding or cutting operation. Do not hang your torch, or electrode holder, on the cylinder.

Whenever you leave the job for any length of time, close the cylinder valves, bleed the system, and release the regulator adjusting screws.

WELDING OR CUTTING CONTAINERS THAT HAVE CONTAINED HAZARDOUS MATERIALS

Empty containers that have contained any type of hazardous mate-

VENTILATION AND WORKING IN CONFINED SPACES

Occasionally a welder must work in confined spaces, such as towers, vessels, or even inside large-diameter pipes and stacks. Working in such places can be dangerous. Never work alone and out of sight. Always have a helper on the outside who is trained in handling an emergency. Be sure there is plenty of ventilation. Never enter a confined space without a safety life line and a preplanned rescue procedure. (See Figure 2D-4.) If there is no means of communication to the outside such as telephone or

rials must be approached with care. Before any hot work can take place the containers must be cleaned thoroughly. After cleaning, the containers should be tested to ascertain if all harmful residues have been removed. If the container is to be entered, an oxygen sufficiency test should also be made.

If the container is not to be entered, another method can be employed. The container can be filled with an inert gas, such as nitrogen, carbon dioxide, or argon. These are very dense gases and are heavier than air. The container should be opened at the top. A hose that will introduce the inert gas into the container should be placed at the bottom of the container. Gas should be allowed to flow so that it forces the fumes out of the opening at the top of the container. When flammable gases have been forced out, you may start to cut or weld.

WELDING FULL CONTAINERS

It is possible to weld on containers, vessels, and pipelines that have flammable liquids in them, but proper procedures are necessary. Moreover, no one should ever attempt to perform this type of welding unless he or she has been trained to do so and has actually worked with another welder who has had experience in **hot taps** as this type of welding is called.

Welding hot taps is not as dangerous as it sounds, but only a skilled welder should attempt it.

The welding of hot taps is a necessary function performed mainly in the petrochemical industry. It is used to add a branch connection to a pipeline or a vessel.

The branch is usually a short piece of pipe with a flange on one end. The other end is prepared to conform to or fit the contour of the pipe or vessel it will be welded to.

The purpose of a hot tap is to add a branch connection while the product is flowing through. This reduces financial loss due to shutting down the operation in order to make a connection.

Most hot taps are performed on piping. The continuous flow of liquid through the pipe cools the area where the arc is. The flowing liquid also displaces any air that might be in the line, reducing the chance of combustion taking place. (See Figure 2D-5.) It takes fuel, ignition, and oxygen to support combustion.

ADDITIONAL SAFETY SUGGESTIONS FOR OXY-FUEL WELDING AND CUTTING

1. Wear the recommended protective clothing and equipment.
2. Check the gas equipment for leaks.
3. Always move cylinders with the safety caps on.
4. Store cylinders secured in the upright position.
5. Wear your flash goggles while welding and cutting.
6. Never open the acetylene cylinder valve more than three-quarters of a turn.
7. Keep flame and sparks as well as falling metal clear of gas hoses.
8. Never allow oil or grease to come in contact with the fittings of the oxy-fuel equipment.
9. Make sure there are no combustibles or flammable liquids in the area before welding or cutting.
10. Shut the cylinder valves, bleed the lines, and remove the adjusting screws when you are going to leave the work area for any period of time (break, lunch, etc.).
11. Never attempt to repair a welding or cutting torch or regulator. This should always be done by a qualified individual, one who not only has the know how and the proper tools, but also has the proper test equipment to insure that the repair is completed safely.
12. Never change or modify any oxy-fuel equipment.
13. Always use the proper sequence when pressurizing or depressurizing an oxy-fuel outfit. The correct sequence is as follows:
    (a) Check to insure that the regulator adjusting screws are turned out.
    (b) Open the oxygen cylinder valve slowly (all the way).

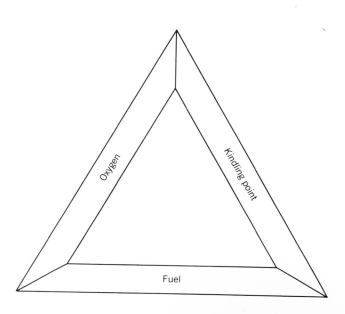

Figure 2D-5

(c) Open the oxygen torch valve (about one turn).
(d) Turn the oxygen adjusting screw in until the correct pressure is reached.
(e) Close the oxygen torch valve.
(f) Open the fuel gas cylinder valve slowly. (In the case of acetylene do not open the valve more than three-quarters of a turn).
(g) Open the fuel gas torch valve (about one turn).
(h) Turn the fuel gas adjusting screw in until the correct working pressure is reached.
(i) Close the fuel gas torch valve. The system is now pressurized and has been bled or purged to insure that there is no air in the system and that there is 100 percent fuel gas and oxygen at the tip of the torch. To insure that there are no leaks, a soap test should be conducted.

14. The following procedure should be followed when depressurizing the outfit.
(a) Close the fuel gas cylinder valve.
(b) Open the fuel gas torch valve.
(c) Close the fuel gas torch valve.
(d) Release the fuel gas adjusting screw and remove.
(e) Close the oxygen cylinder valve.
(f) Open the oxygen torch valve.
(g) Close the oxygen torch valve.
(h) Release the fuel gas adjusting screw and remove. The system has now been secured and the regulators, hoses, and torch bled so that the system is safe.

ADDITIONAL SAFETY THOUGHTS FOR ARC WELDING

There are some additional safety items to consider in the use of arc welding equipment. The following list summarizes some of the items:

1. Welding equipment should only be installed and maintained by qualified personnel. You are not expected to do this.
2. Always check the area for combustibles, flammables, or unsafe conditions. Do not begin welding if any hazards remain to be corrected.
3. Follow all local safety practices.
4. Wear the proper protective clothing and use the recommended safety equipment.
5. Do not work with wet clothing or gloves. Wear boots or use a dry platform to protect yourself from wet ground.
6. Use screens or curtains. Protect others from the rays of the arc.
7. Every once in a while STOP and reevaluate the situation. Take whatever additional precautionary measures are necessary to continue the operation.
8. Keep all equipment in good repair.
9. Never hang your electrode holder over or near the oxy-fuel cylinders.
10. When it is necessary to enter a confined space, be sure to obtain the proper approvals. It is good practice to wear your safety belt with a line attached, and have a reliable trained helper or safety person outside. See that the rescue path is kept clear. Be sure both you and your safety person understand the rescue plan.
11. Welding off of a ladder is dangerous and should be avoided. Use a scaffold.
12. Always make sure that the area you are working in is well ventilated.
13. Discard electrode stubs in the receptacle provided for that purpose.
14. Be sure you understand the job; it's better to be safe than sorry.
15. Never indulge in **horseplay.** It has been the cause of many accidents.

CHAPTER 3
SHIELDED METAL ARC WELDING POWER SOURCES AND EQUIPMENT

Although you do not need to be an electrician, you should have some knowledge of how a welding power source operates. You should know the types of power sources* that are available, as well as their purpose.

This chapter covers the transformer (alternating current) welder along with the rectifier and other direct current welders. Required equipment such as electrode holders and clamps is also discussed.

LESSON 3A
THE TRANSFORMER WELDER

OBJECTIVE
Upon completion of this lesson you should be able to:
1. Explain how high input voltage and low current are changed to low output voltage and high current for welding.
2. Describe the duty **cycle** of a welder.
3. Describe the effects of arc length in the welding process.

INPUT POWER
An arc welder that uses alternating current (AC) from a power line must transform it to a safe, usable welding current. The AC power from the utility company has very high voltage and low amperage. This type of power can be very dangerous to handle and cannot be used for welding.

The utility company power is supplied with a variety of voltages. It ranges from ordinary house voltage of 115 volts on through 230, 380, 460, and even 575 volts AC.

Most shielded metal arc welders come with input taps so that they can be used with several different voltage inputs.

*Sometimes the term *welder* is used instead of power source.

THE STEPDOWN TRANSFORMER
An AC welder is basically a stepdown transformer. A stepdown transformer lowers the input volts and increases the amperes to make the power useful for welding purposes.

For SMA welding power to be considered useful the open circuit voltage must be between 55 and 80 volts. Open circuit voltage means the output voltage when the welder is turned on, but without welding. A voltage measurement taken across the terminals of the welder without a load placed on the **machine** will read within those limits. It is called the no load voltage or **open circuit voltage.** Because ordinary household power is supplied with 115 volts AC, welding power is somewhat safer to handle. However, consider all electrical power as dangerous and able to kill you. It is similar to an automobile engine running with the transmission in neutral.

Closed circuit voltage, on the other hand, is the voltage when welding is taking place. Closed circuit voltage normally fluctuates between 20 to 35 volts. The actual voltage depends upon the arc length you use.

The transformer welder and almost all welders used for shielded metal arc welding are called constant current (CC) machines. Calling them CC machines can be confusing because most of them do not supply perfect constant current. However, over the range of actual welding the current doesn't change too much. Thus, it is called constant current or CC. The most important characteristic of this CC type of welder is that the voltage changes with the length of the arc.

By changing the **arc length** or arc gap you can control the heat input into the workpiece. This ability to change heat input is what makes welding work so well. In manual arc welding there are times when less heat is required to melt the metal, especially in out-of-position welding. At these times, the welder, by skillful manipulation of the electrode, can raise or lower the heat input. Changing the arc length while welding is what gives you control over the heat input.

When a short arc is held, the voltage drops to about 20 volts. This short arc and lower voltage produces a concentrated arc which is a hotter arc. When a long arc is held, the voltage rises to about 35 volts and the arc spreads out. It is cooler and the welder can control the flow of weld metal as well as the size of the puddle. (See Figure 3A-1.)

The ability to control the puddle through shortening and lengthening

SHIELDED METAL ARC WELDING POWER SOURCES AND EQUIPMENT

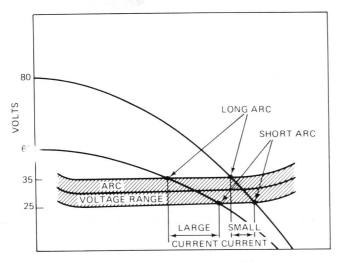

Figure 3A-1 Volt-ampere slope versus welding operation. (Reprinted by permission of Prentice-Hall, Inc., Englewood Cliffs, N.J.)

the arc requires skill. Welding skill is responsible for much of the success of the shielded metal arc process.

HOW THE STEPDOWN TRANSFORMER FUNCTIONS

There are a number of ways that a transformer can be designed to change line power into usable welding power. One simple method uses a movable coil.

In the movable coil transformer two coils are wrapped around a laminated iron core. The laminations are sheets of steel with excellent magnetic properties. (See Figure 3A-2.)

The coils are called the **primary** and the **secondary coils.** The primary coil receives the input current and the secondary coil delivers welding current.

One of these coils stands still while the other moves. The welding current is controlled by moving the coils closer together or farther apart. The distance is usually adjusted mechanically, with a knob or crank, by the welder.

PRIMARY CIRCUIT

The **primary circuit,** into which the input current is connected, consists of a laminated steel core with a wire coil wound around it. The coil must be insulated so there is no possibility of bare wire contacting other metal parts in the welder. Should this happen, anyone touching the welder could be severely shocked and perhaps even killed.

SECONDARY CIRCUIT

The **secondary circuit** is about the same as the primary, except that the secondary coil circuit uses larger diameter wire and has fewer turns than the primary circuit. Also, the secondary circuit is isolated from the primary circuit. It is not physically connected to the input line. When the main switch is turned on, line current enters the primary circuit and a magnetic field induces current in the secondary circuit. This is called transformer action and simply means that the current in one coil circuit induces current in the other, without the coils touching one another.

The induced current leaves the secondary circuit via wires attached to the terminals of the welder. The amount of current is controlled by moving the coils either closer or farther apart. The farther apart the lower the current, and the closer together the higher the current. Usually the primary coil is slightly smaller in diameter than the secondary coil. When the primary coil reaches a position where it is nesting inside of the secondary coil, the maximum **welding current** is obtained. (See Figure 3A-3.)

There are other methods of controlling the welding current, such as the movable shunt, movable core, tapped secondary coil, and saturable reactor. These methods of current control are built into the welding machine. The next lesson discusses the different ways you can adjust the welding current.

TYPE OF WELDING CURRENT

Alternating current welding transformers have high voltage-low current primary coils and low voltage-

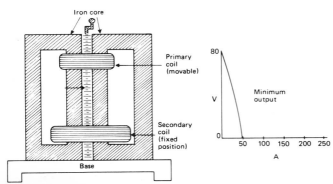

Figure 3A-2 Movable-coil AC power source with coils set for minimum output. (Courtesy American Welding Society)

THE TRANSFORMER WELDER

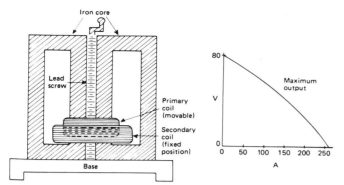

Figure 3A-3 Movable-coil AC power source with coils set for maximum output. (Courtesy American Welding Society)

high current secondary coils. This is the most economical welder sold, but it has several limitations. Only special AC electrodes should be used. On certain weldments and code work, DC electrodes may be specified. In those cases, AC power is unsatisfactory.

Alternating current changes direction each half cycle. It is produced by a generator with an armature that rotates between two magnetic poles. (See Figure 3A-4.)

The electricity produced by the utility companies is known as 60 cycle current. Each turn of the armature makes one cycle of AC. An example is shown in Figure 3A-5.

As you look at the AC wave, notice the crossover points 1, 3, and 1. At these points the current reverses itself. At the crossover point of the wave the current is zero. At points 2 and 4 the current flow is at a maximum. In 60 cycle alternating current the flow reverses itself 120 times a second.

When used for welding one half cycle is called **straight polarity.** The other half cycle is called **reverse polarity.** Although you cannot see it happen, the arc goes out each time the current is reversed. (See Figure 3A-5.) This is the main reason that only certain special electrodes may be used for AC welding. These electrodes contain special materials in the electrode coating that help the arc to restrike after each current reversal.

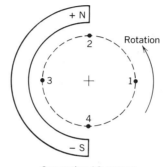

Figure 3A-4 Generating AC current.

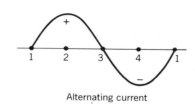

Figure 3A-5 Alternating current sine wave.

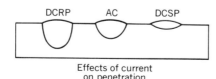

Figure 3A-6 Effects of current on penetration.

Each welding current polarity has its own penetration characteristics. Figure 3A-6 shows the penetration obtained with Direct Current Reverse Polarity **(DCRP),** Alternating Current (AC) and Direct Current Straight Polarity **(DCSP).** Notice that DCRP has deep penetration and DCSP has shallow penetration. Alternating current, which is a combination of the two, has medium penetration.

NAMEPLATE DATA

The following minimum information will usually be found on the power source nameplate.

1. Manufacturers type designation and/or identification number
2. NEMA (National Electrical Manufacturers Association) class designation
3. Maximum open circuit voltage (OCV)
4. Rated load volts
5. Rated load amperes
6. Duty cycle at rated load
7. Maximum speed in rpm at no load (on generator or alternator—not applicable to the AC transformer welder or rectifier)
8. Frequency of power supply input voltage
9. Number of phases of power supply input voltage
10. Input voltage of power supply
11. Amperes input at rated load output

Most of the nameplate information is for the electrician who installs or maintains the equipment. However, the duty cycle, rated load volts, and current rating should be known and understood by the welder.

DUTY CYCLE

You do not actually weld 100 percent of the time. You must stop to change electrodes, to clean the slag before welding the next pass, and for other reasons.

Because you do not weld all of the time, machines are made with different duty cycles. Production welding requires higher duty cycle machines and heavier wiring. By weighing their needs the purchasers can select a machine with a **duty cycle** that will best fit their needs. Just remember high duty cycle machines cost more. The duty cycle rating is based on the number of minutes, in a 10-minute period, a welder can be used at rated load without overheating. Some units have overload protection to prevent welding until the welder has cooled sufficiently.

As an example, a 300-amp welder with a rated duty cycle of 60 percent can safely be operated at 300 amperes for 6 minutes out of every 10 minutes.

A 60 percent duty cycle is recommended for heavy industrial manual welding machines. But just because the welder is rated at 60 percent duty cycle does not mean it can be used at rated load for a steady 36 minutes out of one hour. It means that the welder must be "rested" 4 minutes out of every 10 if it is used at its rated current.

On the other hand, if the welder is used below its rated output, it can be run for a longer time.

LESSON 3B
ALTERNATING CURRENT (AC) AND DIRECT CURRENT (DC) POWER SOURCES

OBJECTIVES
Upon completion of this lesson you should be able to:
1. Explain the different types of power sources available for shielded metal arc welding.
2. Differentiate between engine- and motor-driven welders.
3. Explain the function of rectifiers, capacitors, and reactors.

TRANSFORMER WELDERS
The stepdown transformer welder studied in the previous lesson is also the basis for other types of welders. When other equipment is added, it can be made to produce direct current.

In the last lesson you learned that penetration varies with the type of welding current. By adding a **rectifier** to a transformer it is possible to "straighten out" alternating current so that it flows in one direction. It can be converted to either straight or reverse polarity current depending upon the type of electrodes available.

The ability to select the type of current and control penetration also extends your choice of electrodes. When using direct current, the welder can choose from a wide variety of electrodes and can also use many of the AC electrodes that are available.

A **rectifier** is an electrical device that allows current to flow in only one direction. The rectifier does for electricity what a check valve does for liquid in a pipeline: it prevents reverse flow of current. When properly connected, a rectifier can change into continuous pulses of direct current. Although rectified current flows in one direction it is not steady. It is still quite "wavy." In order to "smooth out" the current a filter can be used. Usually an inductor or capacitor can be used. An inductor stores energy in a magnetic field and can be used to smooth out current flow. A capacitor is a device that has the ability to store energy and release it as needed to smooth the flow of the current.

There are two DC welding circuit connections. When the electrode is connected to the negative terminal, and the workpiece is connected to the positive terminal, the connection is called **straight polarity.** When the electrode is positive and the workpiece is negative, the connection is called **reverse polarity.** Either connection may be used for welding. They are called DCRP and DCSP by many users.

Some electrodes operate better with one polarity than another. As mentioned previously, the depth of penetration can be regulated by the choice of polarity. The electrode melt rate also varies with polarity. Reverse polarity produce welds of deep penetration and straight polarity welds of shallow penetration. Electrodes that can be used on either polarity usually have a higher deposition rate when used with straight polarity.

Some power supplies provide AC, others provide DC, but the AC/DC transformer-rectifier welder can provide both. With the AC/DC welder you can obtain DCSP, DCRP, or AC welding current.

DC MOTOR GENERATORS
Motor generator welders usually produce DC welding current, although there are some that produce AC. The generator (DC), or alternating (AC) is turned by an electric motor. Many people prefer the welding characteristics of generators. They are not as sensitive to line voltage fluctuations as transformer-rectifier units. However, they are more noisy and require more maintenance than transformer-rectifier units.

Generator-type welders have always had great welder appeal. The welding current they supply has less current ripple than rectified current.

The arc is smooth and steady and does not tend to go out as with the transformer welders.

ENGINE-DRIVEN GENERATORS

The generator-type welder can also be powered by internal combustion engines, such as the gasoline or diesel engine. Engine-driven welders are used for fieldwork where electrical power is not readily available.

This type of unit is often found mounted on the back of a truck or portable skid. The engine may be either water cooled or air cooled. An example of an air cooled welding machine is shown in Figure 3B-1.

Some of these welders also have an auxillary 115 volt AC generator (alternator) to provide power for lighting and electric hand tools.

Figure 3B-1 A typical air cooled engine-driven welding generator. (Courtesy Lincoln Electric Company)

LESSON 3C
EQUIPMENT REQUIRED FOR SHIELDED METAL ARC WELDING

OBJECTIVE

Upon completion of this lesson you should be able to:

1. Define the equipment necessary to use the shielded metal arc process.
2. List the types of equipment available and explain both their purpose and use.

SHIELDED METAL ARC WELDING EQUIPMENT

The list of equipment required to weld with the shielded metal arc welding process is a short one. Power supply is the only large expense item. The power cable, electrode holder, and workpiece clamp almost complete the list. You also need your personal safety equipment and clothing.

WELDING CABLE

Make sure the cable is large enough to carry the current and long enough to reach the job without stretching. Stranded copper cable is the most widely used, although aluminum is also used. The cable should be rated for welding duty. It should be tough, abrasion and heat resistant, and fairly flexible. Welding cables are produced in a wide range of sizes, with different coverings to meet the needs of the various applications. Figure 3C-1 shows a table with the recommended sizes of copper welding cable.

Notice that the size of the cable increases with the ampere rating and the distance to the work.

Because of this, a good procedure to follow in selecting cable size is to choose the size that will meet your maximum requirements. Standardize on that one size. If you do this, you may have cable, which because of its size, is not very flexible. To offset this, a short lead of 10 to 15 feet (ft) should be used to connect the heavier cable to the electrode holder. This lighter lead will give you more flexibility of movement and cut down on fatigue.

Power source		Awg cable size for combined length of electrode and ground cables				
Size in amperes	Duty cycle, %	0 to 15 m (0 to 50 ft)	15 to 30 m (50 to 100 ft)	30 to 46 m (100 to 150 ft)	46 to 61 m (150 to 200 ft)	61 to 76 m (200 to 250 ft)
100	20	6	4	3	2	1
180	20-30	4	4	3	2	1
200	60	2	2	2	1	1/0
200	50	3	3	2	1	1/0
250	30					
300	60	1/0	1/0	1/0	2/0	3/0
400	60	2/0	2/0	2/0	3/0	4/0
500	60	2/0	2/0	3/0	3/0	4/0
600	60	2/0	2/0	3/0	4/0	a

a. Use two 3/0 cables in parallel.

Figure 3C-1 Recommended copper welding cable sizes. (Courtesy American Welding Society)

When additional cable is necessary, short lengths may be joined together with special connectors. These connectors must be of low electrical resistance and be covered with an insulation equal to the covering of the cable. Figure 3C-2 shows some of the types of connectors that are available.

Protect the cable covering from damage. It is susceptible to cuts from sharp objects such as falling metal plate and burns from falling sparks, the heat of an open flame, and the welding arc.

WORKPIECE CONNECTORS

A workpiece connector is a device used to attach one of the welding cables to the workpiece. Frequently the workpiece connector is often called the ground connection, but this is a mistake. The workpiece connection is *not* the ground connection. A ground connection is an electrical connection to earth, or another grounded circuit. It is used to reduce the chance of electric shock. The workpiece cable can be connected to a spring clamp or secured to the workpiece by screws and bolts. Sometimes a magnetic device is used to attach the work cable to a steel workpiece.

A securely attached **workpiece lead** is important to the welding circuit. If it is not attached firmly to a clean surface, a poor electrical connection can cause fluctuating current flow and even arcing at the connection point. A poor connection can cause the clamp device to overheat and reduce the amperes delivered to the work.

Many types of clamps are sold commercially. Figure 3C-3 shows some of them.

Some of the connection devices are constructed with a rotating joint. The rotating joint keeps the cable from twisting when the workpiece is rotated. This is illustrated in Figure 3C-4. As the pipe rotates the cable remains in the same positions. This re-

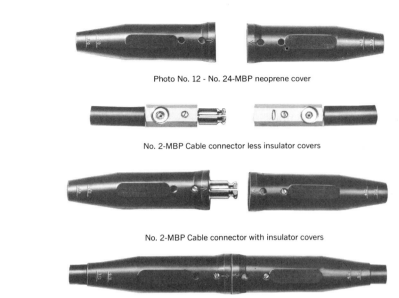

Photo No. 12 - No. 24-MBP neoprene cover

No. 2-MBP Cable connector less insulator covers

No. 2-MBP Cable connector with insulator covers

Figure 3C-2 Typical cable connector. (Courtesy Tweco Company)

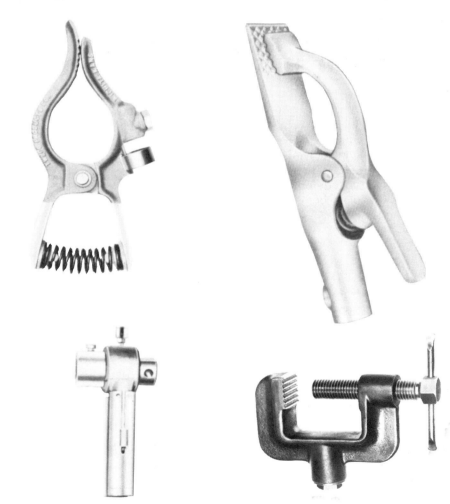

Figure 3C-3 A representative group of work clamps. (Courtesy Tweco Company)

EQUIPMENT REQUIRED FOR SHIELDED METAL ARC WELDING

duces wear and tear on the cable and the cable connection.

ELECTRODE HOLDERS

An **electrode holder** is a device designed to grip the electrode firmly so the welder can work at any angle. There are many satisfactory electrode holders on the market and a popular type is shown in Figure 3C-4.

Figure 3C-4 (Courtesy Tweco Company)

The welder spends a lot of time with it in his or her hand. Because of this, the feel and heft of the holder is as important to the welder as the hammer is to a carpenter. A good rule of thumb would be to use the lightest holder that is practical for a given job.

When selecting an electrode holder, not only the welder's comfort but the current carrying capacity, the availability of spare parts, the ease with which it can be maintained, and how well it is insulated must be taken into consideration.

Some electrode holders work on the leverage principle and clamp the electrode in jaws. Some have a chuck with a quick-acting twist lock action. An examples of this type of holder is shown in Figure 3C-5.

Figure 3C-5 Twist lock type electrode holder. (Courtesy Tweco Company)

Use only electrode holders that have the insulators intact and clean jaws to assure a good electrical contact. The jaws must close tightly around the electrode and hold the electrode firmly. It should not slip or wobble during welding.

CONSUMABLES

Electrodes are the only consumables used in the shielded metal arc welding (SMAW) process. The electrode is part of the welding circuit. It melts down as the welding progresses and provides the joint filler metal.

Electrodes are made in different sizes (diameter) and lengths. Figure 3C-6 indicates the sizes and lengths available as set forth in the **American Welding Society** (AWS) Standards, such as A 5.1-78, the American Welding Society Specification for Carbon Steel Covered Arc Welding Electrodes.

Figure 3C-6 Standard electrode diameters and lengths as specified in AWS A5.1-78 Specifications for Carbon Steel Covered Arc Welding Electrodes. (Courtesy Tweco Company)

Standard sizes, (Core Wire Diameter)[a]		Standard lengths[b,c,d]			
		E6010, E6011, E6012, E6013, E6022, E7014, E7015, E7016 E7018, classifications		E6020, E6027, E7024, E7027, E7028 Classifications	
(in.)	(mm)	(in.)	(mm)	(in.)	(mm)
1/16	1.6[e]	9	230	—	—
5/64	2.0[e]	9 or 12	230 or 300	—	—
3/32	2.4[e]	12 or 14	300 or 350	12 or 14	300 or 350
1/8	3.2	14	350	14	350
5/32	4.0	14	350	14	350
3/16	4.8	14	350	14 or 18	350 or 450
7/32	5.6[e]	14 or 18	350 or 450	18 or 28	450 or 700
1/4	6.4[e]	18	450	18 or 28	450 or 700
5/16	8.0[e]	18	450	18 or 28	450 or 700

[a]Tolerance on the core wire diameter shall be ±0.002 in. (±0.05 mm). Electrodes produced in sizes other than those shown may be classified. See Footnote "k" of Table 8.
[b]Tolerance on the length shall be ¼ in. (±10 mm). See *Note* on page 1.
[c]In all cases, end gripping is standard.
[d]Other lengths are acceptable and shall be as agreed upon by the supplier and the purchaser.
[e]These diameters are not manufactured in all electrode classifications. See Table 8.

Electrodes are produced by many companies under their own trade names and numbering systems. Most electrodes also have AWS numbers because they have the characteristics and mechanical properties recommended by the American Welding Society.

Electrode usage must be considered when estimating the cost of a job. You the welder are the controlling factor in the economical use of the electrodes so it is important that you are not wasteful. Like everything else electrodes are expensive.

Electrodes come in various lengths ranging from 9 in. (230 mm) to 14 in. (350 mm) in the smaller diameters, 14 (350 mm) to 18 in. (450 mm) in the medium diameters, and from 18 in. (450 mm) to 28 in. (700 mm) in the large diameter (production) electrodes.

Electrodes should be used to within 1½ to 2 in. from the end. Train yourself to work in this manner. Use the electrode to the maximum it is meant for. Throwing away unused electrode means you increase the costs of the job. No employer wants this kind of waste, and neither would a responsible welder. If the company you work for is successful, you will be more secure in your job. It is up to you to do those things that help the company to make a profit.

There may come a time when you

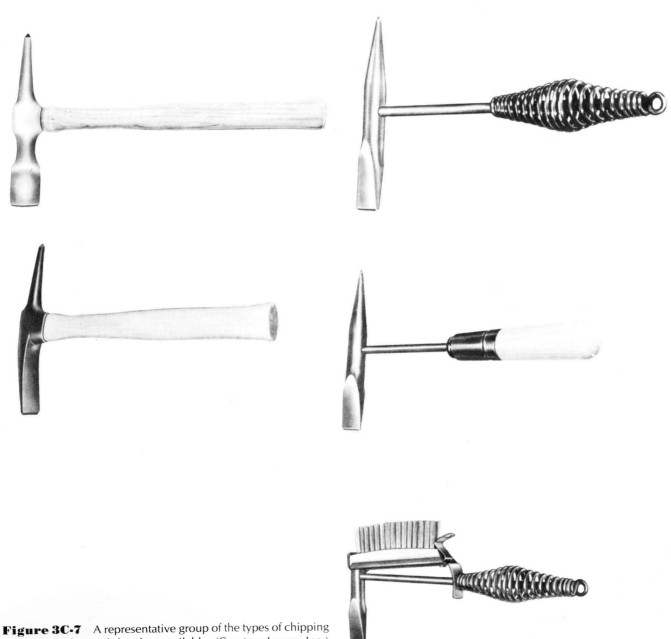

Figure 3C-7 A representative group of the types of chipping hammers and scratch brushes available. (Courtesy Lenco, Inc.)

EQUIPMENT REQUIRED FOR SHIELDED METAL ARC WELDING

will follow in the footsteps of many welders and start a business of your own. The good habits you develop while working for others will aid you, just as bad work habits will cost you time and money.

PERSONAL TOOLS

In most places, especially the smaller shops, you are expected to supply your own hand tools. If you are employed only to join metal, the tools you will need are few. If, on the other hand, you are expected to layout, prepare, and fit up the job, you will need more tools.

For most welding you need a chipping hammer and a wire brush, such as those shown in Figure 3C-7. Sometimes the chipping hammer is called a slag or scaling hammer. They are available with different combinations of ends. As you can see from Figure 3C-7 there are points and chisels, points and cross chisels, chisel and cross chisel. Other selections are also available. Each fits a specific purpose. It is up to the welder to choose the proper hammer for the job. As a suggestion, for removing slag left by the **flame cutting** process, use the cross chisel hammer. When this type of hammer end is used, by striking downward, the slag breaks off easily and cleanly.

The wire brush is used to remove particles of slag that remain after the weld is chipped. Brushes come in different densities and with either long or short handles. Some chipping hammers have a wire brush attached. Wire brushes are available with steel or stainless steel bristles. Clean stainless brushes are used for working with stainless steel or aluminum. Take care of your brushes by keeping them clean. Dirty and oily brushes can leave foreign material in a weld joint, and cause a poor weld.

Two additional homemade tools provide valuable services to the welder. One is a pick. You can purchase a heavy duty ice pick, or make one out of a steel rod by grinding a sharp point on one end. It can be used to reach into sharp corners or undercut areas to remove slag a chipping hammer cannot reach. (See Figure 3C-8.)

The other tool is a piece of hacksaw blade, with a handle covered with tape to protect your hand. It is used on pipe welds. To use the blade, draw it toward you through the joint to remove slag. It is especially good at the toes of the weld on pipe butt joints. This is probably the quickest and easiest method of removing slag from a pipe joint, especially when welding by using the downhill technique. A half-round bastard file is often used in lieu of the saw blade.

One other item the welder should have is a strong leather or canvas bag to carry electrodes or tools, or both. The bag makes it easier to carry things you need when you are working in a place where it would be difficult to carry a toolbox or other container.

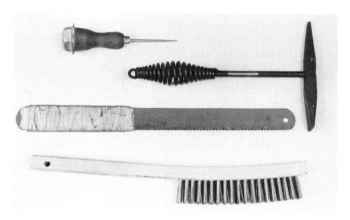

Figure 3C-8 Ice pick, slag, or chipping hammer; hacksaw blade and wire scratch brush.

CHAPTER 4
THE WELDING CIRCUIT, POLARITY, AND ARC BLOW

This chapter deals with the type of **penetration** obtained when different welding currents are used. In direct current welding (DC) there are two polarities, straight polarity (DCSP) and reverse polarity (DCRP). Each produces different amounts of penetration due to the different direction of welding current flow.

It is important that the welder understand the principles involved. This information will aid in selecting the correct electrode for the job at hand.

The effects of **arc blow** will also be covered. An arc is a flexible conductor; it can be bent by a magnetic field. Arc blow is caused by magnetism in the weld zone. It forces the arc to wander from its normal path, which can be very frustrating to the welder. When you understand the causes of arc blow, you will be able to take corrective actions to reduce it. This will make your job easier and the finished weld will be completed better and faster.

LESSON 4A
THE WELDING CIRCUIT

OBJECTIVES
Upon completion of this lesson you should be able to:
1. Draw a diagram and describe alternating current (AC), direct current straight polarity (DCSP), and direct current reverse polarity (DCRP) welding circuits.
2. Describe the type of penetration obtained by the different currents available.

THE ALTERNATING CURRENT (AC) WELDING CIRCUIT
You will remember from Lesson 3A that power from the utility company is transformed by the welding machine to relatively safe welding power. High voltage-low current utility power is changed to low voltage-high current welding power by a stepdown transformer.

From this point on, we will use the term AC, or alternating current, to mean the transformed welding current.

AC welding current changes from negative to positive polarity each half cycle. The half cycles average out each other's effect. The current actually reverses itself 120 times per second when 60 cycle current is used. If any magnetism is present, the arc will be bent or pushed out of position. But with **alternating current** each time the direction is reversed the arc is bent in the opposite direction. The constant reversing of direction reduces the effectiveness of the arc blow. The effect of arc blow will be covered in the next lesson.

With the AC welding circuit the workpiece lead is connected to the terminal of the welding machine marked "Work" or "Ground." The electrode lead is attached to the terminal marked "Electrode," and the leads may never be switched.

THE DIRECT CURRENT (DC) WELDING CIRCUIT
With direct current welding circuits, unlike AC circuits, the welder can choose the direction of welding current flow. The ability to choose the direction of flow allows the welder to choose electrodes that have a wide variety of characteristics.

By choosing wisely the welder can vary the depth of penetration, change the ease of slag removal, and select the rate of filler metal deposition. Rate of filler metal deposition means how much of the electrode is added to the weld in a given period of time.

Direction of flow in direct current welding circuits can be regulated by changing the cable connections.

Sometimes a switch is available to select either Straight or Reverse Polarity.

DIRECT CURRENT, STRAIGHT POLARITY (DCSP)
Direct current straight polarity (DCSP) is the condition where the electrode lead is connected to the negative (−) terminal of the power source and the work lead is connected to the positive (+) terminal. In conventional electrical theory, current flows from positive to negative.

With DCSP the current flows from the workpiece which is positive, to the electrode which is negative. Approximately two-thirds of the total heat is released in the electrode, causing it to melt rapidly. Because most of the heat is released in the electrode the penetration of the workpiece is only moderate. Figure 4A-1 shows the DCSP circuit. This is sometimes called the DCEN (Direct Current Electrode Negative) connection.

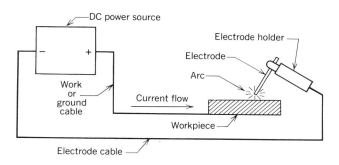

Figure 4A-1 Direct Current Straight Polarity. The current leaves the power source at the positive terminal, travels through the workpiece into the electrode, and returns to the power source via the negative terminal.

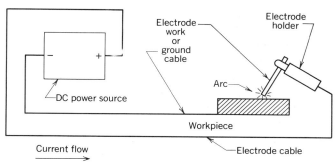

Figure 4A-2 Direct Current Reverse Polarity. The current leaves the power source at the positive terminal, travels through the electrode into the work piece and returns to the power source via the negative terminal.

DIRECT CURRENT, REVERSE POLARITY (DCRP)

In direct current reverse polarity (DCRP) the electrode lead is connected to the positive (+) terminal of the power source and the work lead is connected to the negative (−) terminal. (See Figure 4A-2.)

When welding the DCRP, approximately two-thirds of the heat goes into the workpiece. Compared with DCSP, the DCRP deposition rate is less and the penetration is greater. In Europe this is called DCEP (Direct Current Electrode Positive).

LESSON 4B
ARC BLOW: ITS CAUSES AND HOW TO REDUCE ITS EFFECTS

OBJECTIVES
Upon completion of this lesson you should be able to:
1. Explain the causes of arc blow.
2. Explain the effects of arc blow.
3. List the methods used to reduce arc blow.

ARC BLOW
Arc blow is a condition all welders encounter from time to time. When arc blow occurs, the arc reacts wildly. It wavers and seems to blow away from its normal path. Weld metal can pile up on the trailing edge of the weld bead or be blown ahead of the weld pool. It acts as though it was blown aside by a jet of air, but it is actually due to magnetic deflection.

There are two kinds of arc blow: When the arc is bent forward, in the direction of travel, it is called **forward blow**. When the arc is blown to the rear, it is called **back blow**. (See Figure 4B-1.)

THE CAUSES OF ARC BLOW
Whenever current flows a magnetic

Figure 4B-1 Forward arc blow at the beginning of a V-groove. (Courtesy American Welding Society)

field is created. Magnetic lines of force surround the current carrying conductor. Even current in the plate you are welding can have a magnetic field. When the plates are made of iron or steel, the magnetic field can be very strong. The arc also has its own surrounding magnetic field. When the field in the plate and the field of the arc interact, the arc will bend. The amount of bend is due to the amount of current being used. (See Figure 4B-2.) This condition is sometimes known as the *ground* effect. It is not as noticeable or disturbing to the welder as the end effect discussed below. Most of the time it can be remedied simply by changing the position of the workpiece clamp. When the clamp is moved, the current path in the work is also moved, which, in turn, moves the magnetic field.

In Figure 4B-3 there are examples

THE WELDING CIRCUIT, POLARITY, AND ARC BLOW

of both *ground* and *end* effects. Notice that there is no arc blow near the center of the plate. The flux lines distort at the ends because they are crowded together. The strong, unbalanced field pushes the arc toward the opposite end of the plate.

As you can readily see the flux concentrations at the end are more forceful than in the center of the plate. In large weldments the effect of the work lead connection almost disappears completely. When both concentrations of flux are present at the same time, the stronger one dominates.

In Figures 4B-4 and 4B-5 we see examples of the effect of moving the workpiece connector from one end of the plate to the other. In Figure 4B-4 the direction of travel is away from the workpiece connection. Both the concentration of flux at the plate end and the workpiece connection cause the arc blow. As the arc nears the center of the plate, the end concentration is reduced to a minimum. There is a possibility of some arc blow from the ground effect. As the welding arc approaches the end of the plate, the end concentration becomes forceful, blowing opposite to the direction of travel. The ground

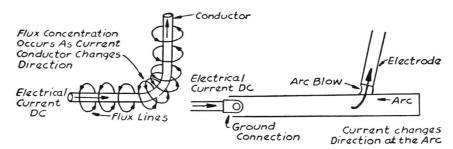

Figure 4B-2 (Courtesy of the Lincoln Electric Company)

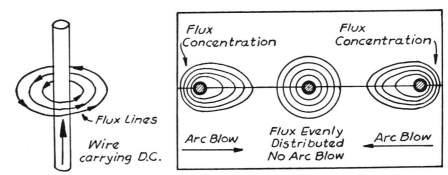

Figure 4B-3 The effects of ground and end effects on arc blow. (Courtesy of the Lincoln Electric Company)

effect exerts force toward the end of the plate opposite the ground connection. Arc blow at this point is strongest in the direction shown.

In Figure 4B-5 the ground connection has been moved to the side of the plate where the weld will end.

The direction of travel will be toward it. By moving the ground connection forward blow is reduced by the ground effect. Any blow is then toward the start of the weld. The combined forces of end concentration and ground effect may cause arc blow at the center of the plate and at the end where the ground connection is.

METHODS OF REDUCING ARC BLOW

Arc blow makes the welder's job more difficult and can cause bad welds. A number of steps can be taken to reduce the effects of arc blow, but they are not always successful.

1. Reduce current. This reduces the strength of the arc blow effect.
2. Move the work lead connection to a position as far as possible from the joint being welded. This reduces the flux density in the plate at the point of welding.
3. Place the workpiece connector at the end of the weld to reduce

Figure 4B-4 and Figure 4B-5 The effects of moving the ground connection from one end of the plate to another. (Courtesy The Lincoln Electric Company.)

ARC BLOW: ITS CAUSES AND HOW TO REDUCE ITS EFFECTS

forward blow. This reduces the strength of the arc blow.

4. Place the ground connector at the start of the weld to reduce back blow.
5. Hold as short an arc as possible. Short arcs are stiffer than long arcs.
6. Set up a magnetic field to help neutralize the one causing the arc blow by wrapping a few turns of the ground lead around the piece being welded.
7. Weld toward a heavy tack or completed section of the weld.
8. Use the back step method of welding on long welds. (Backstepping means to start the welding a short distance from the far end of a joint. After the end of the joint is reached the welder welds another short section traveling toward the one already completed.) Continue to weld in this manner until the joint is finished. (See Figure 4B-6.)
9. Place a steel block across the path of the weld and weld toward it. (See Figure 4B-7.)
10. If none of these remedies is successful and the amount of arc blow is substantial, switch to alternating current. There is practically no arc blow when using an AC arc because the current changes direction 120 times a second.

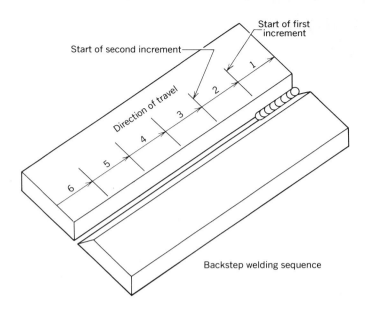

Figure 4B-6 Backstep welding sequence.

Figure 4B-7 The effect that placing a steel block across the path of the weld has on the arc. (Courtesy American Welding Society)

CHAPTER 5
METALS AND WELDING

When metals are joined together by arc welding heat, the structure of the metal is affected. A welder should understand how the metal is affected by heat and how to control the effects.

This chapter discusses the physical and mechanical properties of metal. It also describes how steel is classified, the different structural shapes that are available, how steel can be ordered, and the tests you can make to identify unmarked metal. In addition, the causes of distortion are discussed, as well as the ways to minimize it.

There are many welding applications where metal must be preheated and others where it must be allowed to cool slowly. Sometimes it is necessary to apply heat so the metal takes longer to cool down. Controlled heating and cooling are generally known as heat treating.

The chapter also has a lesson on weld defects. It discusses defects, their causes, and how they can be eliminated.

LESSON 5A
THE MECHANICAL AND PHYSICAL PROPERTIES OF METAL

OBJECTIVE
Upon completion of this lesson you should be able to:
1. Define the term *mechanical properties*.
2. List the mechanical properties that determine metal behavior.
3. State the effects of these properties on metal performance.

DEFINITION OF MECHANICAL PROPERTIES
The mechanical properties of metal are those qualities that determine its behavior when a load is applied. These properties control whether metal will be easy to bend, whether it will be hard, brittle, and so on.

The mechanical properties of metal include: strength, elasticity, ductility, hardness, toughness, and fatigue resistance. Design engineers must consider all of these factors. Determine the choice of metal that will be used to withstand the service load which the product will be exposed to when placed into service.

Sometimes a metal that resists bending will be chosen, whereas in another case ductile metal that can be cold formed readily will be chosen.

The load acting to twist, stretch, bend, or compress the metal is known as **stress.** Unit stress is the load divided by the cross-sectional area of the metal (Figure 5A-1).

During the application of these loads the metal will change shape. This change in the shape is called **strain.** (See Figure 5A-2.)

Metals have an elastic limit. This is the point at which the metal will no longer return to its original shape after the load is removed. If the strain is increased beyond the elastic limit,

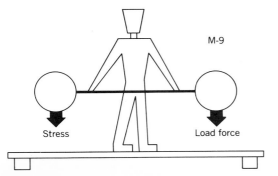

Figure 5A-1 This is an example of stress being exerted on a plank.

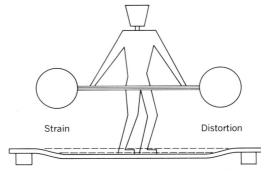

Figure 5A-2 If sufficient stress is applied to metal, it will distort or change its shape.

THE MECHANICAL AND PHYSICAL PROPERTIES OF METAL

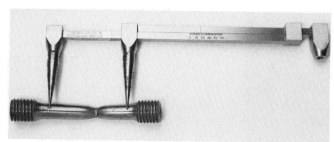

Figure 5A-3 Specimen pulled beyond its yield point. (Courtesy Union County Vocational and Technical School)

the metal will be permanently deformed. An example of permanent deformation is shown in Figure 5A-3 where the specimen has been pulled past its yield point and has "necked down." If the load was continued after this point was reached, the specimen would eventually break.

STRENGTH

Strength is that property of metal which resists forces attempting to pull it apart. It can be measured with a **tensile** tester. An example of this is shown in Figure 5A-4.

The stress required to bring the metal to the breaking point (in pounds per square inch) measures the tensile strength of the metal. This test is a destructive load test. It is commonly used to determine the **tensile strength** of welded joints. Your work will be subjected to tests of this kind many times in your career. A sample of these tests' coupons is shown in Figure 5A-5.

YIELD STRENGTH

During a tensile test the metal begins to "neck down" like a piece of taffy when you try to pull it apart. Yield strength is the load stress just before it begins to deform. The load required to reach the **yield point** is lower than the load required to break the specimen. It may be as little as 50 percent to as high as 85 percent of the tensile strength. The distance the metal stretches before breaking is called **elongation.**

The American Welding Society Specification for Carbon Steel Covered Arc Welding Electrodes (AWS A5.1) lists both the tensile strength and yield strength (in thousands of pounds per square inch), and the elongation and reduction in area (in percent). In some cases which measure toughness, the Charpy V-notch test, is listed. It measures number of foot pounds required, at a given temperature, to break the specimen.

DUCTILITY

Ductility is the property that gives metal the ability to stretch, bend, or twist without breaking or cracking. This quality is measured by bending a specimen in a controlled manner and examining it for cracks or breaks. A specimen that stretches 70 percent or more is considered to be very **ductile.**

Ductility tests are done frequently. You can test your own welds in your school shop. This is an excellent way for you to see the results of your own efforts. You will gain confidence as you see your welds stand up under the tortures of this test. A sample of a "U" bend ductility test specimen is shown in Figure 5A-6. The type of

Figure 5A-4 Specimen mounted in a tensile tester ready for testing. (Reprinted by permission of John Wiley & Sons, Inc.)

Figure 5A-5 A tensile specimen of a ductile material before pull and after pull. (Reprinted by permission of John Wiley & Sons, Inc.)

Figure 5A-6 "U" bend test specimens after testing.

Figure 5A-7 "U" bend test in progress. (Courtesy Union County Vocational and Technical School)

equipment used to perform bend tests is shown in Figure 5A-7.

MALLEABILITY

The ability of a metal to deform permanently when it is compressed, hammered, or rolled into shape is called **malleability.** Most ductile metals are also malleable. Metals such as copper, soft iron, silver, tin, lead, and gold are examples of metals with high malleability.

BRITTLENESS

Brittleness is the opposite of malleability. Brittle metals will not deform easily when placed under load. Metals that possess this property are not only brittle, but they are low in ductility. Cast iron is a good example of a metal that is brittle. Metals that are associated with brittleness crack suddenly when overloaded.

ELASTICITY

Elasticity is the ability of a metal to return to its original shape after the load is released. Spring steel is a good example of a metal that contains this property.

HARDNESS

Hardness is a measure of a metal's ability to resist indentation or penetration by a harder material. There is a need in many applications for metals that possess this quality.

Hardness is measured by pressing a diamond point or hard steel ball into the metal being tested. This makes a small dent in the metal. The size of the dent is then measured and the hardness is calculated with a formula. Figure 5A-8 shows two types of penetrators that are used with a Rockwell hardness tester. Figure 5A-9 shows the **hardness tester** itself.

NOTCH TOUGHNESS

Toughness is the ability of a metal to withstand sudden load without breaking. This property is also called impact resistance. Gears and camshafts require this property.

Toughness is measured by striking a specimen with a sudden force. The specimen is made with a **notch** cut into it, to cause it to break at that point. The energy required to break the specimen determines the toughness of the metal. The piece of equipment most often used for this purpose is the Charpy V-notch tester. It uses a weighted pendulum to strike the specimen. (See Figure 5A-10.)

THE PHYSICAL PROPERTIES OF METALS

Two physical properties of metals are

Figure 5A-8 Brale and Ball. These two penetrators are the basic types used on the Rockwell Hardness Tester. (Note, Brale is a registered trademark of American Chain & Cable Company, Inc., for sphero-conical diamond penetrators.) (Reprinted by permission of John Wiley & Sons, Inc.)

THE MECHANICAL AND PHYSICAL PROPERTIES OF METAL

electrical conductivity and thermal expansion. Both of these properties are present in all metals. Metals are better conductors than nonmetals. The ability to conduct both heat and electricity is related. If metals were not good conductors of electricity, it would be almost impossible to join them with arc welding.

Whether you apply heat to a metal as in shielded metal arc welding, or by some other means, the metal will expand. This property is called thermal expansion. Every welder should understand that as the metal rises in temperature its size increases in all directions. It is possible that the pieces will not fit if "fit up" allowances are not made beforehand.

For example, you may be called on to heat the area surrounding a bronze bearing. The heat expands the surrounding metal so the bearing can be removed. In such cases it is important that you heat only the surrounding area and not the bearing itself. You must heat the surrounding metal as rapidly as possible to prevent the heat from transferring to the bearing; otherwise, the bearing will expand faster than the steel. To correct this both the bearing and the metal around it should be allowed to cool. The bearing may reach its **elastic limit** and will not return to its original size when cooled. This means that the bearing will have a smaller diameter than before and can be easily removed.

Engineers take thermal expansion into consideration in the design of structures such as bridges and other large weldments. They can calculate the effects of thermal expansion.

Thermal expansion can cause all kinds of problems, as when replacing a section in a pipeline, for example. If a section of line is removed in the early morning, the temperature may rise dramatically by early afternoon, and the pipeline will expand. The expansion depends upon the change in temperature and the length of the line. The line may have expanded a few feet, and the replacement section will have to be cut to size before it can be placed in position.

This is true in any **weldment.** Knowledge of how thermal expansion affects metal is useful to a welder.

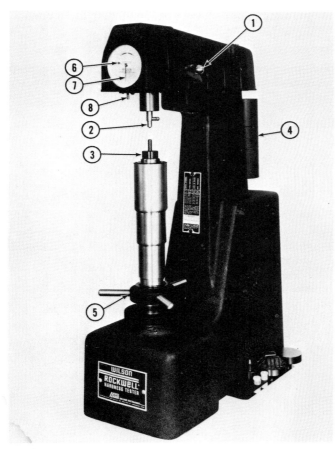

Figure 5A-9 Rockwell Hardness Tester listing the names of parts used in the testing operations (Wilson Instrument Division of Acco). (Reprinted by permission of John Wiley & Sons, Inc.)

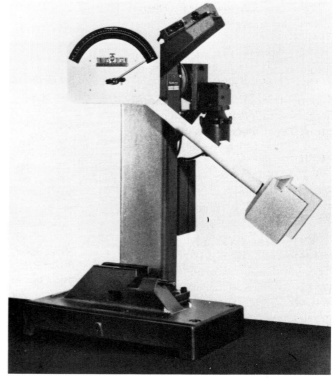

Figure 5A-10 Izod-Charpy testing machine (Tinius Olsen Testing Machine Company, Inc.) (Reprinted by permission of John Wiley & Sons, Inc.)

LESSON 5B
CLASSIFICATION OF IRON AND STEEL

OBJECTIVE
Upon completion of this lesson you should be able to:
1. Explain the use of the AISI-SAE numerical designations of various carbon and alloy steels.
2. Explain the meaning of the number designations.
3. List the percentages of carbon and the predominant element for various steels.

THE IMPORTANCE OF CLASSIFYING STEELS
The majority of the metals joined by the shielded metal arc welding process are **ferrous metal.** Ferrous metals are those that contain iron, such as steels.

It is important that metal composition is known before welding takes place. Some metals need special treatment, such as **preheat, postheat,** controlled **interpass temperatures, travel speeds,** and **rate of heat** input, if the desired results are to be obtained.

There are hundreds of metal compositions available to fit just about every application. There is a need for a classification system that can determine the correct metal to use.

There are several professional societies and trade associations that develop material specifications for steels and other materials. Some of these groups deal only with a particular industry, such as the American Petroleum Institute (API), the Aerospace Materials Specifications (AMS), American Bureau of Shipping, and the Association for American Railroads (AAR).

Others, such as the American National Standards Institute (ANSI), develop specifications that are useful in many industries. The American Society for Testing of Materials (ASTM), the Society of Automotive Engineers (SAE), the American Iron and Steel Institute (AISI) and the American Society of Mechanical Engineers (ASME) are typical organizations.

The AISI and the SAE have developed a system of numbers and prefix letters that identify the type and composition of the steel, as well as the process used in its manufacture. If you work for someone that uses another system, you will have to learn to use that particular system.

Because of its widespread use, the AISI-SAE numerical designation of carbon and alloy steels will be covered in this text. Learning to use this system will help you more easily understand the other systems.

HOW THE AISI-SAE CLASSIFICATION SYSTEM OPERATES
Steels are classified according to their chemical content. To aid in classifying them a four- or five-digit number is used. The first digit indicates the type of steel. 1 denotes carbon steel, 2 nickel steel, 3 nickel-chrome, 4 molybdenum steel, 5 chromium steel, and so forth, as listed in Figure 5B-1.

The second digit indicates the percentage of the main alloying element in the steel. A review of Figure 5B-

Figure 5B-1 SAE-AISI numerical designation of alloy steels (Represents percent of Carbon in hundredths). (Reprinted by permission of John Wiley & Sons, Inc.)

Carbon Steels	
Plain carbon	10xx
Free-cutting, resulfurized	11xx
Manganese Steels	13xx
Nickel Steels	
0.50% nickel	20xx
1.50% nickel	21xx
3.50% nickel	23xx
5.00% nickel	25xx
Nickel-Chromium Steels	
1.25% nickel, 0.65% chromium	31xx
1.75% nickel, 1.00% chromium	32xx
3.50% nickel, 1.57% chromium	33xx
3.00% nickel, 0.80% chromium	34xx
Corrosion and Heat-Resisting Steels	303xx
Molybdenum Steels	
Chromium	41xx
Chromium-nickel	43xx
Nickel	46xx and 48xx
Chromium Steels	
Low-chromium	50xx
Medium-chromium	51xx
High-chromium	52xx
Chromium-Vanadium Steels	6xxx
Tungsten Steels	7xxx
Triple Alloy Steels	8xxx
Silicon-Manganese Steels	9xxx
Leaded Steels	11Lxx (example)

CLASSIFICATION OF IRON AND STEEL

1 will show, in the ASI-SAE number 23XX, the 3 indicates a nickel content of 3.50 percent. That figure is normally the midrange and is not always the actual content. The AISI-SAE numerical classification chart in Figure 5B-2 shows the nickel content for that particular steel can be as little as 3.25 and as much as 3.75 percent.

Carbon is an important element of steel because it hardens steel. The more carbon the steel contains, up to a maximum of 2 percent, the harder and stronger it is. However, a steel with more carbon is less ductile.

Figure 5B-2 AISI-SAE numerical designation of carbon and alloy steels (Source: AISI-SAE) (Reprinted by permission of Prentice-Hall, Inc. Englewood Cliffs, N.J.)

Alloy Steel

AISI Number	C	Mn	P Max	S Max	Si	Ni	Cr	Other	SAE Number
1320	0.18-0.23	1.60-1.90	0.040	0.040	0.20-0.35	–	–	–	1320
1321	0.17-0.22	1.80-2.10	0.050	0.050	0.20-0.35	–	–	–	–
1330	0.28-0.33	1.60-1.90	0.040	0.040	0.20-0.35	–	–	–	1330
1335	0.33-0.38	1.60-1.90	0.040	0.040	0.20-0.35	–	–	–	1335
1340	0.38-0.43	1.60-1.90	0.040	0.040	0.20-0.35	–	–	–	1340
2317	0.15-0.20	0.40-0.60	0.040	0.040	0.20-0.35	3.25-3.75	–	–	2317
2330	0.28-0.33	0.60-0.80	0.040	0.040	0.20-0.35	3.25-3.75	–	–	2330
2335	0.33-0.38	0.60-0.80	0.040	0.040	0.20-0.35	3.25-3.75	–	–	–
2340	0.33-0.43	0.70-0.90	0.040	0.040	0.20-0.35	3.25-3.75	–	–	2340
2345	0.43-0.48	0.70-0.90	0.040	0.040	0.20-0.35	3.25-3.75	–	–	2345
E 2512	0.09-0.14	0.45-0.60	0.025	0.025	0.20-0.35	4.75-5.25	–	–	2512
2515	0.12-0.17	0.40-0.60	0.040	0.040	0.20-0.35	4.75-5.25	–	–	2515
E 2517	0.15-0.20	0.45-0.60	0.025	0.025	0.20-0.35	4.75-5.25	–	–	2517
3115	0.13-0.18	0.40-0.60	0.040	0.040	0.20-0.35	1.10-1.40	0.55-0.75	–	3115
3120	0.17-0.22	0.60-0.80	0.040	0.040	0.20-0.35	1.10-1.40	0.55-0.75	–	3120
3130	0.28-0.33	0.60-0.80	0.040	0.040	0.20-0.35	1.10-1.40	0.55-0.75	–	3130
3135	0.33-0.38	0.60-0.80	0.040	0.040	0.20-0.35	1.10-1.40	0.55-0.75	–	3135
3140	0.38-0.43	0.70-0.90	0.040	0.040	0.20-0.35	1.10-1.40	0.55-0.75	–	3140
3141	0.38-0.43	0.70-0.90	0.040	0.040	0.20-0.35	1.10-1.40	0.70-0.90	–	3141
3145	0.43-0.48	0.70-0.90	0.040	0.040	0.20-0.35	1.10-1.40	0.70-0.90	–	3145
3150	0.48-0.53	0.70-0.90	0.040	0.040	0.20-0.35	1.10-1.40	0.70-0.90	–	3150
E 3310	0.08-0.13	0.45-0.60	0.025	0.025	0.20-0.35	3.25-3.75	1.40-1.75	–	3310
E 3316	0.14-0.19	0.45-0.60	0.025	0.025	0.20-0.35	3.25-3.75	1.40-1.75	–	3316
								Mo	
4017	0.15-0.20	0.70-0.90	0.040	0.040	0.20-0.35	–	–	0.20-0.30	4017
4023	0.20-0.25	0.70-0.90	0.040	0.040	0.20-0.35	–	–	0.20-0.30	4023
4024	0.20-0.25	0.70-0.90	0.040	0.035-0.050	0.20-0.35	–	–	0.20-0.30	4024
4027	0.25-0.30	0.70-0.90	0.040	0.040	0.20-0.35	–	–	0.20-0.30	4027
4028	0.25-0.30	0.70-0.90	0.040	0.035-0.050	0.20-0.35	–	–	0.20-0.30	4028
4032	0.30-0.35	0.70-0.90	0.040	0.040	0.20-0.35	–	–	0.20-0.30	4032
4037	0.35-0.40	0.70-0.90	0.040	0.040	0.20-0.35	–	–	0.20-0.30	4037
4042	0.40-0.45	0.70-0.90	0.040	0.040	0.20-0.35	–	–	0.20-0.30	4042
4047	0.45-0.50	0.70-0.90	0.040	0.040	0.20-0.35	–	–	0.20-0.30	4047
4053	0.50-0.56	0.75-1.00	0.040	0.040	0.20-0.35	–	–	0.20-0.30	4053
4063	0.60-0.67	0.75-1.00	0.040	0.040	0.20-0.35	–	–	0.20-0.30	4063
4068	0.63-0.70	0.75-1.00	0.040	0.040	0.20-0.35	–	–	0.20-0.30	4068
–	0.17-0.22	0.70-0.90	0.040	0.040	0.20-0.35	–	0.40-0.60	0.20-0.30	4119
–	0.23-0.28	0.70-0.90	0.040	0.040	0.20-0.35	–	0.40-0.60	0.20-0.30	4125
4130	0.28-0.33	0.40-0.60	0.040	0.040	0.20-0.35	–	0.80-1.10	0.15-0.25	4130
E 4132	0.30-0.35	0.40-0.60	0.025	0.025	0.20-0.35	–	0.80-1.10	0.18-0.25	–
E 4135	0.33-0.38	0.70-0.90	0.025	0.025	0.20-0.35	–	0.80-1.10	0.18-0.25	–
4137	0.35-0.40	0.70-0.90	0.040	0.040	0.20-0.35	–	0.80-1.10	0.15-0.25	4137
E 4137	0.35-0.40	0.70-0.90	0.025	0.025	0.20-0.35	–	0.80-1.10	0.18-0.25	–
4140	0.38-0.43	0.75-1.00	0.040	0.040	0.20-0.35	–	0.80-1.10	0.18-0.25	4140
4142	0.40-0.45	0.75-1.00	0.040	0.040	0.20-0.35	–	0.80-1.10	0.15-0.25	–
4145	0.43-0.48	0.75-1.00	0.040	0.040	0.20-0.35	–	0.80-1.10	0.15-0.25	4145
4147	0.45-0.50	0.75-1.00	0.040	0.040	0.20-0.35	–	0.80-1.10	0.15-0.25	–
4150	0.48-0.53	0.75-1.00	0.040	0.040	0.20-0.35	–	0.80-1.10	0.15-0.25	4150

The carbon content of the steel is indicated by the last two digits in a four-digit number. The carbon content is expressed in hundreths of 1 percent. Once again the figure given is not the actual carbon content, but somewhere near midrange. As an example, 1020 steel has a carbon content ranging from 0.18 to 0.23 percent. If you look at Figure 5B-2, you will find that holds true throughout the chart. As another example, look up 4140 steel on the AISI-SAE chart to determine its composition. The classification system shows the following:

First digit 4—Chrome molybdenum steel

Second digit 1—One percent molybdenum (actually 0.75 to 1.00 percent)

Third and fourth digits 40—Carbon content (actually 0.38 to 0.43 percent)

Sometimes this is called "40 points" of carbon. Study the chart until you are able to look up information quickly and accurately.

LESSON 5C IDENTIFICATION OF METALS

OBJECTIVES

Upon completion of this lesson you should be able to:

1. List the various tests used in identifying metals.
2. Explain the various testing methods.
3. Use various tests to determine the identity of metals.

Often, it is necessary to identify the metal you are going to weld on prior to the welding operation. Knowing the type of metal is important in any welding operation because the filler metal should be matched with the metal being joined.

In the construction and fabricating industry the welder doesn't have many opportunities to participate in the identification process. The range of metal types ordered is usually limited and their properties are known, so the identifying process is seldom required.

... 'orks for a large
'ification is nor-
inspectors or
owever, there
in mainte-
makes the
e welding
ification

challenge for the welder. Most of the time the type of metal isn't known or the records are lost.

Before a repair can be attempted the metal must be identified. After identification the welding process and the filler metal should be selected. The type of metal, its accessibility, availability of equipment, and skills of the welder must be considered. They all have a bearing on the decision regarding the welding process selection.

CRITERIA FOR IDENTIFICATION

There are many methods that can be used to identify a metal, but they are not always available to a welder. Many methods require special equipment and training. The spectrograph, an instrument that can accurately determine the percentage of each element contained in a metal, is an example of such equipment. Another example is a chemical analysis kit.

There are, however, other test methods that can provide you with information to identify a metal. These other methods are based on the appearance of a metal, its hardness, its reaction to flame or an arc, the pattern of grinding sparks, its reaction to a magnet, its weight, its color and appearance when fractured, the ease with which it fractures, and its reaction to certain chemicals.

Figure 5C-1 lists some base metals and their reaction to some of the tests just mentioned. Use these charts as a quick reference to aid in the identification of metals.

APPEARANCE TEST

Quite often, through practice and experience, we are able to identify a metal part by its color, appearance, shape, and intended use. Castings such as pump housings, engine blocks and manifolds, and other parts with mold markets are easily recognizable. Some parts have grinding marks where the mold marks have obviously been removed. Structural beams and supports are easily recognized by their cross-section shapes and are usually made of hot rolled steel.

Bare hot rolled steel is dark gray and covered by a mill scale from the rolling operation. It may also have an oxidized, or rusty, appearance if it has been exposed to the atmosphere for some time. Cold rolled steel is brighter in color. It has no mill scale and has a smooth surface and square edges. Normally it is free of rust and oxidization, unless its protective coating has been removed.

IDENTIFICATION OF METALS

Figure 5C-1 Summary of identification tests of metals. (Reprinted by permission of Prentice-Hall, Inc., Englewood Cliffs, N.J.)

Base Metal or Alloy	Color	Magnet	Chisel	Fracture	Flame or Torch	Spark
Aluminum & alloys	bluish-white	non-magnetic	easily cut	white	melts wo/col	non-spark.
Brass, navy	yellow or reddish	non-magnetic	easily cut	not used	not used	non-spark.
Bronze, alum. (90Cu-9Al)	reddish yellow	non-magnetic	easily cut	not used	not used	non-spark.
Bronze, phosphor (90Cu-10Sn)	reddish yellow	non-magnetic	easily cut	not used	not used	non-spark.
Bronze, silicon (96Cu-3Si)	reddish yellow	non-magnetic	easily cut	not used	not used	non-spark.
Copper (deoxidized)	red; 1 cent piece	non-magnetic	easily cut	red	not used	non-spark.
Copper nickel (70Cu-30 Ni)	white 5 cent piece	non-magnetic	easily cut	not used	not used	non-spark.
Everdur (96Cu-3Si-1Mn)	gold	non-magnetic	easily cut	not used	not used	non-spark.
Gold	yellow	non-magnetic	easily cut	not used	not used	non-spark.
Inconel (76Ni-16Cr-8Fe)	white	non-magnetic	easily cut	not used	not used	non-spark.
Iron, cast	dull grey	magnetic	not easily chip.	brittle	melts slowly	see text
Iron, wrought	light grey	magnetic	easily cut	bright grey fib.	melts fast	see text
Lead	dark grey	non-magnetic	very soft	white; crystal	melts quick	non-spark.
Magnesium	silvery white	non-magnetic	soft	not used	burns in air	non-spark.
Monel (67 Ni-30Cu)	light grey	slightly magnet.	tough	light grey	not used	non-spark.
Nickel	white	magnetic	easily cut	almost white	not used	see text
Nickel silver	white	non-magnetic	easily chipped	not used	not used	non-spark.
Silver	white-pre 1965 10¢ pc	non-magnetic	not used	not used	not used	non-spark.
Steel, low alloy	blue-grey	magnetic	depends on comp	medium grey	shows color	see text
Steel, high carbon	dark grey	magnetic	hard to chip	very lgt. grey	shows color	see text
Steel, low carbon	dark grey	magnetic	continuous chip	bright grey	shows color	see text
Steel, manganese (14 Mn)	dull	non-magnetic	work hardens	coarse grained	shows color	see text
Steel, medium carbon	dark grey	magnetic	easily cut	very lgt. grey	shows color	see text
Steel, stainless (austentic)	bright silvery	see text	continuous chip	deps. on type	melts fast	see text
Steel, stainless (matensitic)	grey	slightly magnetic	continuous chip	deps. on type	melts fast	see text
Steel, stainless (ferritic)	bright silvery	slightly-magnet.	-	deps. on type	-	see text
Tantalum	grey	non-magnetic	hard to chip	-	high temp.	-
Tin	silvery white	non-magnetic	usually as plating	usually as plating	melts quick	non-spark.
Titanium	steel grey	non-magnetic	hard	not used	not used	see text
Tungsten	steel grey	non-magnetic	hardest metal	brittle	highest temp.	non-spark
Zinc	dark grey	non-magnetic	usually as plating	at R.T.	melts quick	non-spark

Other metals such as aluminum, brass, copper, and stainless steel, have distinctive colors that can be seen when they are cleaned.

MAGNETIC TEST

With the exception of cobalt and nickel most nonferrous metals (metals that do not contain iron) are nonmagnetic. Nonmagnetic metals include aluminum, zinc, magnesium, and copper-based alloys. Iron is strongly ferromagnetic, cobalt is second, and nickel is the least. Iron, because of its low cost, is the most commercially important of the three. Most **ferrous** metals range from strongly magnetic for the steels, to some stainless steels that are slightly magnetic, and some stainless steel alloys that are nonmagnetic.

Practice with a small magnet will provide the welder with the experience necessary to use this test effectively.

HARDNESS TEST

Accurate tests for hardness of a metal must be conducted with laboratory equipment such as a Rockwell or a Brinnel hardness tester.

There are two simple hardness tests that a welder can use to compare a metal sample with an unknown metal. These tests are the scratch and the file test.

The scratch test requires that the corner of a known metal, such as the keystock in Figure 5C-2, be drawn across an unknown metal sample. If the sample is not scratched, use it to try to scratch the key stock, as in

Figure 5C-2 A piece of keystock (mild steel) is scratched across an unknown metal sample. Because the sample is not scratched, it is harder than the keystock and probably is an alloy or tool steel. (Reprinted by permission of John Wiley & Sons, Inc.)

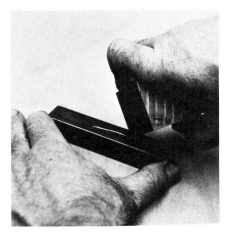

Figure 5C-3 The sample is now scratched against the keystock as a further test and it does scratch the keystock. (Reprinted by permission of John Wiley & Sons, Inc.)

File Test and Hardness Table

File Reaction	Rockwell B	Rockwell C	Brinell	Type Steel
File bites easily into metal	65		100	Mild steel
File bites into metal with pressure		16	212	Medium carbon steel
File does not bite into metal except with difficulty		31	294	High alloy steel / High carbon steel
Metal can only be filed with extreme pressure		42	390	Tool steel
File will mark metal but metal is nearly as hard as the file and filing is impractical		50	481	Hardened tool steel
Metal is as hard as the file		64	739	Case hardened parts

Figure 5C-4 File test and hardness table. (Reprinted by permission of John Wiley & Sons, Inc.)

Figure 5C-3. This test is comparative. It will only tell you which of the metals is the hardest. If you have metal samples of varying hardness, you will be able to match the unknown metal with a sample.

A good file can be used to identify the steels by testing for hardness. Draw the file across the edge of the metal, and note its reaction to the metal. Compare it with the reactions in Figure 5C-4.

CHIP TEST

This test requires a small sharp chisel and a hammer. They are used to attempt to remove a chip from the edge of the sample. The ease or difficulty with which the chip is removed is in direct proportion to the hardness of the metal. It also indicates whether the metal is brittle or ductile. For instance, a continuous chip indicates that the metal is fairly soft and ductile. Mild steel, aluminum, and malleable iron are among the metals that easily make a continuous chip.

When the chips break apart, the metal is brittle. Cast iron is a good example of a brittle metal.

FRACTURED TEST

If the part to be welded is broken or fractured, look at the metal in the break. The appearance of the metal in the fracture will help in the identification. The ease with which it is broken will show the true color of the metal. There won't be any oxidation present until the break has been exposed to the atmosphere for an appreciable time. Samples of metal structures that will be encountered when examining breaks in cast iron are shown in Figures 5C-5 and 5C-6.

The results of the fracture test should

Figure 5C-5 Fracture of gray cast iron. (Reprinted by permission of John Wiley & Sons, Inc.)

Figure 5C-6 Fracture of white cast iron. (Reprinted by permission of John Wiley & Sons, Inc.)

IDENTIFICATION OF METALS

be compared with the chart in Figure 5C-1 as an aid in identifying the metal. When an unknown metal is encountered, it is often possible to remove a small piece from it in order to perform a fracture test.

CHEMICAL TEST

The average welding shop does not have the chemicals necessary to conduct a chemical analysis of metals. Commercial testing kits are available, however. If there is an individual trained in their use, the tests are very valuable in determining the identity of metals.

SPARK TEST

The grinding wheel spark test is a popular test. It provides the welder with the opportunity to separate ferrous and nonferrous metals rapidly. Aluminum, copper, and other nonferrous-based alloys do not give off spark streams like those from ferrous metals.

For example, monel and stainless steel are difficult to tell apart visually. However, stainless steel gives off spark patterns, whereas the monel does not.

Spark tests are conducted by pressing a sample lightly against a grinding wheel as in Figure 5C-7. It is best to view the sparks against a dark background. A piece of black cloth or a board painted black will be fine. The intensity, color, and shape of spark patterns can be used to identify the metal.

METHOD OF CARRYING OUT SPARK TESTS

To perform a spark test correctly it is important to carry it out in an area where the light is subdued, because the color of the sparks is important. A portable or fixed grinding wheel is required that has a surface speed of at least 5000 feet per minute. You can calculate the speed by multiplying the circumference of the wheel, in feet, by the number of revolutions per minute.

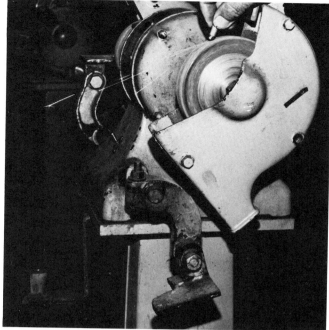

Figure 5C-7 The proper way to make a spark test on a pedestal grinder (Lane Community College). (Reprinted by permission of John Wiley & Sons, Inc.)

Always take proper safety precautions when making this test, especially when using a pedestal grinder as shown in Figure 5C-7. Wear eye protection and watch out for other people.

SPARK PATTERNS OF CERTAIN METALS IN COMMON USE

Wrought iron. Long, straw-colored carrier lines, usually whiter away from the grinding wheel. Carrier lines usually end in spearhead arrows or small forks.

Low carbon or mild steel. Long, straight yellow carrier lines with a few forks or branches and a few carbon bursts or arrowheads.

Low alloy steel (medium carbon). Each alloying element has an effect on the spark appearance and very careful observation is required. Type 4130 steel has carrier lines that often

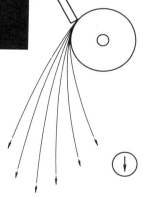

Figure 5C-8 Wrought iron long, straw-colored carrier lines, usually whiter away from the grinding wheel. Carrier lines usually end in spearhead arrows or small forks.

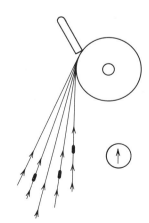

Figure 5C-9 Low carbon steel. Straight carrier lines having a yellowish color with very small amount of branching and very few carbon bursts.

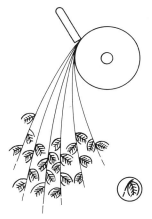

Figure 5C-10 Low alloy steel (medium carbon). Each alloying element has an effect on the spark appearances and very careful observation is required. Type 4130 steel has carrier lines that often end in forks and sharp outer points with few sprigs.

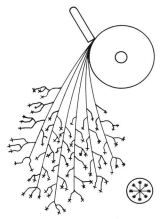

Figure 5C-11 High carbon steel. Short, very white, or light yellow carrier lines with considerable forking, having many starlike bursts. Many of the sparks follow around the wheel.

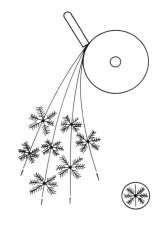

Figure 5C-12 Manganese steel. Bright white carrier lines with fan-shaped bursts.

end in forks and sharp outer points with few sprigs.

High carbon steel. Abundant yellow carrier lines with bright and abundant star bursts.

Manganese steel. Bright white carrier lines with fan-shaped bursts.

Stainless steels. The chrome nickel steels give off short carrier lines, sometimes making a dotted line without buds or sprigs.

Nickel alloys—extremely short spark stream. Carrier lines are orange. There are no forks or sprigs and the sparks may follow the grinding wheel.

Cast iron. Red carrier lines with many bursts that are red near the grinder and orange-yellow farther out. Considerable pressure is required to produce sparks.

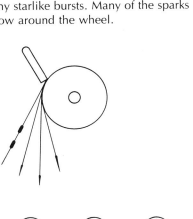

Figure 5C-13 Stainless steel. The chrome nickel steels give off short carrier lines, sometimes making a dotted line without buds or sprigs.

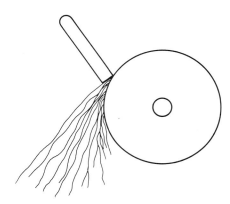

Figure 5C-14 Nickel alloys. Extremely short spark stream. Carrier lines are orange. There are no forks or sprigs and the sparks may follow the grinding wheel.

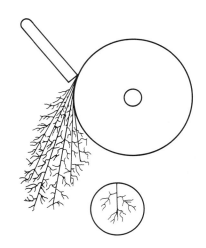

Figure 5C-15 Cast iron. Short carrier lines with many bursts, which are red near the grinder and orange-yellow further out. Considerable pressure is required to produce sparks on cast iron.

LESSON 5D
STRUCTURAL SHAPES

OBJECTIVE
Upon completion of this lesson you should be able to:
1. List the various structural shapes.
2. Develop material lists using the correct procedure for ordering structural shapes.

STRUCTURAL SHAPES
Steel comes in many different sizes and shapes to fit industrial needs. Steel can be purchased from a supplier, or in some cases directly from the mill of the producer. When ordering directly from the producer, the order must be quite substantial. An automobile manufacturer, shipyard, bridge builder, a large construction company requiring special fabricated structural shapes, might order from a steel mill.

In most cases the steel is ordered from the warehouse of a supplier. It is ordered by the purchasing department in large concerns, by the owner or manager of some companies, by a supervisor in many of the smaller shops, and in some cases by the welder.

Steel is sold by the pound. The price depends on the amount of steel ordered at one time. Steel suppliers have guidelines governing the sale of steel.

A supplier will charge a certain amount per pound of steel purchased up to some limit, let us say, 1000 pounds (lb). If the customer wishes to purchase more than the limit, the price per pound is reduced. The price may be reduced even more as the amount of steel purchased increases. There are other reasons for price reductions; they may vary from one company to another. If you are going to purchase an amount that will take a long time to use, you must consider the storage problem. Steel should not be left outside or in a damp area. Store it properly; otherwise it will oxidize or rust. Additional cleaning might be required before it can be used.

Sheet and plate are usually manufactured to the composition standards of the American Society of Testing and Materials (ASTM). Sheet and plate come in widths over 8 in. and are available in a great variety of sizes. Sheet is specified by gauge thickness and is three-sixteenths of an inch or less. Plate, on the other hand, is specified in thickness in excess of three-sixteenths of an inch.

PURCHASING STEEL
Although it is possible to have unusual shapes especially made; Channel, I beam, H beam, wide flange beams, tees, angle bar, Zee bar, octagon bar, hexagon bar, round bar, square and rectangular bar, half round bar, as well as flat bar are normally available. These shapes come in a variety of thicknesses and weights. Because of the wide variety, it is important to give the correct information when placing an order for steel.

To order steel, you must have a working knowledge of the size specifications and how to express your needs to the supplier. For example, by checking the weight per lineal foot of the shape being ordered you can determine the total weight of the order, and calculate the price category of the order.

Often you may find that by ordering a small amount extra you will qualify for a price reduction. Sometimes the extra material is almost free because of the price reduction, making it profitable to purchase in this manner.

When ordering particular shapes, you should check the length that is offered. Some items are sold by specific length, whereas some come in random lengths. This means that all of the pieces may not be the same length.

METHODS OF SPECIFYING SIZES AND SHAPES
There are a number of methods used to specify the size of the shapes being ordered. The best method is to use the terminology in the steel supplier's catalog.

Plate and sheet are normally ordered by giving the thickness of the metal first, then the width, followed by the length and the type of metal.

Structural shapes may be ordered using a number of methods. Some purchasers still use the old symbols to signify the shape they are ordering. This method is being phased out because computers cannot readily interpret symbols. Therefore, letters have been assigned for use as symbols.

The letter designation simplify the ordering procedure. The designations in Figure 5D-1 show the shape, symbol, and the method of ordering the material.

For example: When ordering channel the symbol (C) indicates channel, the first group of numbers (15) the size of the beam, and the last two numbers (50) the weight per lineal foot of the channel.

This is an excellent method of ordering because the different weights of each size of channel depend upon its dimensions.

Figure 5D-2 is representative of the information provided in a supplier's date book for ordering material. Notice that there are two or three different weights of channel available in each size. Also note the dimensions of the channel vary in depth of section, width of the flange, thickness of the flange, and thickness of the web. The size designation, in the example above, gives all of these measurements without writing them out in detail.

The manner in which the dimensions of structural shapes are to be given depends upon the industry in which you work. Because welders work in many industries you should be aware of this difference. For example, in the structural steel fabrication industry all dimensions over 12 in. are given in feet, whereas in most machine shops dimensions under 6 ft are given in inches.

Name of Shape	Shape	Symbol	Example Designation
Wide flange beam	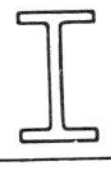	W	W12×16.5
American Standard beam	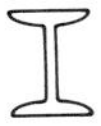	S	S10×35
Channel	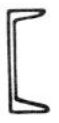	C	C15×50
Angle	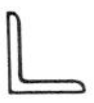	L	L4×3×½
Tees, cut from W shape	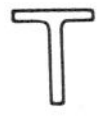	WT	WT5×8.5
Tees, cut from S shape	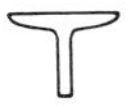	ST	ST6×15.9
Zees	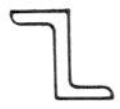	Z	Z5×16.4
Construction pipe	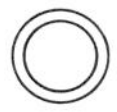	TS	TS4×0.188

Figure 5D-1 Designations for steel shapes. (Reprinted by permission of Prentice-Hall, Inc., Englewood Cliffs, N.J.

C
American Standard Channels

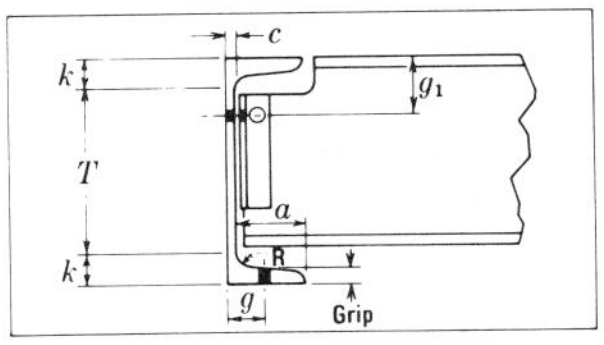

Dimensions for Detailing

Designation and Nominal Size	Weight per Foot	Flange		Web		Distances					Usual Gage g	Grip	Max. Flange Fastener	Fillet Radius R
		Width	Aver. Thickness	Thickness	Half Thickness	a	T	k	g_1	c				
In.	Lbs.	In.	In.	In.	In.	In.	In.	In.	In.	In.	In.	In.	In.	In.
C15 15 x 3⅜	50	3¾	⅝	11/16	⅜	3	12⅛	17/16	2¾	¾	2¼	⅝	1	.50
	40	3½	⅝	½	¼	3	12⅛	17/16	2¾	9/16	2	⅝	1	
	33.9	3⅜	⅝	⅜	3/16	3	12⅛	17/16	2¾	7/16	2	⅝	1	
C12 12 x 3	30	3⅛	½	½	¼	2⅝	9¾	1⅛	2½	9/16	1¾	½	⅞	.38
	25	3	½	⅜	3/16	2⅝	9¾	1⅛	2½	7/16	1¾	½	⅞	
	20.7	3	½	5/16	⅛	2⅝	9¾	1⅛	2½	⅜	1¾	½	⅞	
C10 10 x 2⅝	30	3	7/16	11/16	5/16	2⅜	8	1	2½	¾	1¾	7/16	¾	.34
	25	2⅞	7/16	½	¼	2⅜	8	1	2½	9/16	1¾	7/16	¾	
	20	2¾	7/16	⅜	3/16	2⅜	8	1	2½	7/16	1½	7/16	¾	
	15.3	2⅝	7/16	¼	⅛	2⅜	8	1	2½	5/16	1½	7/16	¾	
C9 9 x 2½	15	2½	7/16	5/16	⅛	2¼	7⅛	15/16	2½	⅜	1⅜	7/16	¾	.33
	13.4	2⅜	7/16	¼	⅛	2¼	7⅛	15/16	2½	5/16	1⅜	7/16	¾	
C8 8 x 2¼	18.75	2½	⅜	½	¼	2	6⅛	15/16	2½	9/16	1½	⅜	¾	.32
	13.75	2⅜	⅜	5/16	⅛	2	6⅛	15/16	2½	⅜	1⅜	⅜	¾	
	11.5	2¼	⅜	¼	⅛	2	6⅛	15/16	2½	5/16	1⅜	⅜	¾	
C7 7 x 2⅛	12.25	2¼	⅜	5/16	3/16	1⅞	5¼	⅞	2½	⅜	1¼	⅜	⅝	.31
	9.8	2⅛	⅜	3/16	⅛	1⅞	5¼	⅞	2½	¼	1¼	⅜	⅝	
C6 6 x 2	13	2⅛	5/16	7/16	3/16	1¾	4⅜	13/16	2¼	½	1⅜	5/16	⅝	.30
	10.5	2	5/16	5/16	3/16	1¾	4⅜	13/16	2¼	⅜	1⅛	⅜	⅝	
	8.2	1⅞	5/16	3/16	⅛	1¾	4⅜	13/16	2¼	¼	1⅛	5/16	⅝	
C5 5 x 1¾	9	1⅞	5/16	5/16	3/16	1½	3½	¾	2¼	⅜	1⅛	5/16	⅝	.29
	6.7	1¾	5/16	3/16	⅛	1½	3½	¾	2¼	¼	...	...	...	
C4 4 x 1⅝	7.25	1¾	5/16	5/16	3/16	1⅜	2⅝	11/16	2	⅜	1	5/16	⅝	.28
	5.4	1⅝	5/16	3/16	1/16	1⅜	2⅝	11/16	2	¼	...	...	...	
C3 3 x 1½	5	1½	¼	¼	⅛	1¼	1⅝	11/16	...	5/16	...	...	...	.27
	4.1	1⅜	¼	3/16	1/16	1¼	1⅝	11/16	...	¼	...	...	...	

Figure 5D-2 (Courtesy United States Steel)

AVAILABLE STRUCTURAL SHAPES

The following examples are based on available structural shapes and size specifications indicated in catalogs. After the size is shown, indicate the type of material such as Hot Rolled Steel (HRS), Cold Rolled Steel (CRS), and so on.

There are a number of ways to express the size specifications, and the method is dependent on the structural shape. Consult any steel distributor's manual for practice in ordering.

LESSON 5E
DISTORTION: ITS CAUSES AND CURES

OBJECTIVES
Upon completion of this lesson you should be able to:
1. Explain distortion and warpage and their causes.
2. Explain how temperature change affects metal.
3. List the problems and suggested remedies for distortion and warpage.

CAUSES OF DISTORTION
Distortion and warpage are the result of the temperature change of metal during welding, cutting, and heating operations. When heat is put into metal, it expands in all directions. As the metal cools it contracts and should return to its original shape. However, heated metal does not always return to its original shape when it cools. The manner and rate at which the heat is applied, the shape of the weld, and other factors affect the final shape after cooling.

There are no hard and fast rules that describe how distortion and warpage can be controlled. However, there are some general guidelines that can be followed. In the end only experience will give you the ability to predict the effects with any degree of accuracy.

HOW HEATING AND COOLING AFFECTS METAL
A good example of the affect that heating and cooling have on metal is shown in Figure 5E-1a. The piece of round bar stock, between the jaws of the vise, is restrained from moving along its length.

After the ends of the bar are secured firmly in the jaws of the vise, the restrained bar should be heated until it is "cherry red" or a little over 1000 degrees Fahrenheit (°F). It should be brought to this temperature by using a neutral oxy-fuel flame. After the bar reaches the required temperature, remove the flame and allow the bar to cool slowly.

As the restrained bar was heated, its diameter increased more than it would have had it not been held in between the vise jaws. Because the bar couldn't expand normally, as it would have if it had been heated while unrestrained, it was "upset." Upsetting is a condition where metal held in restraint expands more than normal in another direction. (See Figure 5E-1b.)

As the bar cools it contracts. Eventually it will fall from between the jaws of the vise. It is shorter than before because it was "upset." It was not allowed to expand by the vise; its length was reduced and the diameter increased. (See Figure 5E-1c.)

Another example of the effects of heat on metal is to tack a piece of ¼-in. bar, 1 to 2 in. wide and about 12 in. or more long, to a plate as indicated in Figure 5E-2a. After tacking one end only, weld toward the free end. As welding progresses the bar will gradually curve toward the side on which the welding is taking place. (See Figure 5E-2b.)

Next, take a thinner piece of metal and stagger the tack welds, as in Figure 5E-2c. Notice how the free end of the bar moves from side to side as the tacks are welded.

THE EFFECT OF WELDING ON A WELDMENT
During the welding procedure the base metal of the weldment and the filler metal are heated to very high temperatures. (Steel melts at approximately 2750° F.) When metal is close to its **melting point** temperature, it has very little strength. The strength

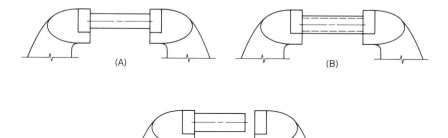

Figure 5E-1 Effects of heating.

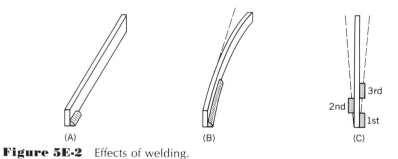

Figure 5E-2 Effects of welding.

increases when the temperature decreases during the cooling cycle.

Heated metal expands and cool metal contracts. A problem arises because weld metal does not heat and cool uniformly. Normally, weld metal solidifies from the root outward to the face of the weld.

As the weld metal freezes it shrinks. The weld is pulled toward the center, causing the weldment to warp.

Figure 5E-3 shows the vertical member of the fillet weld tilting in the direction of the weld. This is due to the shrinkage forces within the weld.

Figure 5E-4 shows a good example of distortion encountered in single V-groove butt joints. The amount of contraction at the root of the weld is small, due to the small amount of weld metal present. Because there is a greater volume of weld metal at the top of the weld, the shrinkage will be greater at that point, and the weldment will distort toward the center line of the weld.

When welding a multipass V-groove butt joint, distortion still occurs, but with slight differences. The first weld pass causes very little, if any, distortion due to the small amount of metal involved. The second pass does not create much distortion because of the first pass, unless the first pass is completely remelted by the second pass.

All passes create some shrinkage. As the number of passes increases the shrinkage forces increase. A multipass weld tends to distort more than a single heavy pass. (See Figure 5E-5.)

DISTORTION
PROBLEMS AND SUGGESTED PREVENTATIVE PROCEDURES
PROBLEM:
The Tee joint in Figure 5E-6a has had its vertical member pulled out of position by shrinkage forces of the fillet weld.

SUGGESTIONS:
No. 1. To offset the shrinkage forces, and avoid the condition in Figure 5E-6a, the vertical member will be pulled upright if the correct prepositioning is used. (See Figure 5E-6b.) The proper amount of prepositioning is usually found by trial and error.

No. 2. Weld the joint from both sides. When the weld on the second side is larger than on the first side, the additional shrinkage may overcome the forces of the first weld. (This procedure is not too popular. The extra time and material increases the cost.)

No. 3. Weld the joint intermittently as shown in Figure 5E-7. With short welds placed on opposite sides of the vertical member, the shrinkage forces tend to neutralize each other. Intermittent welds may be used when a watertight joint is not required or when the strength requirements of the weldment do not call for continuous welds.

No. 4. Weld the fillets with two passes on each side of the vertical member. Notice the welding sequence of the beads in Figure 5E-8. By alternating the placement of weld beads the effect of the shrinkage forces is reduced.

PROBLEM:
Distortion that occurs when welding inside corner fillet joints. When welding an inside corner fillet, the vertical member has a tendency to

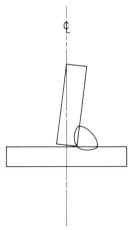

Figure 5E-3 Pulling effect of weld metal.

Figure 5E-4 Distortion due to welding.

Figure 5E-5 Distortion due to welding.

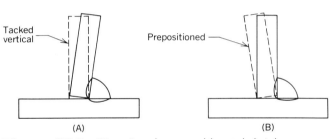

Figure 5E-6 Distortion due to weld metal shrinkage.

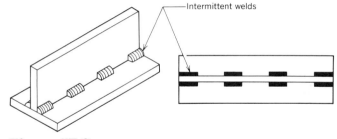

Figure 5E-7

DISTORTION: ITS CAUSES AND CURES

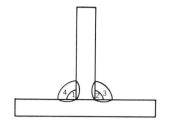

Figure 5E-8 Double fillet weld.

be pulled toward the welded side, as shown in Figure 5E-9a.

SUGGESTIONS:

Preposition the upright member away from the side to be welded. Shrinkage will cause the member to be drawn into the desired position, as in Figure 5E-9b. Another method uses a single bevel on the upright member, as in Figure 5E-9c. Weld the beveled side carefully. Overwelding will cause distortion in that direction.

PROBLEM:

The tendency of single pass groove welds in light metal to pull toward the centerline of the weld. Weld metal shrinkage at the top of the joint causes the sides of the plates to pull upward, as in Figure 5E-10.

SUGGESTIONS:

Change the shape of the weld groove. The more equal the opening at the bottom of the joint is to the opening at the top, the more equal the shrinkage forces will be as the weld metal cools. Use the smallest groove angle possible, within reasonable limits. For example, a square groove butt joint would be excellent on very light-gauge metal. The weld metal would tend to have equal shrinkage forces. (See Figure 5E-11.)

Preposition the plates in the direction opposite the face of the weld, as in Figure 5E-12. Plates bent in this manner will pull into the desired position as the weld metal cools.

PROBLEM:

Distortion in multiple pass V-grooves in heavy plate. Although there is a greater mass, heavier plate sections will also warp. (See Figure 5E-13.)

SUGGESTIONS:

Many small passes greatly increase distortion in groove welds of plate welded from one side. One method of offsetting this distortion is to use larger electrodes and fewer passes. However, the use of larger electrodes and wider pass may not be as applicable for out-of-position welding and when using low hydrogen and alloy electrodes. The best way to cure warpage in heavy plate is to double bevel the joint and weld from both sides, as in Figure 5E-14. Use the double bevel joint on all metal five-eighths of an inch or more in thickness where access can be gained to the second side.

With double groove butt joints it is important to use the proper weld sequence, whether welded with the weave or stringer method. Notice in Figure 5E-14 that the welding alternates sides. Applying the heat evenly throughout the joint equalizes the shrinkage forces. Distortion does not become a problem.

GENERAL SUGGESTIONS FOR REDUCING DISTORTION

1. **Backstep** welding is effective in some instances. It tends to keep distortion to a minimum. When

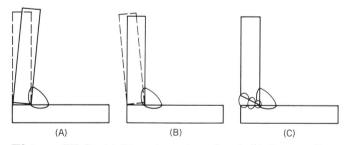

Figure 5E-9 (a) Distortion w/o prebend. (b) Correct alignment w/prebend. (c) Single bevel and fillet.

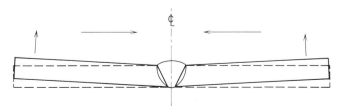

Figure 5E-10 Shrinkage forces pulling plate toward the center line of the weld.

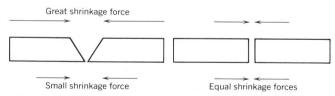

Figure 5E-11 Small shrinkage force, equal shrinkage forces.

Figure 5E-12 Plat prebend before welding.

Figure 5E-13 Multipass single V-butt joint.

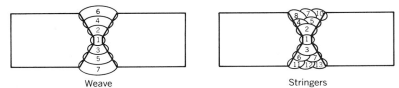 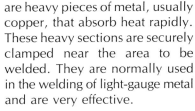

Figure 5E-14 Bead sequence double V-butt joint.

backstepping, the welds are performed in short increments, usually the length that can be obtained with a single electrode. Notice in Figure 5E-15, that although the direction of travel is from left to right, welding starts a short distance to the left of the right-hand side of the plate.

A single bead is run to the end of the plate, then the slag is cleaned from the weld. Another bead is started to the left of the first starting point and welded toward that point. The short welds are continued until the first pass is completed. The following pass is welded in the same manner. Remember to stagger all beads and passes after the first pass, so the starts and stops do not occur in the same place.

2. *Wandering* is simply a method where the welder skips around. Small segments are welded here and there. This method distributes the heat throughout the entire weldment, instead of concentrating it in one place.
3. *Heat sinks* are used in some applications to draw the heat away from the weld metal. Heat sinks are heavy pieces of metal, usually copper, that absorb heat rapidly. These heavy sections are securely clamped near the area to be welded. They are normally used in the welding of light-gauge metal and are very effective.
4. *Restraints* of different types are used to hold the metal in the desired position until the welding is completed and the weldment has cooled down. Heavy tack welds, clamps, and fixtures are used for this purpose.
5. *Peening* is an operation that is also used. Its purpose is to release the locked-in stresses within the weld. **Peening** means hammering the weld metal, usually with an air hammer equipped with a chisel-like tool. As the peening operation is moved over the weld, the metal is reshaped slightly. This relieves most of the stresses. Care must be taken that peening is not overdone. Excessive peening can harden the weld metal. The root or finish pass should never be peened.

SUMMARY: As you read in the beginning of this lesson, there are no set rules that apply to all conditions. The ability to reduce the effects of welding distortion caused by heating and cooling is gained only by trial and error. As you fit and tack the various joints in the welding lessons of this text, you will gradually learn the techniques necessary to control distortion. It will be the same when you enter industry. As you repeat operations you will learn the techniques that produce the desired results.

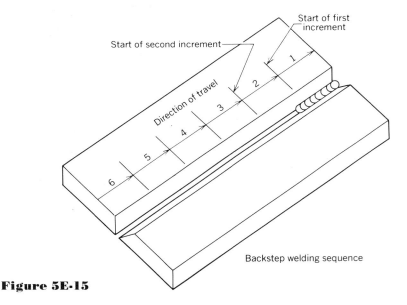

Figure 5E-15

LESSON 5F
PREHEATING AND POSTWELD HEAT TREATMENT

OBJECTIVES
Upon completion of this lesson you should be able to:
1. Explain the need for preheating and postheating in certain weldments and steels.
2. List temperatures at which annealing, stress relieving, and normalizing occur.
3. Explain the importance of maintaining proper interpass temperatures.

DEFINITION OF PREHEATING AND POSTWELD HEAT TREATMENT

Preheating means the act of raising the welding zone temperature by applying heat before welding. Postweld heat treatment, on the other hand, means the act of applying heat after welding. This treatment may be applied anytime prior to the weld being placed in service.

There are a number of methods utilized to heat the weldment. The method selected will be determined by the size and configuration of the weldment, and the availability of facilities for heating purposes.

Small weldments can be placed in an electric or gas furnace. In the case of very large weldments it is sometimes necessary to construct a temporary furnace around them. The heat in most instances can be supplied by a torch or series of torches using natural gas, propane, acetylene, or a similar type of fuel mixed with either air or oxygen.

Electric strip heaters, induction heaters, and radiant heaters are also used for this purpose when applicable.

THE PURPOSE OF PREHEATING

Preheating is used to raise the weldment to a temperature ranging from just below 100 degrees Fahrenheit (100° F) to slightly over 700 degrees Fahrenheit. (700° F) (Specific temperatures are dictated by the qualification welding procedure.)

As you learned in Lesson 5D, stresses are created by heat in the weld area, both in and adjacent to the joint. This heat is relatively localized. There is a sharp difference between the temperature in and near the weld zone. This difference in temperature is known as *temperature differential*.

When a weld is made, especially in a thick piece of metal, the mass of metal adjacent to the weld area quickly draws the heat away from the weld. This is what is called a **quick quench**. (See Figure 5F-1.)

Although it is not as dramatic, quick quench is similar to dousing a welded plate into cold water immediately after welding. Quick quenching causes weld metal to become very hard and brittle. What actually happens in a condition such as this is the grain structure of the metal becomes needlelike, coarse, and extremely sensitive to cracking. (See Figure 5F-2.)

There are two areas of a welded joint that are most vulnerable to cracking as a result of the welding procedure. One is the heat-affected zone (HAZ), which is the area immediately adjacent to the weld. (See Figure 5F-3.) The other is the weld metal itself, which usually is less prone to cracking.

There are three zones involved in a welded joint, and they are shown in Figure 5F-3. Zone A is the area that has been brought to the melting temperature of the metals involved, and includes both the **base metal** and the **filler metals** that make up the weld deposit. Zone B is the **heat-affected zone (HAZ)** which, although it was not heated to the melting point, was raised sufficiently high in temperature to change its original microstructure and physical properties. This is the danger area.

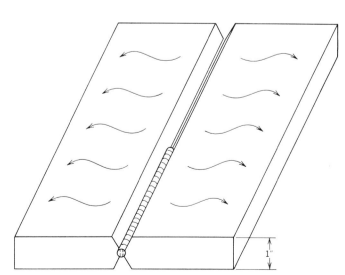

Figure 5F-1 Heat being rapidly drawn away from the weld area by the mass of metal.

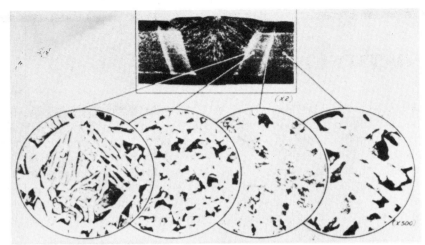

Figure 5F-2 Typical microstructures in the heat-affected zone of a butt weld. (Courtesy United States Steel); (Courtesy Tempil Division, Big Three Industries)

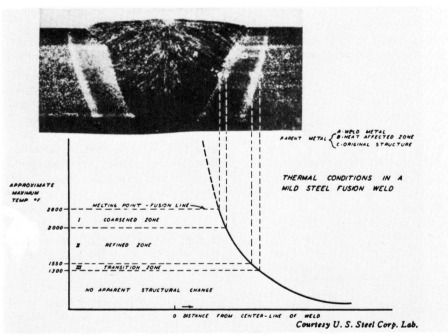

Figure 5F-3 Typical macrostructures and maximum attained temperatures in a butt weld. (Courtesy United States Steel) (Courtesy Tempil Division, Big Three Industries)

Zone C is the area next to the (HAZ). It has not been affected by the heat of the welding operation to the extent that changes have been made in its original microstructure or physical properties.

Notice in Figure 5F-1 the differences in the structure of the metals in the different zones. Welds on steel can cool rapidly, especially when made on larger and heavier sections. Rapid cooling results in the formation of hard, brittle grain structures within the metal. This type of grain structure causes the metal to have low ductility and poor ability to withstand stresses. It is usually found in Zone 1 in Figure 5F-2, which is between the weld metal and the refined structure in Zone 2.

Zone 1 has the greatest danger of cracking. Preheating can eliminate, or at least minimize the danger. Preheating to high temperatures helps to slow the rate of cooling. The slow cooling rate gives time for the desired grain structure to form.

The more carbon in the metal, the greater the chance of brittle structure forming. Longer cooling increases the opportunity for the metal to transform to other structures.

The slower cooling prevents the formation of a brittle structure next to the weld deposit. The weld metal and the adjacent metal will be softer and more ductile with preheating.

Another important reason for preheating is the removal of moisture. Moisture can be a source of hydrogen which can increase the possibility of cracking. Because of the slower cooling rate, distortion and shrinkage stresses are reduced, and hydrogen and other gases have a longer period of time to escape from the metal. Because of its importance preheating is recommended where there is the slightest possibility of any weld defect being present.

Metal subjected to a low ambient temperature should always be preheated. Metal brought into a shop from the outside or metal that is being welded in the field on cold winter days should be preheated before welding.

The need to preheat increases if the weldment fits any of the following categories:

1. If it is heavy or has large mass.
2. If it is cold or damp due to atmospheric conditions.
3. If it is going to be welded with smaller electrode diameters.
4. If it is to be welded at great speed.
5. If it has a complicated shape or design.

PREHEATING AND POST WELD HEAT TREATMENT

6. If the parts to be joined have various thicknesses.
7. If the weldment has a high carbon content.
8. If the metal has a high alloy content.
9. If low hydrogen electrodes are not being used for the welding. Figure 5F-4 is a chart indicating suggested preheating temperatures for a variety of steels in common use.

INTERPASS TEMPERATURES

Preheating steel is only the first step in critical applications. In many cases it is also necessary to maintain a specific temperature throughout the welding procedure. Heating or cooling periods may be needed to maintain the specified temperature. Maintaining the temperature of the weldment between weld passes is known as *maintaining the interpass temperature. Interpass means between the passes.*

On lighter metal it is possible that the welding heat itself will maintain the interpass temperatures; additional heat may not be required.

Heavy sections, on the other hand, may require extra heating to maintain the specified interpass temperatures. If the temperature in the weld area drops below the specified temperature, welding must stop immediately. Apply heat until the prescribed temperature is reached again. Then, and not before, can welding be resumed.

Once welding has started on a weldment requiring special interpass temperatures, it is important to finish with as few interruptions as possible. When it is necessary to stop for a short period, it is important to maintain the temperature level until welding can be resumed. There is one caution to remember: Do not let the interpass temperature exceed the maximum temperature allowable. If it does, stop until the weldment cools down to the prescribed interpass temperature. Then welding can be resumed.

POSTWELD HEAT TREATMENT

Although preheating prevents cracking during and immediately after welding, **postheating** insures crack-free welds much later, before the metal enters service. Postheat also ensures that the grain structure and metallurgical properties are satisfactory.

After a weldment has been completed, it is often heat treated for another reason. Up to this point heat was used to reduce the possibility of cracking in the weld zone. The heat treatments we will discuss now are for stress relieving, annealing, and normalizing.

The chart in Figure 5F-5 helps us understand the various heat treatments and their temperature ranges. It also serves as a color guide to assist us when we are heating metal. It shows the color changes of metal as it rises in temperature. These color changes are rough guides to temperature. They are usable for some applications, but cannot be used when the exact temperature is needed. But for accurate temperature some sort of calibrated device must be used.

STRESS RELIEVING

The most widely used **postweld heat treatment** is for **stress relieving.** It is of great importance to weldments. Normally it is carried out in the 1050 to 1200° F temperature range for carbon steel. Temperatures up to 1400° F are used for alloy steels. Stress relieving is required by many codes and the person responsible for the project should know and understand the code requirements before starting the operation. Stress relieving is used to reduce residual stresses from the welding, casting, or forging process. It improves resistance to corrosion and embrittlement, tempers the metal, and improves the service life. Stress relieving should always be used if the weldment is subject to impact loading or low temperature service.

Stress relieving is accomplished at relatively low temperatures. Temperatures are below that where change in the crystal structure of metal takes place.

ANNEALING

In **annealing,** the metal is heated to approximately 100° F above the point where it begins to change its crystal structure. This range is somewhat between 1675° F and slightly less than 1800° F, as indicated in Figure 5F-5. Usually the weldment is placed in a furnace. Then its temperature is reduced slowly by gradually reducing the temperature of the furnace.

Annealing reduces the stresses present in the weldment. The cooling cycle takes place in the furnace where the weldment was heated. In addition to relieving the stresses, annealing also produces higher ductility and lower strength.

NORMALIZING

Normalizing takes place at the same temperatures as annealing. The weldments are held at temperature for similar periods, but there are some differences.

In annealing the weldment is cooled in the controlled atmosphere of a furnace. In normalizing the weldment is cooled in still air, and therefore the cooling rate will be somewhat faster. The increased rate of cooling reduces ductility, but the weldment has greater strength than an annealed weld.

Both annealing and normalizing relieve more stresses than stress relieving, but they require higher heat input. They can create heavy scale and under certain circumstances can cause changes in weldment dimensions.

No heat treatment should be car-

Table 4—Preheat Temperatures for Metals and Alloys

METAL GROUP	METAL DESIGNATION	APPROXIMATE COMPOSITION — PERCENT							METAL DESIGNATION	RECOMMENDED PREHEAT
		C.	Mn.	Si.	Cr.	Ni.	Mo.	Cu.		
PLAIN CARBON STEELS	PLAIN CARBON STEEL	BELOW .20							PLAIN CARBON STEEL	UP TO 200°F
	PLAIN CARBON STEEL	.20-.30							PLAIN CARBON STEEL	200°F-300°F
	PLAIN CARBON STEEL	.30-.45							PLAIN CARBON STEEL	300°F-500°F
	PLAIN CARBON STEEL	.45-.80							PLAIN CARBON STEEL	500°F-800°F
CARBON MOLY STEELS	CARBON MOLY STEEL	.10-.20					.50		CARBON MOLY STEEL	300°F-500°F
	CARBON MOLY STEEL	.20-.30					.50		CARBON MOLY STEEL	400°F-600°F
	CARBON MOLY STEEL	.30-.35					.50		CARBON MOLY STEEL	500°F-800°F
MANGANESE STEELS	SILICON STRUCTURAL STEEL	.35	.80	.25					SILICON STRUCTURAL STEEL	300°F-500°F
	MEDIUM MANGANESE STEEL	.20-.25	1.0-1.75						MEDIUM MANGANESE STEEL	300°F-500°F
	SAE T 1330 STEEL	.30	1.75						SAE T 1330 STEEL	400°F-600°F
	SAE T 1340 STEEL	.40	1.75						SAE T 1340 STEEL	500°F-800°F
	SAE T 1350 STEEL	.50	1.75						SAE T 1350 STEEL	600°F-900°F
	12% MANGANESE STEEL	1.25	12.0						12% MANGANESE STEEL	USUALLY NOT REQUIRED
HIGH TENSILE STEELS (SEE ALSO STEELS BELOW)	MANGANESE MOLY STEEL	.20	1.65	.20			.35		MANGANESE MOLY STEEL	300°F-500°F
	JALTEN STEEL	.35 MAX.	1.50	.30				.40	JALTEN STEEL	400°F-600°F
	MANTEN STEEL	.30 MAX.	1.35	.30				.20	MANTEN STEEL	400°F-600°F
	ARMCO HIGH TENSILE STEEL	.12 MAX				.50 MIN.	.05 MIN.	.35 MIN.	ARMCO HIGH TENSILE STEEL	UP TO 200°F
	DOUBLE STRENGTH #1 STEEL	.12 MAX.	.75			.50-1.25	.10 MIN.	.50-1.50	DOUBLE STRENGTH #1 STEEL	300°F-600°F
	DOUBLE STRENGTH #1A STEEL	.30 MAX.	.75			.50-1.25	.10 MIN.	.50-1.50	DOUBLE STRENGTH #1A STEEL	400°F-700°F
	MAYARI R STEEL	.12 MAX.	.75	.35	.2-1.0	.25-.75		.60	MAYARI R STEEL	UP TO 300°F
	OTISCOLOY STEEL	.12 MAX.	1.25	.10 MAX.	.10 MAX.			.50 MAX.	OTISCOLOY STEEL	200°F-400°F
	NAX HIGH TENSILE STEEL	.15-.25		.75	.60	.17	.15 MAX.	.25 MAX. Zr. .12	NAX HIGH TENSILE STEEL	UP TO 300°F
	CROMANSIL STEEL	.14 MAX.	1.25	.75	.50				CROMANSIL STEEL	300°F-400°F
	A. W. DYN-EL STEEL	.11-.14					.40		A. W. DYN-EL STEEL	UP TO 300°F
	CORTEN STEEL	.12 MAX.		.25-1.0	.5-1.5	.55 MAX.		.40	CORTEN STEEL	200°F-400°F
	CHROME COPPER NICKEL STEEL	.12 MAX.	.75		.75	.75		.55	CHROME COPPER NICKEL STEEL	200°F-400°F
	CHROME MANGANESE STEEL	.40	.90		.40				CHROME MANGANESE STEEL	400°F-600°F
	YOLOY STEEL	.05-.35	.3-1.0			1.75		1.0	YOLOY STEEL	200°F-600°F
	HI-STEEL	.12 MAX.	.6	.3 MAX.		.55		.9-1.25	HI-STEEL	200°F-500°F
NICKEL STEELS	SAE 2015 STEEL	.10-.20				.50			SAE 2015 STEEL	UP TO 300°F
	SAE 2115 STEEL	.10-.20				1.50			SAE 2115 STEEL	200°F-300°F
	2½% NICKEL STEEL	.10-.20				2.50			2½% NICKEL STEEL	200°F-400°F
	SAE 2315 STEEL	.15				3.50			SAE 2315 STEEL	200°F-500°F
	SAE 2320 STEEL	.20				3.50			SAE 2320 STEEL	200°F-500°F
	SAE 2330 STEEL	.30				3.50			SAE 2330 STEEL	300°F-600°F
	SAE 2340 STEEL	.40				3.50			SAE 2340 STEEL	400°F-700°F
MEDIUM NICKEL CHROMIUM STEELS	SAE 3115 STEEL	.15			.60	1.25			SAE 3115 STEEL	200°F-400°F
	SAE 3125 STEEL	.25			.60	1.25			SAE 3125 STEEL	300°F-500°F
	SAE 3130 STEEL	.30			.60	1.25			SAE 3130 STEEL	400°F-700°F
	SAE 3140 STEEL	.40			.60	1.25			SAE 3140 STEEL	500°F-800°F
	SAE 3150 STEEL	.50			.60	1.25			SAE 3150 STEEL	600°F-900°F
	SAE 3215 STEEL	.15			1.00	1.75			SAE 3215 STEEL	300°F-500°F
	SAE 3230 STEEL	.30			1.00	1.75			SAE 3230 STEEL	500°F-700°F
	SAE 3240 STEEL	.40			1.00	1.75			SAE 3240 STEEL	700°F-1000°F
	SAE 3250 STEEL	.50			1.00	1.75			SAE 3250 STEEL	900°F-1100°F
	SAE 3315 STEEL	.15			1.50	3.50			SAE 3315 STEEL	500°F-700°F
	SAE 3325 STEEL	.25			1.50	3.50			SAE 3325 STEEL	900°F-1100°F
	SAE 3435 STEEL	.35			.75	3.00			SAE 3435 STEEL	900°F-1100°F
	SAE 3450 STEEL	.50			.75	3.00			SAE 3450 STEEL	900°F-1100°F
MOLY BEARING CHROMIUM AND CHROMIUM NICKEL STEELS	SAE 4140 STEEL	.40			.95		.20		SAE 4140 STEEL	600°F-800°F
	SAE 4340 STEEL	.40			.65	1.75	.35		SAE 4340 STEEL	700°F-900°F
	SAE 4615 STEEL	.15				1.80	.25		SAE 4615 STEEL	400°F-600°F
	SAE 4630 STEEL	.30				1.80	.25		SAE 4630 STEEL	500°F-700°F
	SAE 4640 STEEL	.40				1.80	.25		SAE 4640 STEEL	600°F-800°F
	SAE 4820 STEEL	.20				3.50	.25		SAE 4820 STEEL	600°F-800°F
LOW CHROME MOLY STEELS	2% Cr.-½% Mo. STEEL	UP TO .15			2.0		0.5		2% Cr.-½% Mo. STEEL	400°F-600°F
	2% Cr.-½% Mo. STEEL	.15-.25			2.0		0.5		2% Cr.-½% Mo. STEEL	500°F-800°F
	2% Cr.-1% Mo. STEEL	UP TO .15			2.0		1.0		2% Cr.-1% Mo. STEEL	500°F-700°F
	2% Cr.-1% Mo. STEEL	.15-.25			2.0		1.0		2% Cr.-1% Mo. STEEL	600°F-800°F
MEDIUM CHROME MOLY STEELS	5% Cr.-½% Mo. STEEL	UP TO .15			5.0		0.5		5% Cr.-½% Mo. STEEL	500°F-800°F
	5% Cr.-½% Mo. STEEL	.15-.25			5.0		0.5		5% Cr.-½% Mo. STEEL	600°F-900°F
	8% Cr.-1% Mo. STEEL	.15 MAX.			8.0		1.0		8% Cr.-1% Mo. STEEL	600°F-900°F
PLAIN HIGH CHROMIUM STEELS	12-14% Cr. TYPE 410	.10			13.0				12-14% Cr. TYPE 410	300°F-500°F
	16-18% Cr. TYPE 430	.10			17.0				16-18% Cr. TYPE 430	300°F-500°F
	23-30% Cr. TYPE 446	.10			26.0				23-30% Cr. TYPE 446	300°F-500°F
HIGH CHROME NICKEL STAINLESS STEELS	18% Cr. 8% Ni. TYPE 304	.07			18.0	8.0			18% Cr. 8% Ni. TYPE 304	USUALLY DO NOT REQUIRE PREHEAT BUT IT MAY BE DESIRABLE TO REMOVE CHILL
	25-12 TYPE 309	.07			25.0	12.0			25-12 TYPE 309	
	25-20 TYPE 310	.10			25.0	20.0			25-20 TYPE 310	
	18-8 Cb. TYPE 347	.07			18.0	8.0		Cb. 10XC	18-8 Cb. TYPE 347	
	18-8 Mo. TYPE 316	.07			18.0	8.0	2.5		18-8 Mo. TYPE 316	
	18-8 Mo. TYPE 317	.07			18.0	8.0	3.5		18-8 Mo. TYPE 317	

Figure 5F-4 Preheat temperatures of metals and alloys. (Courtesy United States Steel); (Courtesy Tempil Division, Big Three Industries)

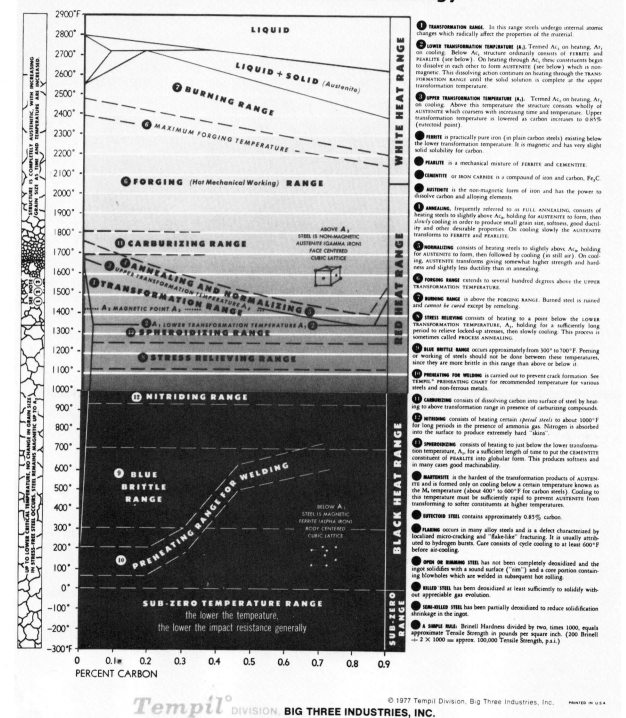

Figure 5F-5 Preheating for welding. (Courtesy United States Steel); (Courtesy Tempil Division, Big Three Industries)

ried out in an indiscriminant manner. Follow the specifications in the number of degrees the weldment can be increased in temperature each hour and how long it must remain at these temperatures. It is important to raise and lower the temperature according to the specifications. Otherwise the procedure will fail.

TEMPERATURE CONTROL

Temperatures can be monitored through the use of meters in conjunction with thermocouples. Heat input can be controlled by the settings on induction and resistance heating equipment when used.

When heat is being supplied by other devices by open flame, temperature can be determined by one of the fine, commercially available temperature indicators. Figure 5F-6 shows a number of the products available.

Probably the most popular and widely used is the temperature indicating stick. These crayons, or

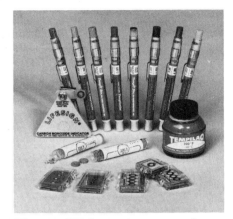

Figure 5F-6 Various types of temperature indicators that are available. (Courtesy Tempil Division of Big Three Industries Inc.)

sticks, are available in over 100 different temperature ranges, from 100° F to 2500° F. The systematically spaced temperature ratings of the crayons are accurate within ± 1 percent of their rating.

They are simple to use. Choose the appropriate crayon for the job at hand and mark the workpiece with it. The stick will leave a dry opaque mark until the metal reaches the required temperature. When the desired temperature is reached, the mark will melt and become a liquid smear. When working with temperatures below 700° F, the mark can be applied to the surface before heating begins and the mark will melt when the proper temperature is reached. You can also stroke hot metal with a crayon as the metal is being heated. Other items, such as pellets, stick-on labels, and liquids are also available for indicating temperatures.

LESSON 5G
WELDMENT DEFECTS

OBJECTIVES
Upon completion of this lesson you should be able to:
1. Define the term **structural discontinuity**.
2. List the types and causes of **discontinuities**.
3. Define the term **dimensional discrepancy**.

DEFINITION
A weld discontinuity is an imperfection in the weldment. It can be a space that is not filled with metal of

Some of the illustrations in this lesson are reproductions of American Welding Society material. The author gratefully acknowledges the American Welding Society.

the same metallurgical structure as the rest of the weld. Such imperfections can affect the weld performance to a point where the weld is unacceptable or will fail in service.

CAUSE OF DISCONTINUITIES
Overlap, undercut, lack of penetration, slag inclusions, foreign matter, cracks, **porosity,** tungsten inclusion, lack of fusion, poor weld profiles, and arc strikes are all discontinuities and may affect the quality of a weld.

In most cases *welder* error is the major cause of discontinuities. The error can be due to things such as lack of ability, fatigue, and poor judgment or carelessness on the part of the welder.

Some weld discontinuities, their causes and suggestions that will help

eliminate them, are discussed in the following paragraphs.

POROSITY
Gas pockets or voids in the weld area are called porosity. These voids are caused by gases that are released by chemical reactions of the welding process. The gas is trapped as the weld metals cool and solidify. Porosity may be localized or found uniformly throughout the weld. (See Figure 5G-1.)

Excessive heat can cause porosity. The welder must be careful and not

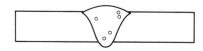

Figure 5G-1 Porosity

WELDMENT DEFECTS

weld with too high an amperage setting. High travel speed is another factor that causes porosity. When the rate of travel is high, the weld puddle tends to cool rapidly causing the gases to be trapped. You have seen the effervescent bubbles rising to the surface when a glass of soda is poured. Think of them as the gases attempting to escape from a weld and visualize the effect should the soda freeze instantaneously. The bubbles would be trapped, as are the gases in the welding process.

SLAG INCLUSION

During welding, oxides and other materials can be trapped in the weld metal and the molten base metal. (See Figure 5G-2.) Often, this material comes from the electrode coating and foreign matter that has not been properly removed between passes. The number of **slag inclusions** in the completed weld can be reduced by cleaning thoroughly between beads and passes. Do not just clean the base metal of all rust, scale, paint, oil, or other foreign matter before welding. Use the correct amount of arc heat; make sure you are not welding with too low an amperage.

TUNGSTEN INCLUSION

These discontinuities are *found only* in the **Gas Tungsten Arc process (GTAW)**. Particles of the **tungsten electrode** can be deposited in the weld area. They leave unwanted hard spots that can cause a weld to fail. Tungsten is the hardest metal known and has a high melting point. It does not fuse into the weld metal.

Too much arc current will cause the tungsten to overheat. Tiny particles can melt and fall into the puddle, where they remain as the weld solidifies. Touch starting the electrode or dipping it into a molten pool will also cause the problem. Prevent tungsten contamination by use of the correct amperage for the tungsten used. Do not allow the tungsten to touch the molten pool at any time.

INCOMPLETE FUSION

Incomplete fusion or lack of fusion is due to the failure to melt and join the base metal and the weld metal. Failure to properly join a bead to one previously deposited can also cause lack of fusion. (See Figures 5G-3 and 5G-4.)

Normally, incomplete fusion is caused because of too little heat. It is necessary to raise the temperature of the metal to its melting point. The presence of foreign material can sometimes cause incomplete fusion. High travel speed can also cause lack of fusion. Use of the proper cleaning, the correct amperage, and the proper rate of travel can eliminate this problem. (See Figure 5G-5.)

INADEQUATE PENETRATION

Inadequate penetration, or lack of penetration, is the failure to fuse the joint at the weld root. One way to check for good penetration is to see that a specified amount of weld metal protrudes out the other side of the joint. (See Figure 5G-6.)

Failure to penetrate properly can be caused by: poor joint design, wrong choice of electrode, wrong size electrode for the root opening, too small a root opening, root face too large, travel speed too fast, amperage too low, electrode angle or manipulation faulty.

A good welder makes sure the joint is prepared correctly and uses welding conditions that give complete penetration.

ARC STRIKES

Arc strikes are small weld spots, usually outside the weld area, where the welder has accidently struck an arc. They can occur wherever the electrode touches the base metal or where a work cable has been improperly connected to the workpiece. The melting usually takes place in a very small area. The mass of metal surrounding the melted spot tends to cool it rapidly. The rapid cooling creates a hard spot which can cause cracking in that area. The spot can also create **undercut,** which reduces the thickness of the metal at that point. Avoid arc strikes; they can cause weld failure and are very expensive to repair. They should be avoided at all times.

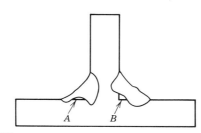

Figure 5G-3 Incomplete fusion in fillet welding "B" is often called bridging.

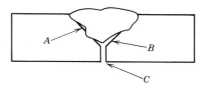

Figure 5G-4 Incomplete fusion in a groove weld.

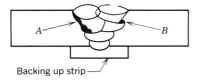

Figure 5G-2 Slag inclusions between passes at *A* and undercut at *B*.

Figure 5G-5 Incomplete fusion from oxide or dross at the center of a joint especially on aluminum.

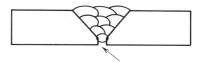

Figure 5G-6 Inadequate penetration.

Only strike the arc in places that will be rewelded. The new weld will cover the arc spot.

UNDERCUT

A groove melted into the base metal next to a weld is called undercut. (See Figure 5G-7 and 5G-8.)

Undercut can occur during the finish pass or during one of the fill passes in a multipass joint. Undercut of a joint sidewall creates slag cleaning problems because the slag can be trapped in the undercut. It is difficult for the welder to completely "burn out" an undercut area and obtain a good weld. Trapped slag is very difficult to remove and, if left in, can cause slag inclusion.

To reduce the possibility of getting undercut, hold a medium to short arc. Do not use too high an amperage and pause briefly at the edges of the weld to allow the metal to fill in the low spots. Also, whip the electrode toward the center of the weld if necessary.

OVERLAP

Overlap is a condition where the weld metal spills over the edge of the weld bead without being fused to the base metal at the point. (See Figure 5G-9.)

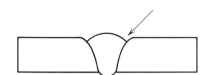

Figure 5G-7 Undercut at the toes of a groove weld.

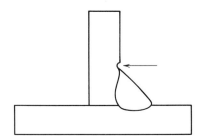

Figure 5G-8 Undercut in a fillet weld.

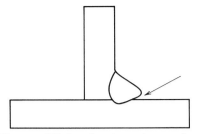

Figure 5G-9 Overlap—An overflow of weld beyond the end of fusion.

Overlap, because of its nature, is a **notch.** It can concentrate stress and reduce the effective size of a fillet weld. Its effect is similar to undercut. (A notch is a place where the section of a weld has been reduced. This place can become a starting point for a crack. Once started, a crack can work its way through a weld, causing it to fail.)

Overlap is caused by excessive heat from using too much current, slow travel, improper electrode angles, or manipulation on the part of the welder.

WELD PROFILE

The term *weld profile* refers to the shape of a weld when viewed from one end. Figure 5G-10 shows the desired profile for both a groove and a fillet joint. Figure 5G-11 shows profiles that are not acceptable.

It is important to maintain the proper profile throughout the weld if it is to be acceptable. For example, a weld with a throat that is too small may fail in service. The correct throat size is necessary if weld strength is to be maintained.

If the welder uses the proper welding procedures and techniques, there should be no problem in producing acceptable welds.

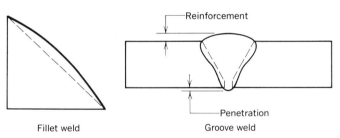

Figure 5G-10

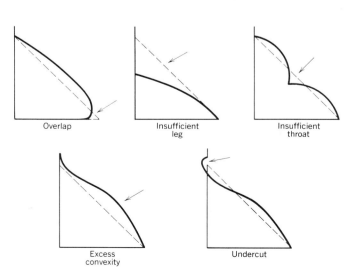

Figure 5G-11

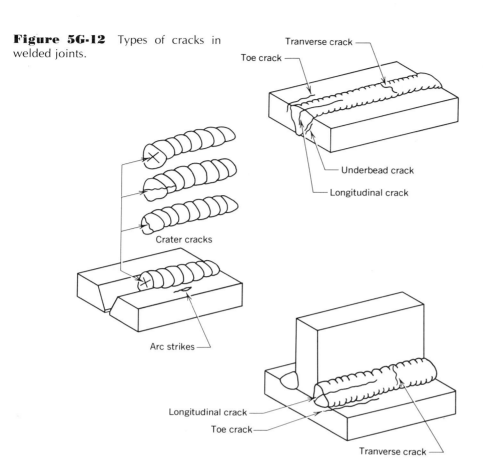

Figure 5G-12 Types of cracks in welded joints.

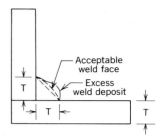

Figure 5G-13 Overwelded fillet.

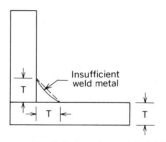

Figure 5G-14 Underwelded fillet.

CRACKS

There are two types of cracking; they are hot cracking and cold cracking. (See Figure 5G-12.) Hot cracking depends on the composition of the metal. There is little the welder can do except to follow the prescribed procedure. This includes preheating the workpiece, and keeping the preheat and interpass temperatures at the lower end of the operating range. (Each electrode has a recommended amperage range. The welder should use the lower portion of the range.)

Cold cracking occurs in steels that have low ductility or hardenability and have hydrogen present. To prevent cold cracking use low hydrogen electrodes. Also use the correct preheat procedure.

DIMENSIONAL DISCREPANCIES

Another type of weld defect is known as a **dimensional discrepancy**. Included in this category are improper weld shape, lack of penetration, failure to fill the joint, and any change in the size of the weld joint.

OVERWELDING

Overwelding means depositing more filler metal than specified. It adds nothing of value to the weld. (See Figure 5G-13.) Overwelding not only wastes time and material, but creates great stress at the toes of the weld and in the adjacent metal. Overwelding should be avoided at all times.

UNDERWELDING

Underwelding is the opposite of overwelding. (See Figure 5G-14.) The welder deposits less filler metal than specified, which produces a weld without the required design strength. Underwelding is not acceptable.

MISMATCH

Mismatch is a term used to describe the offset of two pieces. For example, between two sections of a butt joint. In Figure 5G-15 the stresses of the welded condition are very high. They can be reduced by grinding provided the design section thickness is not reduced. A far better solution, if there is no way to eliminate the mismatch, is to grind the joint before welding.

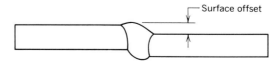

Figure 5G-15 Mismatch.

CHAPTER 6
SHIELDED METAL ARC WELDING ELECTRODES

Every welder should have some knowledge of electrode coatings, as well as electrode selection and electrode classification.

Four electrode groups are described in this chapter. The American Welding Society Classification system for covered electrodes, in the carbon and low-alloy steel categories, is also described. This information will be invaluable to you.

Learn to use the classification system. It will help you select the correct electrode for the job at hand. When it is important to do a job quickly, and maintain quality, proper electrodes are essential.

LESSON 6A
ELECTRODE COATING FUNDAMENTALS

OBJECTIVES
Upon completion of this lesson you should be able to:
1. Explain the functions of the electrode coating.
2. Describe the proper care of electrodes.

PURPOSE OF ELECTRODE COATING
Before good coatings were developed it was difficult to constantly produce acceptable welds. Many of those earlier good welds might not meet today's high standards.

Now, the chemistry of coatings is complex. Good chemistry gives us the ability to produce high quality welds.

Some coating functions follow:

1. Protect the molten metal and shield the arc from the atmosphere. Oxygen and nitrogen, present in the atmosphere, can cause weld porosity.
2. Stabilize the arc and help direct the arc stream.
3. Reduce impurities.
4. Control bead shape.
5. Refine the weld metal.
6. Increase the deposition rate.
7. Control the strength and composition of the weld metal.
8. Make it possible to use AC current.

CARE OF ELECTRODES
Take care of your electrodes if you want them to work right. Protect the coating from damage. Store electrodes in a secure, dry area. Don't allow excessive temperature and humidity fluctuations. Store electrodes as recommended by the manufacturer. Electrode storage ovens are available, and many have controls which can recondition electrodes that have absorbed moisture. Ovens similiar to the one in Figure 6A-1 can be used for storage and reconditioning.

Electrodes are reconditioned by baking. Baking at elevated temperatures removes the moisture, and it restores the electrodes to normal. They become usable again. The chart in Figure 6A-2 provides information to store and recondition some typical electrodes.

Figure 6A-1 Type 300 dryrod electrode oven. (Courtesy Phoenix Products Company, Inc.)

ELECTRODE COATING FUNDAMENTALS

Figure 6A-2 Revised guide to flux and electrode stabilization. (Temperatures shown are not *guaranteed* to be correct or safe.) (Courtesy Phoenix Products Company, Inc.)

Type (AWS)	Air Conditioned Storage Before Opening RH = Relative Humidity	Dry Rod Oven Holding After Opening	After Exposure to Moisture a Sufficient Time to Affect Weld Quality	
			Recondition Step #1	Rebake Step #2
Standard EXX10 EXX11 EXX12 EXX13	Keep Dry @ Room Temp. 40°–120° F 60% (±10%) RH	100° F (±25°)	Not Required NEVER STORE ABOVE 130° OR BELOW 50% RH	Not Required
EXX20 EXX30 **Iron Powder** EXX14 EXX24 EXX27	90° F (±20°) 50% Max. RH	150°–200° F	250°-300° F ONE HOUR ───────── TWO HOUR TOTAL	350° F (±25°) ONE HOUR
Iron Powder- Lo-Hydrogen EXX18 EXX28 **Lo-Hydrogen** EXX15 EXX16	90° F (±20°) 50% Max. RH	300° F (±50°)	500°–600° F ONE HOUR ───────── ONE & ONE-HALF HOUR TOTAL	700° F (±50°) ONE-HALF HOUR
Lo-Hydrogen- Hi-Tensile EXXX15 EXXX16 EXXX18	90° F (±20°) 50% Max. RH	250°–450° F	500°–600° F ONE HOUR ───────── ONE & ONE-HALF HOUR TOTAL	650° F (±50°) ONE-HALF HOUR
Stainless Inconel[a] Monel[a] Nickel Hard Surfacing	Keep Dry @ Room Temp. 40°–120° F 60% (±10%) RH	180° F (±25°)	350° F (±25°) ONE HOUR ───────── TWO HOUR TOTAL	500°–600° F ONE HOUR
Brasses Bronzes Special Alloys	Keep Dry @ Room Temp. 40°–120° F 60% (±10%) RH	150°–200°F	Not Required	Not Required
Granulated or Agglomerated Flux	Keep Dry @ Room Temp. 40°–120° F 60% (±10%) RH	250° F (±50°)	Not Required	700° F (±100°) TWO HOURS
Flux-Core Coils	Keep Dry @ Room Temp. 40°–120° F 60% (±10%) RH	250° F (±50°)	400° F (±50°) ONE HOUR	Not Required

Note: Some HTS, Stainless electrode groups, and 15 & 16 type coatings may require higher or lower temperatures for rebaking than those shown. These can be determined by special request to the particular manufacturer involved.

[a]Registered Trade Marks of International Nickel Co., Inc.

Figure 6A-3 illustrates still another type of electrode storage container. This type of container is used at the welding site. Oven protection is mandatory with electrodes such as low hydrogen, iron powder, stainless steel. and certain other alloys.

Figure 6A-3 Welder using electrodes directly from a portable electrode oven. *Note:* Leather-gauntlet type gloves would be a better choice, especially for out-of-position welding. (Photo courtesy Phoenix Products Company, Inc.)

LESSON 6B
ELECTRODE SELECTION

OBJECTIVES
Upon completion of this lesson you should be able to:
1. Explain the importance of correct group selection.
2. List the four factors involved in selecting the correct electrode group.
3. List the AWS specifications governing mild carbon steel and **low-alloy** steel covered electrodes.
4. List the characteristics of each electrode group.

THE IMPORTANCE OF ELECTRODE GROUP SELECTION
What electrode should be used for the job assigned to you? Someone must select the electrode to use. The selection is important to both you and your employer. The wrong choice will make the job harder; it will increase welding time, as well as cost.

Your employer must run a profitable business. Otherwise it won't be long before you are looking for another job. Even if excessive cost is not a problem, the wrong electrode will make welding very frustrating.

FACTORS INVOLVED IN SELECTING AN ELECTRODE GROUP
The shielded metal arc welding electrode not only adds filler metal to the weld, it also helps control weld shape. With the correct electrode and welding technique you have complete control over the bead shape.

The core wires of low alloy steel electrodes are similar. Alloying elements are added by the electrode coating. Other coating materials provide for ease of slag removal, reduced porosity and spatter, easier arc starting control of penetration, and ease of operation.

The materials in the coating can be used to classify electrodes into four groups, according to their intended use. Take these groups into consideration when you wish to select the best electrode for the job.

Other selection factors include:

1. Position of the weld
2. Quality requirements

3. Rate of deposition

The position of the weldment has some influence on the deposition rate. Out-of-position weld joints may prevent you from using high deposition rate electrodes.

GROUP CHARACTERISTICS

American Welding Society specification A 5.1 Specifications for Carbon Steel Covered Arc Welding Electrodes governs the mild carbon steel electrodes. Specification A 5.5 Low-Alloy Covered Arc Welding Electrodes governs low alloy steel covered electrodes.

F numbers are assigned to groupings of electrodes based on their usability characteristics. These characteristics determine the ability of welders to make satisfactory welds with a given filler metal.

The characteristics of the electrodes vary from one group to another. The electrodes are placed in groups according to their ability to: deposit filler metal, freeze or solidify rapidly, penetrate moderately or deeply, and weld hard to weld steels. Each group has its own particular characteristics, although the electrodes from more than a single group can often be used for a given weldment

Group F-1

Fast fill, high deposition electrodes: E-6020, E-6027, E-7024, and E-7028.

These electrodes have a high deposition rate. They have a lot of iron powder, or iron oxide, in their coatings, in some cases as much as 50 percent.

The iron powder melts as the electrode is consumed. Most of it becomes part of the weld puddle. You can deposit up to 50 percent more weld metal with these electrodes.

Electrodes in this group do not provide a forceful arc. Because of their high deposition rate they can *only* be used for welds in the *flat* (downhand) position, or for horizontal fillets. (See Figure 6B-1.)

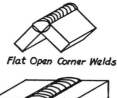

Figure 6B-1 Group F-1 fast fill electrodes. (Courtesy Lincoln Electric Company)

The characteristics of electrodes in this group are as follows:

1. The weld puddle is very fluid. It solidifies slowly.
2. The slag is heavy and solidifies before the weld metal. Because of this, the weld metal takes on the shape of the solidified slag. It acts similar to lead or wax poured into a mold.
3. Slag is easily removed.
4. The arc is soft with little spatter.
5. Penetration is shallow to moderate.
6. Bead appearance is excellent, smooth, and free of large ripples.
7. Possible to use on **crack-sensitive** metals, when low hydrogen (Group F-4) electrodes are not available.
8. X-ray quality welds are possible.

Group F-2

Fast Follow Mild Penetration Electrodes: E-6012, E-6013, and E-7014.

The deposition rate of these electrodes is somewhat less than those in Group F-1. The puddle is less fluid than those in Group F-1. The puddle tends to "freeze" rapidly. They are an excellent choice for out-of-position welding. (See Figure 6B-2.) The characteristics of the electrodes in this group are as follows:

1. The weld pool is slightly fluid. It solidifies more rapidly than Group F-1 electrodes.
2. The arc is slightly stiffer than with Group F-1. It is fairly quiet with minimal spatter.
3. Penetration is shallow to moderate.
4. Weld quality is good.
5. It is possible to use the "Drag" technique to good advantage.
6. Excellent for use on sheet metal, structural steel, and general fabrication.
7. The slag is easily removed from a flat surface, but is sometimes difficult in fillet welds.

Figure 6B-2 Group F-2 fast follow electrodes. (Courtesy Lincoln Electric Company)

Group F-3

Fast Freeze, Deep Penetration Electrodes: E-6010, E-6011.

These electrodes give a deep digging arc. This group is good for butt welds in plate, pipe, and vessels. It gives excellent melt through on root passes.

The puddle is deep and highly agitated. The quick-freezing puddle is easy to control out of position. (See Figure 6B-3.) The electrodes are good for downhill welding. In widespread use for out-of-position butt joints on pipe.

The characteristics of these electrodes are:

1. Harsh digging arc.
2. Deep penetration. Excellent melt through on root passes.
3. Fast freezing weld pool.
4. Fairly rough weld deposit. (Because of fast freeze show every movement the electrode made.)
5. Excellent mechanical properties. (X-ray quality possible.)
6. High level of spatter.
7. Requires a high degree of operator skill.
8. Excellent for all types of open root joints.

Group F-4

Hard to Weld Steels or Low Hydrogen Electrodes, E-6015, E-6016 and E-7018.

Low hydrogen electrodes are known for their ability to weld the "hard to weld" steels. These include free-machining, high sulphur bearing steels, high carbon and low alloy abrasion resistant steels, and armor plate (See Figure 6B-4.)

These electrodes are less crack sensitive than the Group F-3 electrodes. In some instances, such as welding pipe at extremely low temperatures, F-4 electrodes are better.

These electrodes are the obvious choice when the best mechanical properties and X-ray quality is desired.

The characteristics of the electrodes in this group are as follows:

1. Always store at the temperature recommended by the manufacturer. This reduces the possibility of moisture pick-up by the coating.
2. Medium or moderate arc force.
3. Penetration is medium to moderate. It is difficult to obtain excellent melt through on *open root* passes.
4. Moderate deposition rate.
5. Less spatter than the Group F-3 electrodes.
6. Weld metal has superior mechanical properties.
7. Can be used to weld steels not readily welded by the other groups.

The groups are listed in the order of difficulty of use. It follows then that those in Group F-1 are less difficult to weld with than those in Group F-4.

When a welder takes a certification test he or she is qualified to weld the joint configuration used in the test, in the position the test was welded, in using the same electrodes used in the test or electrodes of a lower group number. Example: If the electrodes used in the test were from Group F-4, the welder would be qualified to use all of the electrodes in the remaining groups.

Overhead Butt and Fillet Welds

Vertical Butt and Fillet Welds

Sheet metal Edge, corner and Butt welds

Figure 6B-3 Group F-3 fast freeze electrodes. (Courtesy Lincoln Electric Company)

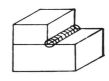

Welds in all positions

Figure 6B-4 Group F-14 electrodes for hard-to-weld steels. (Courtesy Lincoln Electric Company)

LESSON 6C
ELECTRODE CLASSIFICATION

OBJECTIVES
Upon completion of this lesson you should be able to:
1. Explain the reason for classifying electrodes.
2. Name two organizations that set up specifications for covered electrodes.
3. List the mechanical property tests used to determine covered steel electrode specifications.
4. Pass a test demonstrating your ability to use a classification chart.

PURPOSE OF CLASSIFYING ELECTRODES
Many companies manufacture similar electrodes under a variety of trade names. Their operating characteristics may differ slightly, but their mechanical properties should be similar. They are designed to meet published code requirements. Codes, or specifications, are published by the American Welding Society (AWS), the American Bureau of Shipping (ABS) the United States Coast Guard (USCG), the American Society of Mechanical Engineers (ASME), the Federal Bureau of Roads, Military Specifications (Mil Specs), and the American Petroleum Institute (API) and other agencies.

These codes tend to agree or conform with the AWS code. They insure the product will produce weld deposits in which the mechanical properties equal or exceed the base metal properties.

The American Welding Society has set up an Electrode Classification system. It sets standards that must be met when an electrode is to be used on a weldment that must conform to special requirements.

TESTS USED TO DETERMINE MECHANICAL PROPERTIES
The standards mentioned above deal with the minimum mechanical and radiographic (X-ray) requirements. They described the electrode characteristics with regard to:

Tensile strength
Yield point
Elongation
Brittleness

Tests have been devised to measure these characteristics. A description of these terms follows:

Tensile strength. Pressure needed to pull a metal specimen to the breaking point. Typically measured in pounds per square inch.

Yield point. The point reached, during the pulling of a specimen, when the metal no longer resists. It will continue to stretch with no increase in the pulling force. If you have ever pulled a piece of taffy apart, you may remember how it "necked-down". When the taffy reached a certain point, you no longer had to pull harder. The taffy continued to stretch until finally it broke; that was the yield point of the taffy.

Elongation. The distance the specimen stretches is elongation. Before the test, two center punch marks are placed on the specimen. The distance between the marks is recorded before the specimen is pulled apart. Afterward the distance is measured again. The percent of stretch, or elongation, is calculated from the two measurements.

Brittleness. The charpy V-notch is one method of determining the brittleness of a specimen. It is a measure of the impact resistance of the material. This is determined by one of two methods. In each of the methods a notch is cut in the specimen. Then sudden force is applied, which causes the specimen to break at the notch. The energy used to make the break is measured. It is used to determine brittleness.

THE AWS COVERED ELECTRODE CLASSIFICATION SYSTEM
Shielded metal arc welding electrodes for steel are governed by two separate American Welding Society specifications. Specification AWS A5.1 governs mild steel covered arc welding electrodes. Specification AWS A5.5 governs low-alloy steel covered arc welding electrodes.

Classification numbers consist of the four parts shown in Figure 6C-1. As you can see, the classification number contains a great deal of information. All mild steel and low-alloy electrodes have the prefix "E" which means Electric arc welding electrode.

The first two digits in a four-digit number, or the first three digits in a five-digit number, indicate the minimum tensile strength of the filler metal in thousands of pounds per square inch.

In the example, 60 indicates the tensile strength is 60,000 pounds per

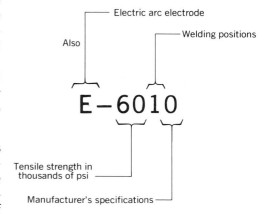

Figure 6C-1 Meaning of each digit or group of digits.

square inch (60,000 psi.) The 60 multiplied by 1,000 equals 60,000.

EXAMPLE
E-6010 → 60 × 1000 = 60,000 psi
E-10018 → 100 × 1000 = 100,000 psi

The second digit from the *right* indicates the position of welding use. The digit one, in E-60*1*0 for example, means the electrode can be used in *all positions*. All position welding means flat, horizontal, vertical, and overhead.

The welding position digits are defined as follows:

Second Digit from the Right	Welding Positions
E-XX*1*X	All positions; flat, horizontal, vertical, and overhead
E-XX*2*X	Flat position and horizontal fillets
E-XX*3*X	Flat only
E-XX*22*	Single-pass horizontal fillets and flat only
E-XX*18-1*	Like E-7018 except Charpy V notch @ −50 degrees F
E-XX*48*	Like E-7018 except also suitable for 3G vertical position (Downhill method)

The third and fourth digits taken together indicate the manufacturer's specifications. It includes the type of coating, recommended welding current, type arc, penetration, and weld quality. Figure 6C-2 shows the operating characteristics for electrodes in all four groups.

Figure 6C-2 Operating characteristics of mild steel and low-alloy steel electrodes.

Group Classification	Type of Covering	Current	Penetration	Welding Positions	Iron Powder Content
F-3 E-XX10	Cellulose Sodium	DCRP (electrode positive)	Deep	All	0–10%
F-3 E-XX11	Cellulose Potassium	AC or DCRP (electrode positive)	Deep	All	0
F-2 E-XX12	Titania Sodium	AC or DCSP (electrode negative)	Medium	All	0–10%
F-2 E-XX13	Titania Potassium	AC or DCSP or DCRP (electrode negative)	Shallow	All	0–10%
F-2 E-XX14	Iron powder Titania	DCRP or DCSP or AC	Medium	All	25–40%
F-4 E-XX15	Low-hydrogen Sodium	DCRP (electrode positive)	Medium	All	0
F-4 E-XX16	Low-hydrogen Potassium	AC or DCRP (electrode positive)	Medium	All	0
F-4 E-XX18	Iron powder Low-hydrogen	AC or DCRP (electrode positive)	Medium	All	25–45%
F-1 E-XX20	High iron oxide	DCSP (electrode negative) or AC for horizontal fillets and DC either polarity or AC for flat position welding	Medium	H-fillets and flat	0
F-1 E-XX24	Iron powder Titania	AC or DC either polarity	Shallow	H-fillets and flat	50%
F-1 E-XX27	Iron powder Iron oxide	DCSP (electrode negative) or AC for horizontal fillets and DC either polarity or AC for flat position	Medium	H-fillets and flat	50%
F-1 E-XX28	Iron powder Low-hydrogen	AC or DCRP (electrode positive)	Shallow	H-fillets and flat	50%

Note: In the figure above, the X's are used so only the last two digits are present to make the chart easier to read.

LESSON 6D
STAINLESS STEEL ELECTRODES

OBJECTIVES
Upon completion of this lesson you should be able to:
1. Explain the relationship between the American Iron and Steel Institute classification system and that of the American Welding Society.
2. Explain the reason for the corrosion resistance of the stainless steels.
3. Pass a test on the classification of electrodes for the welding of stainless steel.
4. Explain the purpose of the suffixes in stainless steel electrode classification.
5. List at least four methods of minimizing the problems of carbide precipitation and distortion.

The previous lessons explained the classification of mild steel and low alloy steel covered electrodes. This lesson discusses stainless steel electrodes.

These electrodes are used to weld steels with good corrosion resistance. Many chemicals, liquids, and gases will attack ordinary steel. The corrosion resistance of stainless steel is provided by addition of chromium. Up to 30 percent (30%) chromium is added to these electrodes.

The chromium forms an oxide coating on the surface of the metal. This coating resists further oxidation, rust, or corrosion. Because of this ability to resist corrosion these steels are in great demand. Industries such as the petrochemical, pharmaceutical, and food industry, use stainless steel.

Stainless steels are classified by the American Iron and Steel Institute (AISI) and the American Welding Society (AWS).

THE STAINLESS STEELS
The American Iron and Steel Institute (AISI) classifies stainless steels in five groups.

These are:
1. Series 200, Chromium Nickel Magnesium-Austenitic (nonhardenable)
2. Series 300, Chromium Nickel Austenitic (nonhardenable except by cold working)
3. Series 400, Chromium-Martensitic (hardenable)
4. Series 400, Chromium-Ferritic (nonhardenable)
5. Series 500, Chromium-Molybdenum-Martensetic

The series indicates the major elements in the electrode core wire. Note that only the 400 series, Chromium-Martensetic, can be hardened. This means the finished weldment can be heat treated to increase the hardness.

The martensetic and ferritic steels (Series 400 and 500) are magnetic. The austenitic steels (Series 200 and 300) are not magnetic when properly annealed. This means they have been heated and cooled under controlled conditions.

CLASSIFICATION OF STAINLESS STEEL ELECTRODES
The American Welding Society specification A 5.4 describes corrosion-resisting chromium and chromium-nickel steel covered welding electrodes.

The American Welding Society classification of stainless steel electrodes closely follows the AISI. The AWS system places S and E before the number. The E indicates an electric arc welding electrode. The AWS system also adds a suffix following the number. The suffix will always be a "15" or "16." The 15 indicates the electrode is lime coated. It may be used with direct current reverse polarity.

The 16 means the electrode is titania coated. It can be used with either alternating current or direct current reverse polarity.

On out-of-position welding, lime-coated electrodes operate better than titania-coated electrodes. In addition, they are less crack sensitive. The titania coated electrodes (suffix 16) produce a smoother bead in the flat or downhand position. They are slightly more difficult to weld with, in out-of-position work.

In Figure 6D-1 classification numbers also have a letter suffix in ad-

Figure 6D-1 Stainless steel filler metal alloys per AWS A5.4. (Reprinted by permission of Prentice-Hall, Inc., Englewood Cliffs, N.J.

AWS Class	Typical Composition %						
	C	Cr	Ni	Mo	Mn	Si	Others
E308	0.08	19.5	10.5	—	2.5	0.90	—
E308L	0.04	19.5	10.5	—	2.5	0.90	—
E309	0.15	23.5	13.5	—	2.5	0.90	—
E309Cb	0.12	23.5	13.5	—	2.5	0.90	Cb + Ti —0.85
E309Mo	0.12	23.5	13.5	2.5	2.5	0.90	—
E310	0.20	26.5	21.5	—	2.5	0.75	—
E310Cb	0.12	26.5	21.5	—	2.5	0.75	Cb + Ti —0.85
E310Mo	0.12	26.5	21.5	2.5	2.5	0.75	—
E312	0.15	30.0	9.0	—	2.5	0.90	—
E316	0.08	18.5	12.5	2.5	2.5	0.90	—
E316L	0.04	18.5	12.5	2.5	2.5	0.90	—
E317	0.08	19.5	13.0	3.5	2.5	0.90	—
E318	0.08	18.5	12.5	2.5	2.5	0.90	—
E320	0.07	20.0	34.0	2.5	2.5	0.60	—
E330	0.25	15.5	35.0	—	2.5	0.90	—
E347	0.08	19.5	10.0	—	2.5	0.90	—
E410	0.12	12.5	0.60	—	1.0	0.90	—
E430	0.10	16.5	0.60	—	1.0	0.90	—

Note: Remainder is iron.

dition to the 15 and 16. These letters indicate the presence of alloying elements.

EXAMPLE

Suffix	Meaning
L	Low carbon
ELC	Extra low carbon
Cb	Columbium
Mo	Molybdenum

In Figure 6D-2 the following also holds true:

E	Electric arc welding electrode
First three digits	American Welding Society designation for the filler metal
Fourth digit and Fifth digit	Welding position and Type of coating and current

WELDING STAINLESS STEEL —DIFFERENT NOT DIFFICULT

Some welders consider stainless steel difficult to weld, but this is not so. Any welder who can weld in all positions with low hydrogen electrodes will have little, if any, difficulty with stainless steel electrodes.

This is not to say there are no differences. However, they are easily mastered. Some differences between stainless and mild steel electrodes are.

1. Less amperage required than with same diameter mild steel electrodes.
2. Higher electrical resistance.
3. Higher thermal expansion (stainless steel expands more than the carbon steels).
4. Lower thermal conductivity (the metal holds the heat longer than the carbon steels).
5. Lower melting point.

When welding stainless steels, these differences must be taken into consideration. The lower melting temperature means you have to choose electrode sizes carefully, especially when welding open root butt

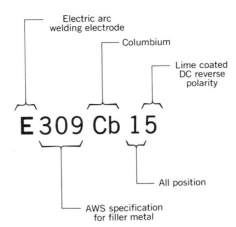

Figure 6D-2

joints. The puddle is slightly more fluid than with low hydrogen electrodes. Also, the root face has a tendency to melt away rapidly. Sometimes you may have to use a 3/32 in. (2.4 mm) dia. electrode in place of a 1/8 in. (3.2 mm) electrode, to obtain the desired root pass quality.

Excessive heat causes distortion. Heat from welding causes stainless steels to expand. Cooling causes shrinkage in the weld and the immediate area. Because of their high thermal expansion, as little heat as possible should be put into the welding zone of stainless steel.

SUGGESTIONS FOR WELDING STAINLESS STEELS

Heat input can be kept to a minimum by using lower amperage than with carbon steels. Another problem caused by excessive heat input is carbide precipitation.

The austenitic chrome-nickel steels contain small amounts of carbon. During the heating cycle undesirable chromium carbide tends to form. In order to minimize this possibility use the following preventative measures.

1. Use the lowest amperage possible.
2. Use an electrode designated as ELC, for extra low carbon.
3. Use an electrode containing Columbium (Cb).
4. Utilize **back-step** welding, sometimes called the cascade or skip method of welding. Back-step welding means to weld a small section at what would normally be the end of a joint. Then weld another section the same length. The second weld should butt the first. Repeat as indicated in Figure 6D-3.
5. Keep the width of the weld bead from 2 to 2½ times the diameter of the core wire. Move rapidly.

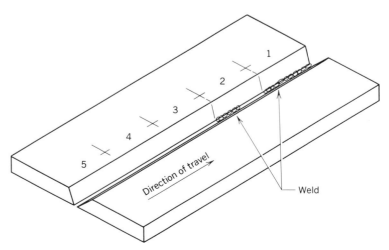

Figure 6D-3

CHAPTER 7
PROCEDURES AND PREPARATION FOR PLATE WELDING

You should be aware of the procedures and preparations for the type of joint you are to make before you start to weld.

It is important that you understand the proper procedures, especially those for preparing, fitting, and tacking the plates before welding. In addition, you should be aware of the economic factors involved in the welding of joints.

There are many types of joints and they may be welded in many positions. Also, it is important that you prepare the plates properly before welding. Proper preparation is needed to meet the standards of the industry or a particular code.

Follow the prescribed details and learn to apply the proper preparatory procedures. Do not deviate or make any changes. If you follow the procedures, you will have no trouble in producing acceptable welds. This is especially true when you work with similar joint designs.

LESSON 7A
TYPES OF JOINT CONFIGURATIONS

OBJECTIVES
Upon completion of this lesson you should be able to:
1. Explain the term *joint configuration* and describe the five basic weld joints.
2. List the parts of a Tee joint fillet and a groove weld.
3. Explain why plates in excess of ⅝ in. thick (15.9 mm) are welded from both sides.
4. Explain how joint design affects weld economics.
5. Describe effective throat of a fillet weld.
6. Explain distortion.

WELD JOINT TYPES
Configuration is a term used to define the shape of an object. The configuration of a weld joint simply means how the parts are placed, or the shape of the joint.

TYPES OF JOINTS
There are five basic joints for welding plates. They are the butt, lap, T, corner, and edge joints.

All plate or sheet joints may be joined by bead, groove, or fillet, welds. However, flat sheet and plate may also be joined by plug or slot welds. (See Figure 7A-1.) At times a slot weld is used instead of a plug

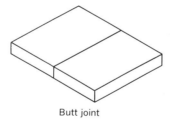
Butt joint

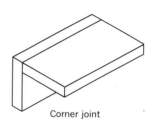
Corner joint

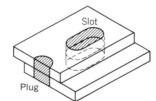
Slot
Plug

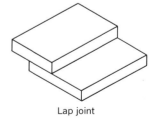
Lap joint

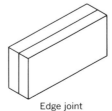
Edge joint

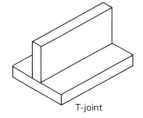
T-joint

Figure 7A-1 Basic weld joints.

Some of these illustrations in this lessons are reproductions of American Welding Society materials. The author gratefully acknowledges the American Welding Society.

weld in order to obtain more holding surface.

In some cases combinations of joint preparations are used on a single weldment (pieces to be joined or welded). For example; the single bevel T joint is a variation of the basic T. (See Figure 7A-2.) The difference is that the vertical plate is beveled. With this type of joint the beveled area is filled with weld metal. Then fillet welds are added to the sides. This type of joint has maximum strength because the entire cross section is weld metal. There are no voids or spaces. Heavy plate may be beveled from both sides. (See Figure 7A-2.)

JOINT DESIGN

Joint design is very important in the fabricating industry, especially where cost is important. Low cost is the main reason that T, lap, and to some extent the square groove butt joints, are popular designs. These joints require the least preparation, and this tends to keep welding costs down.

Some other money and time savers are the single and double "J" and "U" groove joints. (See Figure 7A-3.) These joints are designed so that large diameter electrodes can get deep into the joint. This lets you deposit metal at a high rate of speed and increase your work output.

JOINTS WELDED FROM BOTH SIDES

Notice that some of the joints such as the double V and the double U, are grooved from both sides. (See Figure 7A-3.) This type of design reduces distortion and warping. Experience has shown that plate 5/8 in. (15.9 mm) thick and over, tends to distort toward the welding side. The distortion stresses are equalized and bending is reduced by welding from both sides, and staggering or alternating the sides being welded. Figure 7A-4 shows how the proper **bead sequence** will achieve the best results.

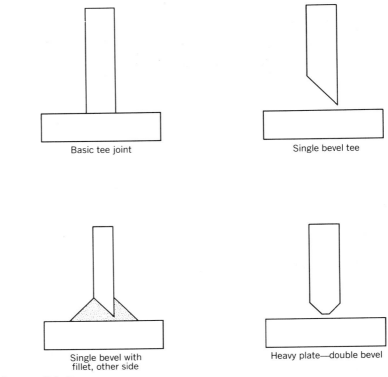

Figure 7A-2

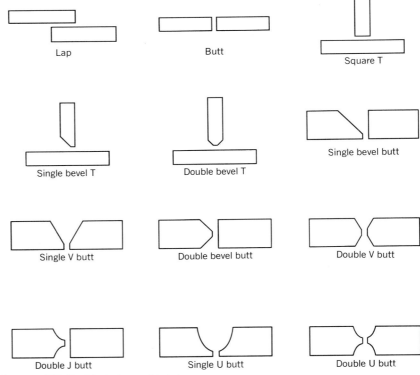

Figure 7A-3 Variations of joints.

TYPES OF JOINT CONFIGURATIONS

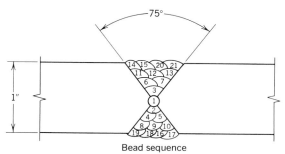

Figure 7A-4 Bead sequence double V-butt.

PARTS OF JOINTS AND WELDS

The T, lap, and V groove butt joints can be used to learn the parts of joints and welds. It is important that you know these parts. When you do, you can understand your instructor throughout your training period.

The lap joint in Figure 7A-5 has a heel, two legs, and two toes. The **fillet weld** used to make this joint has two legs, two toes, and a root at the same point as the heel. The surface of the completed weld is called the face. The weld **throat** is the distance from the heel or root to the surface of the face. Notice that there are two throats indicated. The throat is the distance from the root to the face. The **effective throat** is the distance from the root to a line running at a 45° angle or from toe to toe. It is used in determining the strength of the weld.

The V-groove has a face and a root. (See Figure 7A-6.) The root is somewhat different from the T joint. With the T both parts should be fused to a depth sufficient to join them without leaving any voids at the heel.

The root of a V-groove weld must not only be fused, but the filler metal must penetrate through. It should form a slightly convex (outwardly curved) bead on the side of the joint opposite the face. A V-groove that lacks penetration will fail a test and may also fail in service.

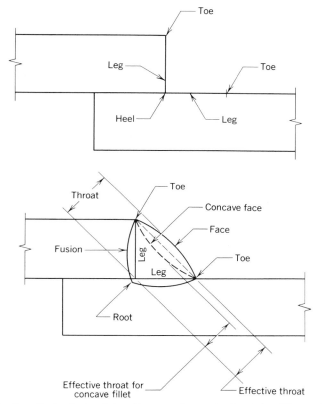

Figure 7A-5 Top: Parts of a lap joint. Bottom: Parts of a lap joint fillet weld.

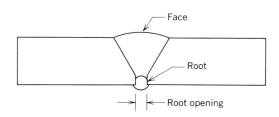

Figure 7A-6

LESSON 7B
WELD POSITIONS FOR TEST PURPOSES

OBJECTIVES
Upon completion of this lesson you should be able to:
1. List the test position using the correct AWS designation.
2. Describe the most commonly used test positions.

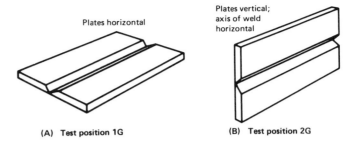

WELD POSITIONS
Employers who manufacture products to meet code requirements must test their welders. They are tested on the joints to be welded, in the same positions as on the product. Various agencies and groups have set standards for these welding positions. Some of them are shown in Figure 7B-1.

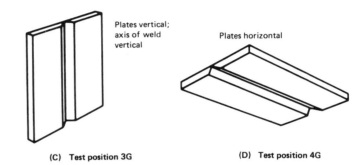

LIMITS FOR TEST PURPOSES
These diagrams show the limits of the weld axis. The axis is the line that passes through the center of the weld, from end to end. The diagrams also show the limit the joint can be rolled or turned around the axis.

For example, in Figure 7B-2 a groove weld in the **vertical position** can be tilted forward. It can move from the perpendicular (straight up), forward 10°, and rotate 360° according to E. But if it is tilted forward from 15 to 80° according to axis limits for D, it can only be rotated from 80 to 280°.

Look closely at the Tabulation of Positions in Figure 7B-2 and check

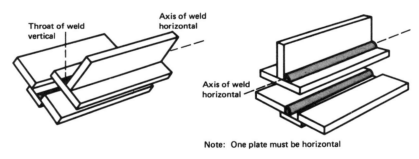

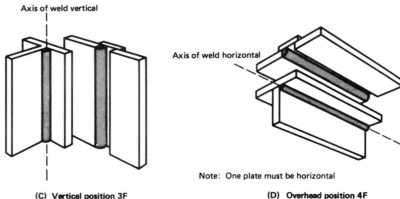

Figure 7B-1 Positions of test plates for groove welds. (a) Test position 1G. (b) Test position 2G. (c) Test position 3G. (d) Test position 4G. (Courtesy American Welding Society) Positions of test plates for fillet welds. (a) Flat position 1F. (b) Horizontal position 2F. (c) Vertical position 3F. (d) Overhead position 4F. (Courtesy American Welding Society)

PRACTICE AND TEST PLATE PROCEDURES

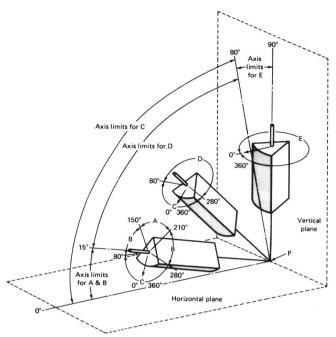

Figure 7B-2 Positions of groove welds. (Courtesy American Welding Society)

Tabulation of positions of groove welds			
Position	Diagram reference	Inclination of axis	Rotation of face
Flat	A	0° to 15°	150° to 210°
Horizontal	B	0° to 15°	80° to 150° 210° to 280°
Overhead	C	0° to 80°	0° to 80° 280° to 360°
Vertical	D	15° to 80°	80° to 280°
	E	80° to 90°	0° to 360°

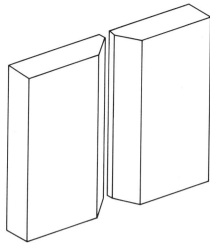

Figure 7B-3 V-groove butt 3G test position.

out Overhead C. Note the weld must be lying face down and between 0 to 80° to be considered **overhead**. Now look at the **axis** indicated by C, on the rotational line around the axis, on the lowest weld on the drawing.

Note the face of the weld can be turned only from 0 to 125° or from 235 to 360°. Basically the face of the weld points 120° on either side of a line perpendicular from the bottom or face of the weld.

One of the most commonly used test positions for plate is shown in Figure 7B-3. When the test is begun, the weldment may not be moved, not even for inspection or cleaning.

Some test positions, such as the 1G and 1F, qualify the welder only for that position. Other tests can qualify the welder for positions requiring less skill and ability. For example, according to D.1.1 of the AWS structural code the 3G position qualifies the welder to weld in the 1G–2G as well as the 3G position. The weld thickness limit plate is described in the various codes. This is done for each type of joint and welding position.

LESSON 7C
PRACTICE AND TEST PLATE PROCEDURES

OBJECTIVES
Upon completion of this lesson you should be able to:
1. Name the major organizations that provide the various welding codes.
2. List the parts of a test piece.
3. List the various names of the root face.
4. Define the word **variable** and explain the reason for maintaining dimensions.

This is an important unit if you wish to become proficient in arc welding.

It is common knowledge that proper joint preparation completes half the job.

There are many variables to be concerned with in plate preparation. A good welder uses these variables as standards. A good welder follows a pattern and does not vary methods.

By doing things the same way each time problems are eliminated. Joints with the same shape and constant dimensions are produced. Because of proper preparation, joints can be welded with a minimum of effort.

WELDING VARIABLES
The variables you will find are:

1. *Bevel angle.* The amount of **bevel**, in degrees, that is required on each plate. (See Figure 7C-1.)
2. *Root face size.* The flat surface on the leading edge of the bevel. (This flat surface is also called a nose, shoulder, chamfer, or land.) (See Figure 7C-1.)
3. *Root opening.* The distance between the two plates at the **root face**. (See Figure 7C-1.)
4. *Preposition amount.* The amount the plates are bent to reduce the effects of heat distortion. (See Figure 7C-2.)

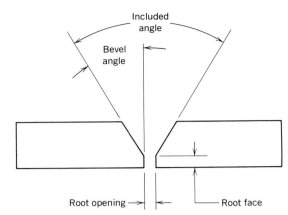

Figure 7C-1

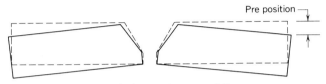

Figure 7C-2 Prepositioning of plates prior to welding.

Follow the procedures faithfully while you work on practice plates. It will develop your test plate preparation ability. If you conform with the respective codes, it will not be difficult.

Before beginning to weld two pieces of metal to be joined, they must be properly prepared. These pieces of metal may be called by names such as: the workpiece, base metal, parent metal, practice piece, the weldment, or test piece.

A student should develop good habits in the preparation of the plates. Such a student can develop the ability to pass the required **welder certification** tests.

CODES AND AGENCIES
Today all critical welding performed is governed by codes. These codes have been developed over the years by such organizations as the American Welding Society (AWS), the American Society of Mechanical Engineers (ASME), and the American Petroleum Institute (API), to name a few. They produce some of the most important and widely used codes.

These organizations set the standards for welding on all types of pressure vessels, bridges, buildings, and oil and gas transmission lines.

Section IX Boiler and Pressure Vessel Code of the American Society of Mechanical Engineers governs the design, fabrication, construction, and inspection, during construction, of boilers and unfired pressure vessels.

AWS D1.1 Structural Welding Code-Steel of the American Welding Society covers welding requirements applicable to any type of welded structure. It may be used in conjunction with any complementary code or specification for the design and construction of steel structures.

The API code of the American Petroleum Institute deals with the petrochemical industry and pressure piping, especially in the area of oil and gas transmission lines.

LESSON 7D
PLATE PREPARATION FOR TEST PURPOSES

OBJECTIVES
Upon completion of this lesson you should be able to:
1. List the steps to properly and safely prepare a set of plates according to Section IV of the ASME code.
2. List the method used to dimension the bevel angle, **root face**, and **root opening**.
3. Explain the main causes of **porosity** and its effects on the completed weld.

PLATE PREPARATION
Proper plate preparation is necessary if the joint is to fit up correctly and conform to code specifications.

The next step is to remove all slag, if any is left by the flame cutting pro-

PLATE PREPARATION FOR TEST PURPOSES

cess. Slag, or oxidized metal, may be removed with a chipping hammer or a hammer and chisel. It can also be removed by sanding or grinding or by a combination of any of these methods. (See Figure 7D-1.)

Mill scale (the hard, smooth material that coats the surface of the metal during the steel-making process) must be removed in the weld area. Scale, rust, paint, or other foreign matter left on the surface of the metal may cause porosity.

Porosity is the name for the tiny bubbles, about the size of a pinhole or slightly larger, in the weld metal. It can cause weld failure. Also, porosity is one reason that welds fail to pass a radiographic (X-ray) examination. (See Figure 7D-2.)

The face of any **flame cut** beveled surface should be sanded or filed to remove any oxide (coating). It is best to use a vise to hold the plates while this is being done.

After the plate faces have been cleaned, the land, nose, root face, and so on, should be filed to the proper dimension or size. (See Figure 7D-3.)

Be sure to run the file along the length of the land, on both the bevel side and the root side, to remove any rough metal.

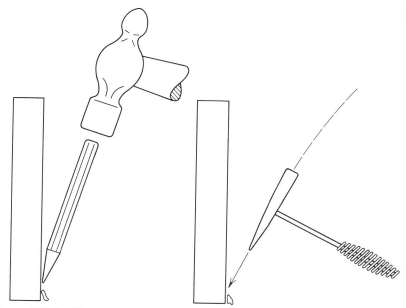

Figure 7D-1 The correct manner of removing slag with a chisel and ballpeen or chipping hammer

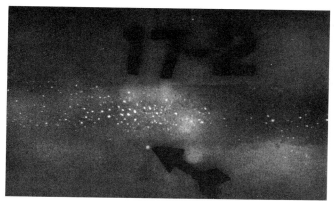

Figure 7D-2 Radiograph (X-ray) photo showing cluster porosity. (Courtesy American Welding Society)

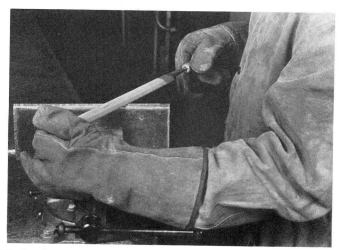

Figure 7D-3 Correct way to insure that you have the required root face

LESSON 7E
PROCEDURES FOR TACK WELDING PRACTICE AND TEST PIECES

OBJECTIVES
Upon completion of this lesson you should be able to:
1. List the proper tacking operations, according to prescribed procedures.
2. List the problems encountered with T- and V- groove joints.
3. Explain the importance of spacers in the tacking procedure.
4. Explain why the dimensions of the **root face** and **root opening** should be equal.

Figure 7E-1 Pad of stringers.

PREPARATORY PROCEDURES
After the plates have been cleaned and prepared, as you learned in Lesson 7D, they must be fit up and tacked before welding.

If you are simply going to weld a pad of stringer beads, you are ready to weld after the cleaning operation. (See Figure 7E-1.) However, if you are going to weld two plates together, to form a T, a lap joint fillet, or possibly a V- groove butt joint, you have more preparatory work.

FILLET WELDS
In a T- joint a plate is placed perpendicular (at a 90° angle) to another plate. Its cross section looks like the letter "T." (See Figure 7E-2.) It is important that the bearing surfaces (the point where the two plates touch) fit well. A poor fit with a too large, or an uneven, opening between the plates will make welding difficult. There should be little or no space between the pieces. Also, the space should be uniform.

Tack weld the vertical, or perpendicular, plate while it is held in place. Tacks, at least 1 in. long, should be placed at each end, on one side of the joint. The side opposite those tacks should be welded first. After the first weld, on the other side, the tacked side may be welded. Tacking on one side and then welding the other side reduces the chance of distortion. (Distortion means the bending out of shape or position of any weldment.)

The lap joint fillet (Figure 7E-3), shows one plate overlapping another. To conserve metal do not overlap the practice plates more than ¾ of an inch. Remember, it is important that the welding surfaces of the plates must be free of rust, scale, paint, and so on.

When both fillets of the lap joint are to be welded, follow the same procedure as used for the T-joint.

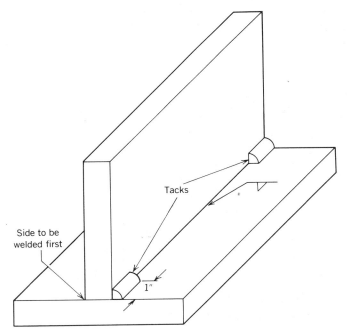

Figure 7E-2 T-joint fillet.

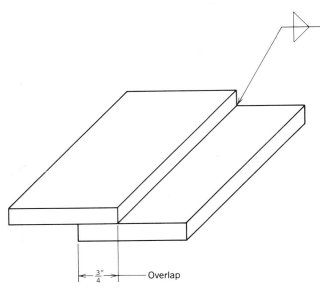

Figure 7E-3 Lap joint fillet.

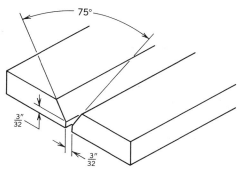

Figure 7E-4 Single bevel and single V-grooves

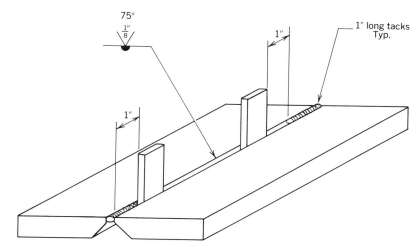

Figure 7E-5 V- groove butt weld with spacers in place

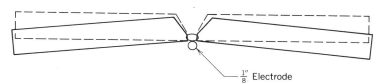

Figure 7E-6 Prepositioning of the plates by bending slightly in the direction of the root

V-GROOVE BUTT JOINTS

The last and most difficult joint to prepare is the V- groove butt joint.

First, the prescribed **bevel angle** must be either flame cut, machined, or ground on the two edges to be joined. The area next to the bevel, on both sides of the plate, must be sanded or ground clean. Next, a root face, shoulder, or land must be sanded or filed on the feather edge of the bevel. This is shown in Figure 7E-4. The proper and most common term used is the **root face**.

When both plates have been properly prepared with equal size root faces, they are ready to be tack welded. First, the plates should be placed face down as in Figure 7E-5. Then spacers of the same thickness as the root face should be placed between the plates 1 in. in from each end. Hold the plates in place and weld 1 in. long tacks at each end. (Weld from the end toward the center.)

After tacking remove the spacers. If they are removed before welding, the heat will pull the plates together unevenly. This causes distortion and difficulty in welding. On the job it causes distortion and poor fit up for the next joint.

The final step before welding can be started is prepositioning of the plate. With a V- groove test piece, a good procedure is to bend the plates slightly in the direction of the root (see Figure 7E-6). The heat of welding will cause the plates to straighten up, and the finished surface will be flat.

To preposition plates, bend the plates until a ⅛ in. (3.2 mm) electrode can be placed between a flat surface and the root of the joint, when the weldment is lying with the root side down. (See Figure 7E-6.) Weld metal shrinks as it solidifies and cools. This causes the metal to distort from its original shape. The distortion is reduced by prepositioning the plate. Prepositioning reduces, or even eliminates distortion because weld metal contraction pulls the plates into position. This is shown by the dashed line in the figure.

LESSON 7F
THE USE OF BACK-UP BARS AND BACKING RINGS

OBJECTIVES
Upon completion of this lesson you should be able to:

1. Describe backing rings.
2. Describe consumable inserts.
3. Explain the differences in the usage of backing rings and consumable inserts.
4. Describe the method used in welding a first pass without removal of and with removal of the "nubs."
5. Describe back-up bars and their use.

BACK-UP BARS AND BACKING RINGS

All welders should be familiar with the use of back-up bars and backing rings. They are an important factor in the proper fit up of plate and pipe.

Both items have wide industrial use for a number of reasons. Back-up bars can simplify plate preparation. Depending on the welding position they allow the use of larger and more economical electrodes. In addition to shielded metal arc welding, back-up bars are also used with the automatic and semiautomatic welding process.

BACK-UP BARS

Back-up bars or **backing strip** are simply bars or strips of metal. Their composition is similar to that of the plate being joined. They are placed across the **root gap** as shown in Figure 7F-1. These bars are tacked in place to help in maintaining the correct root opening. In addition, they hold the plates in position, and in some cases they help contain the weld metal. Occasionally, the back-up bar is removed after the joint is welded, especially on test pieces. Joints prepared with back-up bars let you weld with large diameter electrodes, right from the start. There is no root opening to require extra care in welding. You do not have to use less current and smaller electrodes.

Back-up bars are also known as **backing straps** or backing strips.

BACKING RINGS

A **backing ring** is a narrow strip of metal that has been rolled into a cylinder. Nubs of various lengths, selected according to ring use, stick out from the ring, as shown in Figure 7F-2.

Backing rings are used extensively to weld piping, where restrictions of the fluid flow are an important factor.

Notice that the outside of the ring is machined smooth and flat. It fits snugly to the inside wall of the pipe. The inside of the ring is beveled on the outer edges to provide a smooth

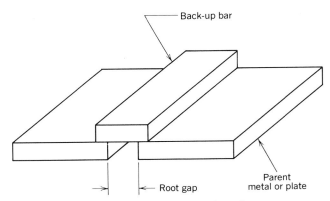

Figure 7F-1 Square groove butt joint with back-up bar

transition for the fluid flowing through the pipe. Backing rings also set the gap for the root pass automatically, eliminate the "icicles" of excessive melt through, reduce fit up time, and help insure correct joint fit up.

The rings have from 3 to 21 nubs, depending on the diameter of the ring. Backing rings are made for use with pipe of from three quarters of an inch, to over 64 in. in diameter. They are available in carbon steel, wrought iron, stainless steel, aluminum, and various alloys. The rings are available in either solid or split, shapes. Solid rings are machined to fit the inside diameter of the pipe. Split rings have an opening to allow for some unevenness or out of roundness of the pipe. Figure 7F-3 shows solid and split rings.

USE OF BACKING RINGS

Backing rings may be used in a number of methods. Some manufacturers of backing rings feel that it is not necessary to tack a joint. They recommend that you skip around the joint when welding. This keeps the pipe from being pulled out of alignment. Good welding practices dictate no joint should ever be welded unless it is tacked in at least four places.

When using rings with long nubs, the nubs are removed by striking with a hammer, but only after the spaces between the nubs have been properly welded. (See Figure 7F-4.) Weld the rings with short punched up nubs in the same manner as with long nubs.

Figure 7F-2 Backing rings with three different nub types. (Courtesy Robvon Backing Ring Co.)

Weld over the short nubs. When welding over nubs, be careful. Never stop welding close to the nub. Also be sure the amperage is high enough to properly fuse the pipe, the backing ring, and the nubs.

When the pipe is ready to receive the backing ring, the axis of the pipe should be in the **horizontal position**. Insert the ring with the split at the bottom of the joint. Do *not* start or stop at the split in the ring when making the root pass. Direct the arc so that it penetrates the inner edges (root faces) of both pieces of pipe and the ring. The molten metal should penetrate approximately one-sixteenth

THE USE OF BACK-UP BARS AND BACKING RINGS

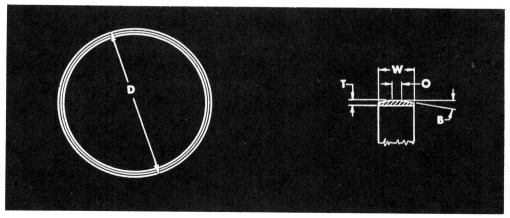

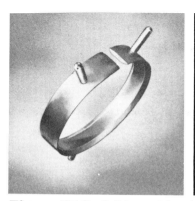
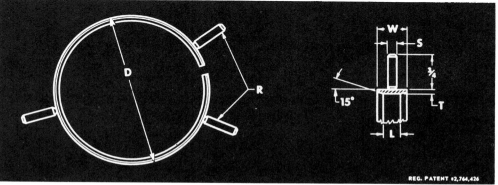

Figure 7F-3 Solid type (above) and split type backing rings (below). (Courtesy Robvon Backing Ring Co.)

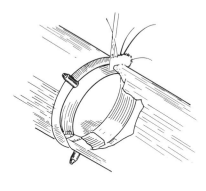

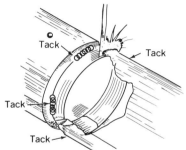

Figure 7F-4 Pipe joint with backing ring tack welded in four places. (Courtesy Robvon Backing Ring Co.)

of an inch into the ring, thereby forming a well-fused joint.

When properly welded, a joint with backing rings can pass radiographic examination.

CONSUMABLE INSERT RINGS

"Consumable insert" ring means just that. The ring is intended to be consumed during gas tungsten arc welding. The ring is placed between the root faces of the pipe. As the welding arc advances the ring becomes the filler for the root metal. The ring acts as filler metal in much the same manner as the edges of the closed corner joint, when you first started your oxyacetylene welding training.

Consumable inserts are frequently used in fossil fuel and nuclear fuel powerhouse welding application. Figure 7F-5 shows typical end preparations for pipe to be welded with consumable inserts.

Figure 7F-6 shows a cut-out section of a joint welded with a con-

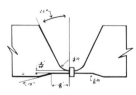
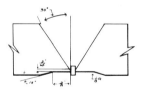

Figure 7F-5 Typical end preparations. (Courtesy Robvon Backing Ring Co.)

sumable insert. Notice the fused insert has become the "melt through" of the first pass.

It must be noted that cleanliness is of great importance when welding with consumable inserts. Care must be taken if consistent quality welds are expected. Follow the procedures suggested by the manufacturer of the rings or the specific welding code you are following.

Figure 7F-6 Cut away view of a consumable insert that has become part of the root bead during the welding process. (Courtesy Robvon Backing Ring Co.)

LESSON 7G
PROCEDURE AND WELDER QUALIFICATION TESTS

OBJECTIVES
Upon completion of this lesson you should be able to:
1. Explain the differences between procedure and welder qualification tests.
2. List the variables involved.
3. Define certification.

WELDER QUALIFICATION TESTS
The main purpose of **welder qualification** tests is to demonstrate the welder's ability. The test welds meet the codes and standards established by the responsible agency. Many codes are in use today. They cover pressure vessels, buildings, bridges, oil and gas transmission lines, and other structures. They are published by the American Welding Society (AWS), The American Society of Mechanical Engineers (ASME), and the American Petroleum Institute (API), to name a few.

Although each code has its own distinct character, they have similar requirements. These requirements must be met; the qualification tests help to ensure this is done.

Very little is left to chance. This is shown by the information in the welder qualification and procedure qualification forms that are illustrated in Figures 7G-1 to 7G-4. These forms spell out the procedures to be followed when welding according to a particular code.

PROCEDURE QUALIFICATION
Procedures set up to produce acceptable welds are described in the **Procedure Qualification**. Follow the procedures to obtain consistently acceptable welds. Follow the procedures on the procedure qualification sheet when you take a welder qualification test. This information may be passed on to you verbally or you may be told to work from the procedure sheet itself.

Compare the Procedure Qualification sheets with the Welder Qualification sheets. You will find the **procedure** sheets contain some information not included in the Welder Qualification sheets. For example, Figure 7G-1 contains the following information not found in the Welder Qualification sheet. (See Figure 7G-2.)

1. The groove design and information regarding bevel angles, root gap, and backing, if any.
2. Preheat and/or postheat treatment.
3. Welding progression or technique.

In summary, all the information the welder needs to complete the test is contained in the procedure qualification sheet.

CERTIFICATION
A welder passes the qualification test and the individual in charge of test-

WELDING PROCEDURE QUALIFICATION TEST RECORD

PROCEDURE SPECIFICATION

Material specification _____
Welding process _____
Manual or machine _____
Position of welding _____
Filler metal specification _____
Filler metal classification _____
Weld metal grade* _____
Shielding gas _____ Flow rate _____
Single or multiple pass _____
Single or multiple arc _____
Welding current _____
Welding progression _____
Preheat temperature _____
Postheat treatment _____
Welder's name _____

*Applicable when filler metal has no AWS classification.

GROOVE WELD TEST RESULTS

Reduced-section tension tests
Tensile strength, psi

1. _____
2. _____

Guided-bend tests (2 root-, 2 face-, or 4 side-bend)

Root | Face
1. _____ | 1. _____
2. _____ | 2. _____

Radiographic-ultrasonic examination

FILLET WELD TEST RESULTS

Minimum size multiple pass Macroetch | Maximum size single pass Macroetch

1. ____ 3. ____ | 1. ____ 3. ____
2. ____ | 2. ____

All-weld-metal tension test

Tensile strength, psi _____
Yield point, psi _____
Elongation in 2 in., % _____

Laboratory test no. _____

WELDING PROCEDURE

Pass No.	Elect. size	Welding current		Speed of travel	Joint detail
		Amperes	Volts		

We, the undersigned, certify that the statements in this record are correct and that the test welds were prepared, welded, and tested in accordance with the requirements of 5B of AWS D1.1, Structural Welding Code.

Procedure no. _____ Manufacturer or contractor _____

Revision no. _____ Authorized by _____

Date _____

Form E-2 © 1978 by American Welding Society. All rights reserved.

Figure 7G-1 (Courtesy American Welding Society)

WELDER AND WELDING OPERATOR QUALIFICATION TEST RECORD

Welder or welding operator's name _____ Identification no. _____

Welding process _____ Manual _____ Semiautomatic _____ Machine _____

Position _____
(Flat, horizontal, overhead or vertical – if vertical, state whether upward or downward)

In accordance with procedure specification no. _____

Material specification _____

Diameter and wall thickness (if pipe) – otherwise, joint thickness _____

Thickness range this qualifies _____

FILLER METAL

Specification no. _____ Classification _____ F no. _____

Describe filler metal (if not covered by AWS specification) _____

Is backing strip used? _____

Filler metal diameter and trade name _____ Flux for submerged arc or gas for gas metal arc or flux cored arc welding _____

Guided Bend Test Results

Type	Result	Type	Result

Test conducted by _____ Laboratory test no. _____
per _____

Fillet Test Results

Appearance _____ Fillet size _____

Fracture test root penetration _____ Macroetch _____

(Describe the location, nature, and size of any crack or tearing of the specimen.)

Test conducted by _____ Laboratory test no. _____
per _____

RADIOGRAPHIC TEST RESULTS

Film identification	Results	Remarks	Film identification	Results	Remarks

Test witnessed by _____ Test no. _____
per _____

We, the undersigned, certify that the statements in this record are correct and that the welds were prepared and tested in accordance with the requirements of 5C or D of AWS D1.1, Structural Welding Code.

Manufacturer or contractor _____

Authorized by _____

Date _____

Form E-4

© 1978 by American Welding Society. All rights reserved.

Figure 7G-2 (Courtesy American Welding Society)

QW-483 SECTION IX – PART QW WELDING

QW-483 Procedure Qualification Record (PQR)
(See QW-201.2)

(COMPANY NAME)

PROCEDURE QUALIFICATION RECORD NO. _____ DATE _____

WPS NO. _____

WELDING PROCESS(ES) _____ TYPES _____
(Manual, Automatic, Semi-Aut.)

JOINTS (QW-402)

Groove Design Used

BASE METALS (QW-403)

Material Spec. _____

Type or Grade _____

P No. _____ to P No. _____

Thickness _____

Diameter _____

Other _____

FILLER METALS (QW-404)

Weld Metal Analysis A No. _____

Size of Electrode _____

Filler Metal F No. _____

SFA Specification _____

AWS Classification _____

Other _____

POSITION (QW-405)

Position of Groove _____

Weld Progression _____
(Uphill, Downhill)

Other _____

PREHEAT (QW-406)

Preheat Temp. _____

Interpass Temp. _____

Other _____

POSTWELD HEAT TREATMENT (QW-407)

Temperature _____

Time _____

Other _____

GAS (QW-408)

Type of Gas or Gases _____

Composition of Gas Mixture _____

Other _____

ELECTRICAL CHARACTERISTICS (QW-409)

Current _____

Polarity _____

Amps. _____ Volts _____

Travel Speed _____

Other _____

TECHNIQUE (QW-410)

String or Weave Bead _____

Oscillation _____

Multipass or Single Pass _____
(per side)

Single or Multiple Electrodes _____

4-8-4

Figure 7G-3 (Courtesy American Society of Mechanical Engineers)

QW-484 SECTION IX – PART QW WELDING

QW-484 Manufacturer's Record of Welder or Welding Operator Qualification Tests
(See QW-301)

Welder Name _____ Check No. _____ Stamp No. _____
Welding Process _____ Type _____
In accordance with Welding Procedure Specification (WPS) _____
Backing (QW-402) _____
Material (QW-403) Spec. _____ to _____ of P No. _____ to P No. _____
          Thickness Range _____ Dia. Range _____
Filler Metal (QW-404) Spec. No. _____ F No. _____
          Other _____
Position (QW-405) _____
      (1G, 2G, 6G)
Electrical Characteristics (QW-409) Current _____ Polarity _____
Weld Progression (QW-410) _____

FOR INFORMATION ONLY

Filler Metal Diameter and Trade Name _____ Flux for Submerged Arc or Gas for Inert Gas Shielded Arc
_____ Welding _____

GUIDED BEND TEST RESULTS QW-462.2(a), QW-462.3(a), QW-462.3(b)

Type and Figure No.	Result	Type and Figure No.	Result

Radiographic Results: For alternative qualification of groove welds by radiography in accordance with QW-304 and QW-305 _____
Test Conducted by _____ Laboratory–Test No. _____
      per _____

FILLET WELD TEST RESULTS (See QW-462.4(a), QW-462.4(b))

Fracture Test _____
    (Describe the location, nature and size of any crack or tearing of the specimen)
Length and Per Cent of Defects _____ inches _____ %
Macro Test–Fusion _____
Appearance–Fillet Size _____ in. X _____ in. Convexity or Concavity _____ in.
Test Conducted by _____ Laboratory–Test No. _____
      per _____

We certify that the statements in this record are correct and that the test welds were prepared, welded and tested in accordance with the requirements of Sections IX of the ASME code.

                              Signed _____
                                        (Organization)
Date _____ By _____

(Detail of record of tests are illustrative only and may be modified to conform to the type and number of tests required by the Code.)
NOTE: Any essential variables in addition to those above shall be recorded.

This Form is obtainable from the Order Dept., ASME, 345 E. 47th St., New York, N. Y. 10017

Figure 7G-4 (Courtesy American Society of Mechanical Engineers)

ing signs the document. The signed document certifies that the welder can perform welding operations described in the Procedure Qualifications. The certification is limited; it applies only to weldments made under the conditions of the test, including the weld position. The certification is also limited to the company for which the test was taken. However, it normally is valid indefinitely unless:

1. The welder does not use the qualified process within six months or
2. There is some specific reason, such as the failure of a weld to pass a nondestructive test or sloppy workmanship.

The Welder Qualification test is a record of the person's performance. It is also proof of a welder's ability to perform certain operations. The Procedure Qualification, on the other hand, contains the information needed for the Procedure Qualification and the Welder Qualification test. It provides a permanent record regarding the methods used to produce satisfactory welds.

In summary, the procedure form shows the procedure is workable and can produce acceptable welds. The Welder Qualification test proves the welder has the ability to produce those acceptable welds, using the procedure described.

A manufacturer can use a qualified procedure as long as none of the essential variables, such as root opening, bevel angle, type of filler metal, size of the electrode, position of the weldment, and so on, are changed.

CHAPTER 8
WELDING SYMBOLS

It is important for a welder to understand the meaning of the welding symbols. Welding symbols present a great deal of information in a limited space. They can be used to describe the process, type of welds, weld sizes, and types of filler metal, as well as the testing methods to be used in the manufacture of a product.

Welding symbols are used on blueprints to conserve space. They give the fabricators all the information they require.

The ability to understand welding symbols can be a great advantage. You can develop this ability with a reasonable amount of study and practice.

LESSON 8A
CORRECT USAGE OF WELDING SYMBOLS

OBJECTIVES
Upon completion of this lesson you should be able to:
1. Explain the purpose of the various welding symbol elements.
2. List those symbols common to the shielded metal arc welding process.
3. Draw the desired weld as indicated by the various weld symbols.
4. Explain the information provided by welding symbol elements.

PURPOSE OF THE SYMBOLS
Symbols are used to provide information such as:
1. Type of welds
2. Weld position
3. Process to be used
4. Joint preparation
5. Type and method of finish
6. Dimensions
7. Filler metal, etc., on a design drawing

It is poor drafting practice to write out all the information on the drawing because the drawing becomes crowded and difficult to read.

Symbols are a **shorthand** method used to deliver information from the designer to the welder.

DISTINCTION BETWEEN WELD SYMBOL AND WELDING SYMBOL
The weld symbol is a part of the complete welding symbol. The weld symbol indicates only the type of weld required. (See Figure 8A-1.)

The complete welding symbol passes on the rest of the information needed by the welder. In addition to the weld type, it includes all the other information required.

THE WELDING SYMBOL
The complete welding symbol shown in Figure 8A-2, is made up of the following parts:

Reference line
Arrow
Basic weld symbols
Dimensions and other data
Supplementary symbols
Finish symbols
Tail
Specifications, process, or other reference.

The position of the arrow or the tail may be reversed. But the rest of the elements must always be placed as shown.

Letters, such as the "F" at top center of Figure 8A-2, indicate the position where the method of finishing will always be found. The letter "A" for angle means the groove angle will be found at that point. The same holds true for the other letters on the welding symbol.

Supplementary symbols, such as field weld, weld all around, and contour symbol, are shown in their proper location.

The welding symbol can be just a reference line, with a weld symbol and arrow, or it can contain many of the elements shown. When there is no information to be placed in the tail, the tail is omitted.

When you learn the standard information locations, you will understand the meaning of the various symbols.

WELD LOCATION
There are two sides to the reference line. They are known as "arrow side" and "other side." The side of the line closest to the reader is called the arrow side. The side farthest from the reader is called the other side. (See Figure 8A-3.)

The arrow points to a particular place on the drawing; it indicates the weld location. The position of the weld symbol shows which side of the joint will be welded. (See Figure 8A-4.)

Notice that the weld symbol for a fillet weld is on the side closest to the reader (the arrow side). The weld should be placed as shown. If the weld symbol had been on the side

CORRECT USAGE OF WELDING SYMBOLS

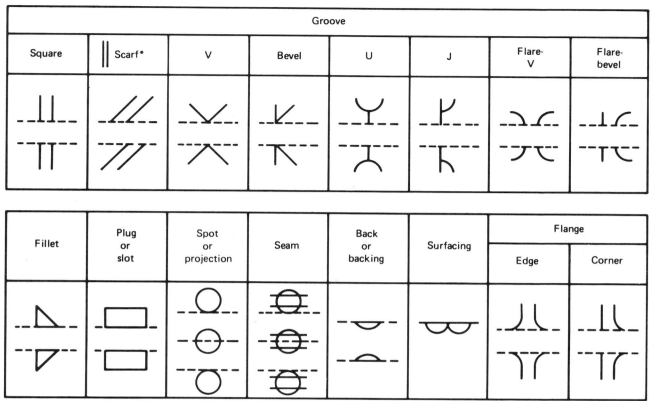

Basic weld symbols

Figure 8A-1 Basic Weld Symbols. (Courtesy American Welding Society)

farthest from the reader, the weld would have been placed on the other side of the vertical member.

In Figure 8A-4 there are actually two joints shown. If the weld was meant to be placed on joint "B", the weld symbol would have to be pointed at the joint below the horizontal part.

When the weld symbol for a V-groove butt joint is above the reference line, the desired weld is on the side opposite from where the arrow is pointing. (See Figure 8A-5B.) Desired welds for the symbol below and on both sides at the reference line are shown in Figures 8A-5A and 8A-5C.

A break in the arrow line, as in Figure 8A-6A&B, indicates that the

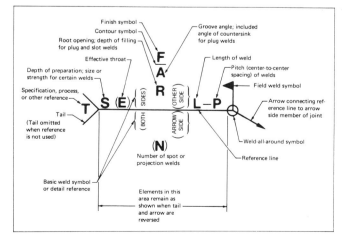

Figure 8A-2 Standard location of elements of a welding symbol.

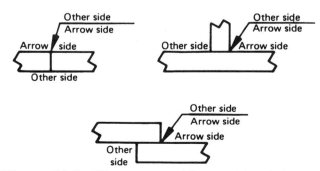

Figure 8A-3 Fillet, groove, and flange weld symbols. (Courtesy American Welding Society)

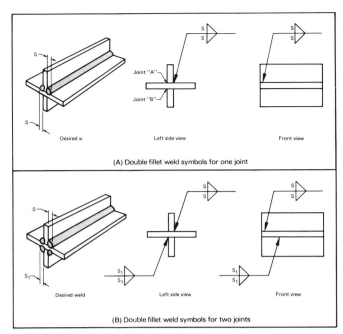

Figure 8A-4 Application of fillet weld symbols. (Courtesy American Welding Society)

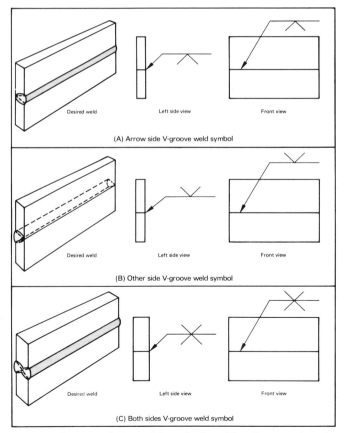

Figure 8A-5 Application of arrow side and other side. (Courtesy American Welding Society)

part the arrow points to is to be beveled, grooved, and so on. In this case the weld symbol calls for a single bevel groove weld. Its position below the reference line indicates the weld is to be placed where the arrow touches the line. The broken arrow line points up, toward the top member of the weldment, indicating that the top member is the one to be beveled.

Some welds must be welded from both sides. Such welds are indicated by placing the weld symbol on both sides of the reference line. (See Figure 8A-6C.) Other examples of this are shown in Figure 8A-7.

SYMBOLS WITH AND WITHOUT REFERENCES

When a specification, process, or other reference is used with a welding symbol, the reference shall be placed in the tail. (See Figure 8A-8.) The meaning of Specification A-2 is shown elsewhere on the drawing. It may even be on a separate sheet that shows drawings for the project.

The GTAW reference is self-explanatory. GTAW is the abbreviation of Gas Tungsten Arc Welding. It is the arc welding process which the engineer specified for that particular weld.

Symbols may be used without references if a note appears on the drawing. For example, "UNLESS otherwise designated all welds ARE TO BE MADE IN ACCORDANCE WITH SPECIFICATION NO. _____ _____ ." References are not needed where the welding procedure is described in detail elsewhere. These descriptions may be found in shop instructions and process sheets.

Notes such as "Unless otherwise indicated, all fillet welds shall be 5/16 inch in size", may also be placed on the drawing.

SUPPLEMENTARY SYMBOLS

Other symbols such as weld all-around, field weld, melt through, backing, and spacer material, as well

CORRECT USAGE OF WELDING SYMBOLS

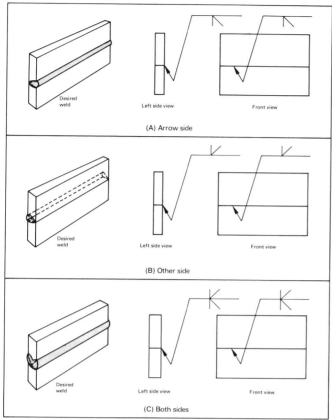

Figure 8A-6 Application of break in arrow of welding symbol. (bevel-groove weld). (Courtesy American Welding Society)

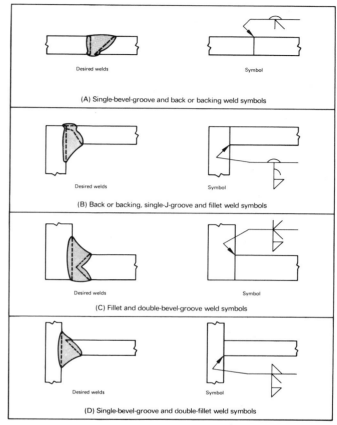

Figure 8A-7 Combination of weld symbols showing break in arrow. (Courtesy American Welding Society)

as contour symbols, give more information about the desired weld. (See Figure 8A-9.) The weld all-around symbol means to weld completely around the joint. The weld all-around symbol is a circle around the junction of the reference line and the arrow line.

Figure 8A-8 Symbols with and without references.

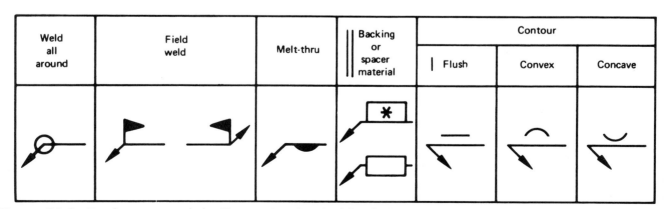

Figure 8A-9 Supplementary symbols. (Courtesy American Welding Society)

In Figure 8A-10 the symbol below the reference line indicates that a fillet weld is desired. It is positioned on the side closest to the reader. The weld should be where the arrow touches the drawing. The circle indicates weld all-around.

Field welds are indicated by a small flag that extends upward from the junction of the reference line and the arrow line. It indicates the particular weld will be made, but at the final job site.

Sometimes, due to the size of a weldment or for ease of handling, a large, cumbersome weldment may be fabricated in sections. It is welded together at the final destination.

The melt-through symbol is used to indicate 100 percent penetration, plus reinforcement, is required. This symbol is normally used on joints welded from one side, such as pipe welds.

Melt through is weld metal at the root that sticks out past the face of the material being welded. Notice that the melt-through symbol in Figure 8A-11 calls for a 3/32 in. reinforcement on the other side.

Contour symbols denote the desired shape for the finished weld. When the symbols are as shown in Figure 8A-9, it means that the welds are to remain in an "**as welded**" condition. This means that nothing is done to the welds except for cleaning.

Other methods are sometimes called for to obtain the desired finish. When one of these methods is to be used, a letter, denoting the method of finishing, is placed over the finish symbol. This is shown in Figure 8A-12. The letters and methods of finishing are as follows:

C Chipping   G Grinding
M Machining   R Rolling
H Hammering

MULTIPLE REFERENCE LINES

When necessary, more than one reference line may be used. The additional lines are used to indicate a sequence of operations. This is shown in Figure 8A-13. Multiple lines can be used to show the type of welds, the welding sequence and, if a tail is added, the process required.

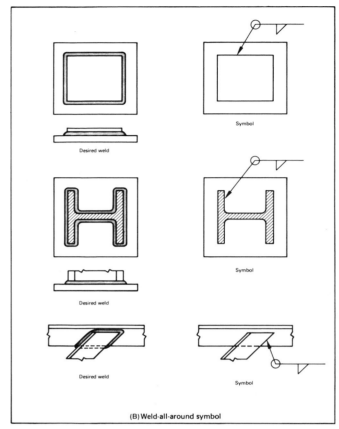

Figure 8A-10 Designation of extent of welding. (Courtesy American Welding Society)

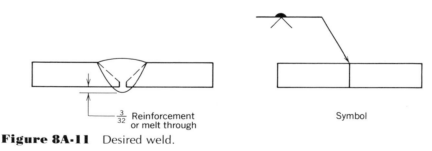

Figure 8A-11 Desired weld.

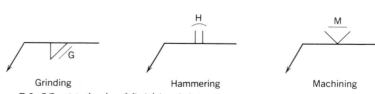

Figure 8A-12 Methods of finishing joints.

Figure 8A-14A indicates a U-groove field weld on the arrow side in the first operation. The second operation will be a field back gouge of the other side and then field weld the joint.

CORRECT USAGE OF WELDING SYMBOLS

Also, Figure 8A-14B indicates an **all-around** groove on the arrow side and ultrasonic testing **all around**.

OTHER INFORMATION FOUND ON THE WELDING SYMBOL

The depth of joint preparation and effective throat size can also be shown. In Figure 8A-15 the depth of preparation is shown by figures to the extreme left of the weld symbol. The effective throat size is shown next, to the right in parentheses.

Sometimes it is difficult to describe the desired weld with weld symbols. In such cases the weld can be shown by the use of a drawing or other necessary information. This is shown in Figure 8A-16.

DIMENSIONING FILLET WELDS

Unless there is a general note on the drawing, such as "Unless otherwise noted, all fillet welds 5/16 in. in size", all fillet welds must be dimensioned.

Weld size is shown directly to the left of the weld symbol, except when the fillet has unequal legs. Then, the dimension of each leg shall be given in parentheses as shown in the unequal leg fillet in Figure 8A-15.

EXTENT OF FILLET WELDING

Fillet weld lengths may be indicated by symbols and dimension lines, as in Figure 8A-16. In the drawing, note that the structural member lines have been broken to show more length. The overall dimension gives the true length. Also note, only 24 in. is welded at each end.

Another way is to show the size and type of weld on the reference line and point the arrow at a shaded area on the drawing. This is shown in Figure 8A-17.

INTERMITTENT FILLET WELDS

Intermittent welds on only one side of a member are shown by a symbol as in Figure 8A-18A. This weld symbol shows the welding will be on the arrow side, and the welds will be 2 in. long. The pitch, or distance from center to center, will be 4 in.

In all intermittent welding, the end welds should be longer, so that they terminate at the end of the joint.

Figure 8A-18B demonstrates the weld symbol for chain intermittent welding. *Chain* intermittent welding places the increments directly opposite each other on the weldment. Notice that the welds are 2 in. long

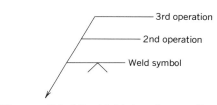

Figure 8A-13 Multiple reference lines

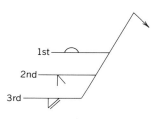

Figure 8A-14 Supplementary symbols

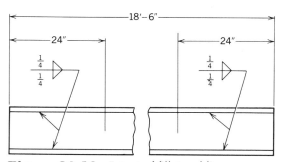

Figure 8A-16 Extent of fillet welding

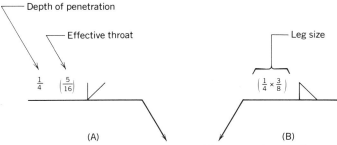

Figure 8A-15 Dimensioning welds (a) Depth and effective throat of a groove weld. (b) Unequal leg fillet.

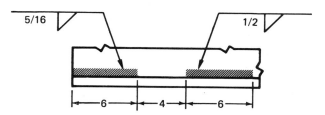

Figure 8A-17 Extent of fillet welding-hatching. (Courtesy American Welding Society)

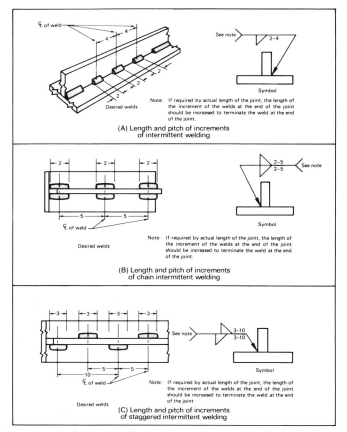

Figure 8A-18 Application of dimensions to intermittent fillet weld symbols. (Courtesy American Welding Society)

and 5 in. from center to center.

Figure 8A-18C demonstrates the weld symbol for *staggered* intermittent welding. Staggered intermittent welding places the weld increments opposite the center of the spaces on the other side of the member. In the case shown, the distances between the increments in the lower part of the drawing are 10 in. The center point of the space between the increments would be 5 in. The 3 in. weld increments on the opposite side are centered on this point.

TYPES AND DIMENSIONING OF GROOVE WELDS

A group of groove weld symbols was shown in Figure 8A-1. It is important that you understand their meaning and the methods used to dimension them.

All information concerning a groove weld is placed on the same side of the reference line as the weld symbol. Figure 8A-19 indicates the positioning of groove weld information.

A figure in the groove symbol, such as "1/8," indicates the root opening. A figure above it, given in degrees, indicates the included bevel angle of both plates. The figure to the far left signifies the depth of the groove preparation, whereas the figure in parentheses signifies the effective throat.

When the dimensions of groove welds do not appear in a general note on the drawing, they will be shown on the welding symbol.

When depth of preparation and effective throat are not shown on the welding symbol, complete penetration is required.

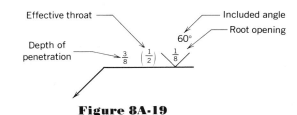

Figure 8A-19

SOME APPLICATIONS OF GROOVE WELD SYMBOLS AND DIMENSIONS

There are a great many variations of groove welds. Figure 8A-20 shows groove weld symbols where the required effective throat is indicated, but the depth of penetration is not. In such cases either the supervisor or the welder determines the joint preparation.

Figure 8A-21 shows variations between the depth of preparation and the effective throat of the weld.

(A) Indicates that the depth of preparation and the effective throat are the same.

(B) Shows the depth of preparation is less than the effective throat. The groove is only 5/8 in. deep. The weld must penetrate deeper, so the effective throat is 3/4 in.

(C) Indicates the depth of preparation to be more than the effective throat. This means a small space at the bottom of the groove does not have to be filled with weld metal.

(D) Indicates a weld without preparation, but with an effective throat of 3/4 in.

Figure 8A-22 indicates both the preparation and the effective throat. Notice the effective throat in all three drawings exceeds the depth of preparation. Also look at the bottom right of the drawing for the note in the tail of the welding symbol. It states this is a complete penetration joint, welded from both sides.

Figure 8A-23 indicates the effective throat, without preparation information. Notice the notes indicate

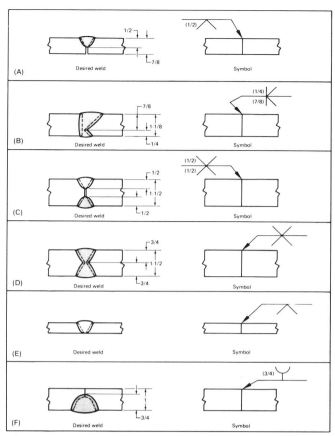

Figure 8A-20 Designation of effective throat of groove welds, depth of preparation not specified. (Courtesy American Welding Society)

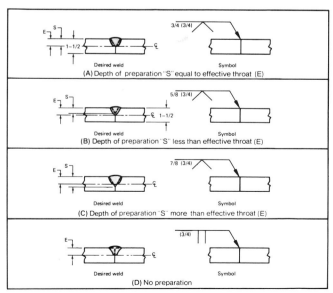

Figure 8A-21 Examples of different relationships between depth of preparation "S" and effective throat (E). (Courtesy American Welding Society)

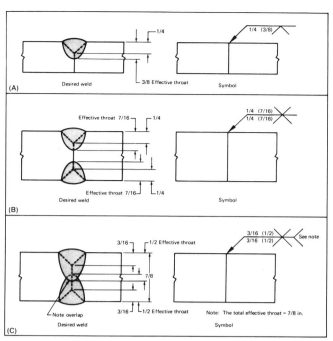

Figure 8A-22 Designation of effective throat of groove welds with specified depth of preparation. (Courtesy American Welding Society)

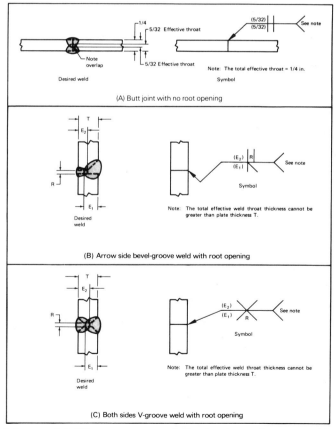

Figure 8A-23 Designation of effective throat of groove welds without specified depth of preparation. (Courtesy American Welding Society)

that the effective throat is never greater than the plate thickness.

Figure 8A-24 shows how root openings are designated. It is important to adhere to root opening dimensions. Too small a root opening can cause lack of complete penetration. Too large a root opening will require an excessive amount of filler metal.

Figure 8A-25 shows the methods of dimensioning groove angles. The correct groove angle combined with the correct root opening produces the proper size joint.

APPLICATION OF BACK OR BACKING WELD SYMBOLS

Figure 8A-26 shows how the back or backing weld symbol is applied to groove welds.

In drawing (A) the note indicates the single pass back weld is to be made after the groove weld is completed.

Drawing (B) indicates the backing weld will be made first, then the groove weld will be welded.

Drawing (C) uses multiple reference lines to show the sequence of operations. The arrow side plate is to be beveled to a depth of 3/8 in., with a root gap of 1/8 in. After the arrow side is welded with an effective throat depth of 9/16 in., the other side is to be back gouged to a depth of 1/2 in., then welded with an effective throat of 9/16 in. (Back gouging is used to remove the metal from the other side. It is indicated by the information in the tail of the welding symbol.)

In drawing (D) the backing weld is used with a root opening and without gouging. Either the arrow side or the other side may be welded first, but the welder must be sure the joint is carefully cleaned to reduce the chance of slag inclusions.

SURFACING WELDS

The surfacing weld symbol is used to indicate surfaces that require buildup by welding. The buildup may be in a single layer or with multiple layers, depending on the thickness desired. The thickness shall be indicated by figures to the left of the weld symbol. When no specific height is indicated, the buildup will consist of one layer.

If only a portion of a surface is to be built up, the size and area loca-

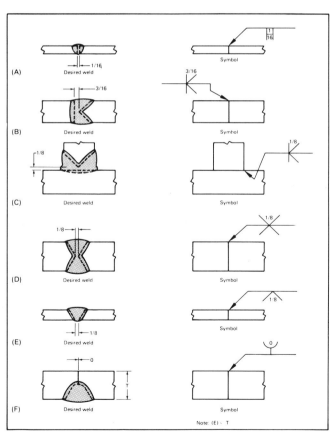

Figure 8A-24 Designation of root opening of groove welds. (Courtesy American Welding Society)

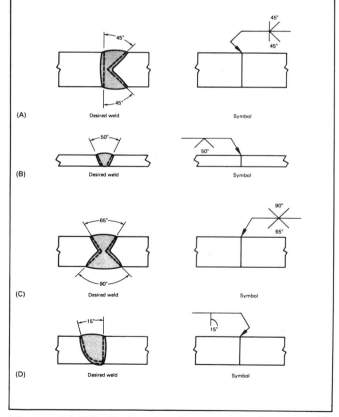

Figure 8A-25 Designation of groove angle of groove welds. (Courtesy American Welding Society)

CORRECT USAGE OF WELDING SYMBOLS

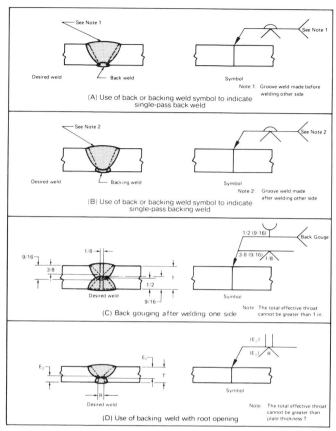

Figure 8A-26 Application of back or backing weld symbol with and without gouging. (Courtesy American Welding Society)

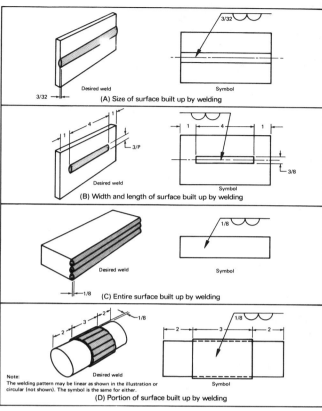

Figure 8A-27 Application of surfacing weld symbol to indicate surfaces built up by welding. (Courtesy American Welding Society)

tion information shall be shown as in Figure 8A-27.

This book does not cover the many other symbol applications that do not apply to the shielded metal arc welding. Only those symbols related to it have been covered.

JOINTS WITH BACKING

Some butt joints call for backing. The square groove butt, especially when heavy plate is being used, requires backing in the form of a bar or strip. A backing symbol is used to provide the welder with information regarding the backing material, its size, and sometimes information regarding its removal.

In Figure 8A-28 the top drawing shows a V-groove butt with a backing bar on the root side. The tail note indicates that the backing material should be the same as the welded plate. The "M" in the symbol also indicates the backing is not to be removed after the joint is completed.

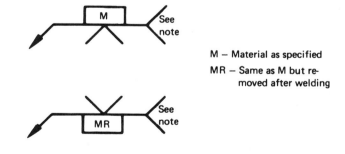

M – Material as specified
MR – Same as M but removed after welding

Note: Material and dimensions of backing as specified

Figure 8A-28 (Courtesy American Welding Society)

The lower drawing shows the type of material and its dimensions will be found either in the note or elsewhere on the drawing. The "R" in the symbol indicates the backing is to be removed after the weld is completed.

CHAPTER 9
WELDING INSPECTION AND QUALITY CONTROL

Welding inspection is of great importance to the manufacturer. It helps to keep costs down by finding unacceptable welds before they are shipped. The shipping destination may be some distance away. The cost of repairs made in the field is quite high. It might be necessary to send a factory crew to the customer's plant. The cost of travel, lodging, meals, and premium pay can add up to a sizable amount.

A good quality assurance program is essential. In many companies it is carried out by the inspection department. Manufacturers should hire only qualified welders and welding inspectors. In addition, there should be a good working relationship between the welders and the welding inspectors.

LESSON 9A
THE IMPORTANCE OF WELDING INSPECTION

OBJECTIVES
Upon completion of this lesson you should be able to:
1. Explain the economic importance of welding inspection.
2. List the steps the welder must take in the weld-inspection process.
3. Describe the duties of the welding inspector.
4. Explain the purpose of a Performance Qualification Test.

THE WELDER'S PART IN WELDING INSPECTION AND QUALITY CONTROL

Welders should visually inspect each and every weld they produce. The inspection should be automatic. It should be done between passes, as well as after the weld is completed.

Between passes the welder should check for undercut and the possibility of slag inclusion. Also, check for proper bead contour and smooth connections.

Completed welds should be checked carefully for things such as even bead appearance, uniform width and height, size of legs on fillet welds, as well as weld face contour. Also check for undercut, surface porosity, and slag inclusions.

The connections should be sound and conform with the appearance of the rest of the weld.

The experienced welder should constantly inspect the weld during its formation. The welder can see if the penetration and fusion are proper. Although large gas pockets that can be seen form, the welder has no way of detecting fine porosity.

In addition to these visual inspection steps, the welder may take part in the measurement of welds.

ROLE OF THE WELDING INSPECTOR

The welding inspector is the watchdog of the welding industry. The inspector makes sure that only high quality weldments are shipped. A wide range of methods and equipment is available for the use of the inspector. These range from visual inspection and weld measurement to more complicated **destructive** and **nondestructive** test methods. Proper use of tests can determine if the welds are strong enough.

Welding inspectors are trained to use specialized inspection equipment. Sometimes these people are called weld technicians or quality control personnel; it all depends on the company they work for. Sometimes, with proper training, a welder can become an inspector.

Although welding inspection is an additional cost, it can save money in the long run. It can eliminate or at least reduce the necessity for field repairs. Common sense tells you it is better to find and repair defects before the product leaves the shop, or is placed in service. Sometimes the welding inspector administers welder performance tests. Almost all employers require a welder to pass a qualification test. The test shows the welder has the ability to weld at the quality level required by the employer.

The test annoys some welders. However, a good welder need not fear the test or the welding inspector. Cooperate with inspectors; they know requirements of:

1. Metal preparation
2. Fit up
3. Root faces and root openings

4. Maximum and minimum amperages
5. Electrode size
6. Welding position
7. Type of bead required
8. Bead sequence
9. Method of cleaning allowed
10. Type of metal removal allowed
11. Preheat and postheat
12. Time limits for the test
13. Special rules governing the test

Some welders think the inspector is supposed to make the job more difficult, but this is not true. They only want you to weld according to regulations. It is better for everyone if you pass the test. Tests are expensive and qualified welders are hard to find.

THE PERFORMANCE QUALIFICATION TEST AND THE WELDER

Do not be nervous when you have to take a performance qualification test. The test is an opportunity to prove your ability to weld with competence. In most trades an apprenticeship of four to five years is required. This includes attending school and on-the-job training. Welders can usually advance themselves by demonstrating their ability to make quality welds. Also, qualified welders are paid fairly. Wages are usually based on welding skill, not only on the length of service for a company.

The welder performance qualification test is just one of many methods used to assure weld quality. Periodic testing is a must in most of the welding industry. There are some types of work that require inspection of every weld produced. Powerhouse work is an example of where such inspection and testing is done. Each weld is routinely tested. Radiographic, ultrasonic, or other nondestructive tests are frequently used. It is not unusual for entire pressure vessels to be X-rayed.

The constant testing and inspection help to prevent the development of bad habits. It also reduces the amount of shoddy workmanship and unacceptable welds.

In summary, the welding inspector tests the welder's skills, and high quality welding is maintained.

LESSON 9B
VISUAL INSPECTION OF WELDS

OBJECTIVES
Upon completion of this lesson you should be able to:
1. List the items to be checked in making a visual inspection of a weld.
2. Make visual inspection of welds and list any imperfections or discontinuities.
3. Use a measuring device to determine the size of fillet welds.

DEFINITION OF VISUAL INSPECTION

Eyesight plays a major role in visual inspections. An extensive knowledge of welding is also important. You should know the qualities to look for in a finished weld. It is important to understand the causes of all faults that may be discovered. Knowledge of how to repair the defects as well as their prevention is also important.

Visual inspection can help determine if the weld meets the required standards. However, visual inspection is limited; only surface weld defects can be determined.

THE EXTENT OF VISUAL INSPECTION

We are limited by what we can see in visual inspection of welds. Often, however, the information we gain is enough to pass judgment. The weld can pass, be rejected, or recommended for rework. There are times when the inspection is for appearance only. It determines if the welds are smooth enough to be pleasing to the eye. Use visual inspection where weld strength is not important. There are many instances such as with sheet metal products, where strength is not a major factor. For instance, where the weld does not have to be watertight and will not be loaded, visual inspection may be sufficient.

The welder should visually inspect the weld while it is being made. Repairs, or changes in the welding operation, should be made as needed. There is probably no better qualified person to make an accurate visual inspection than a welder. A good welder should have pride in the finished product.

A visual inspection should answer the following questions:

1. Are the width and height uniform for the entire length of the finish pass?
2. Does the weld appear neat and uniform throughout?
3. Are all connections smooth? Do they conform with the surface of the weld?
4. Does the weld contour conform to the weld indicated by the weld symbol?
5. Is the surface of the weld free of porosity marks?
6. Are the surface and **toes** of the weld free of any slag inclusions?
7. Is the weld free of undercut?

8. Is the metal next to the weld free of arc strikes?

Fillet welds have another important inspection item; that is the size of the legs of the weld. Most fillet welds are expected to have equal legs. Some unusual fillet welds have unequal legs. This simply means that the vertical toe distance from the heel of the weld is not the same as the horizontal toe from the heel of the weld. (See Figure 9B-1.)

The fillet weld contour must match the weld symbol description. A special contour measuring device is made for that purpose. It is fairly simple to measure fillet welds.

For example, there is the Navy gauge and the standard fillet weld gauge. (See Figures 9B-2 and 9B-3.)

The standard fillet weld gauge ends are shaped differently. Both ends are used in measuring fillet welds. The right side of Figure 9B-4 shows a fillet weld vertical leg measurement. The blade ends on the gauges are carefully shaped and marked. They can be used on various sizes of welds. They can also determine if the weld profile is flat, convex, or concave. This is illustrated in Figure 9B-4.

The size of a fillet weld is determined by the length of the shorter leg. In Figure 9B-5 the weld gauge is being used properly to measure the fillet on the left. The weld gauge on the right indicates the length of the longer leg, but fails to determine the true size of the weld.

In Figure 9B-6 the opposite end of the weld gauge is being used. It will determine if the weld has sufficient throat to meet the specifications.

The wrong end of the fillet gauge is being used on the weld on the left. It shows that the leg of the weld is 5/16 of an inch, but it does not determine if that size isosceles right triangle can be inscribed within the weld cross section. On the other hand, if the specifications called for a ¼ inch concave fillet weld on the right side, that weld would meet that specification. The fillet weld gauge ends indicate that both legs of the fillet meet the specifications. The center

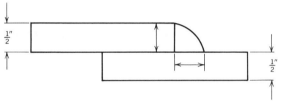

Figure 9B-1 Equal leg fillet weld: both the vertical and horizontal legs of the joint are the same length.

Figure 9B-3 Use of standard fillet gauge. (Reprinted by permission of Prentice-Hall, Inc.)

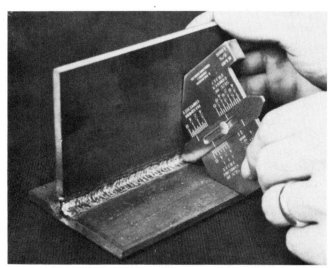

Figure 9B-2 Navy type weld gauge. (Reprinted by permission of Prentice-Hall, Inc.)

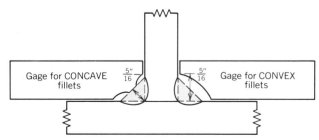

Figure 9B-4 Two types of fillet gauges. Convex fillets may be measured with a gauge like the one shown on the right. In this case it is a measurement of the leg sizes. Concave fillets are measured with a gauge like the one on the left. In this case it measures the throat of the weld. (Courtesy The Lincoln Electric Company)

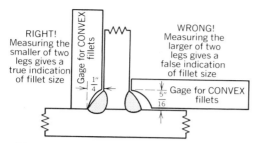

Figure 9B-5 Measuring convex fillets. Notice that the largest isosceles right triangle which can be inscribed within the cross section of fillet is determined by the dimension of the shorter leg. (Courtesy The Lincoln Electric Company)

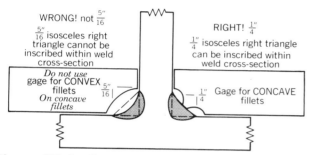

Figure 9B-6 Showing the right and wrong method of gauging a concave fillet. (Courtesy The Lincoln Electric Company)

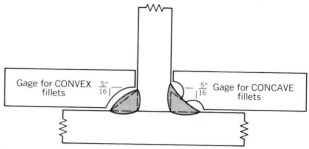

Figure 9B-7 With equal legged 45° fillets, either type gauge (concave or convex) may be used. Both will indicate the same size fillet providing it is not excessively concave. (Courtesy The Lincoln Electric Company)

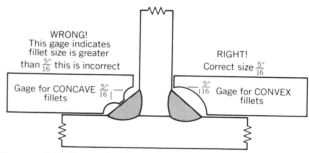

Figure 9B-8 With the fillet shown above, it may not be readily apparent whether it is flat, slightly concave or convex. But if double-checked with both types of gauges, it would then be apparent that the vertical leg is smaller than the bottom leg and that this is the true fillet size. The concave gauge would give the impression that the fillet is larger than 5/16 in. and this would be incorrect. (Courtesy The Lincoln Electric Company)

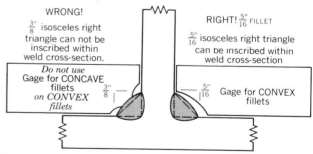

Figure 9B-9 Showing the right and wrong method of gauging a convex fillet. (Courtesy The Lincoln Electric Company)

portion of the gauge just touches the face of the weld at the throat. This indicates that that throat of the weld is also sufficient to meet the specifications.

Concave fillet welds are often specified. Their leg lengths are sometimes lengthened to increase the strength of the weld.

Figures 9B-7 demonstrates that both types of weld gauges may be used on equal leg fillet welds. Figure 9B-8, on the other hand, demonstrates that double checking is often the prudent thing to do. In this case the gauge used on the right side weld gives you the true weld size. The one on the left is somewhat confusing.

Finally, in Figure 9B-9 you see an example of the correct weld gauge to use for convex fillet welds. Welders should be able to use this type of gauge to check their welds to see if they meet the specifications.

LESSON 9C
DESTRUCTIVE TESTING METHODS

OBJECTIVES

Upon completion of this lesson you should be able to:

1. List the various destructive weld tests.
2. Discuss the purpose for which they are used.
3. Describe the various test methods.
4. Describe the preparation of test samples.

DEFINITION OF DESTRUCTIVE TESTING

In **destructive testing** the sample is destroyed during the test. Before that, the weld was destroyed when the sample was obtained. This type of test provides us with important information about the mechanical properties of steel. These properties were discussed in Lesson 5A.

Tensile tests, **bend tests**, chemical analysis, hardness tests, V-notch tests, micro- and macroscopic tests, and in some cases hydrostatic tests are used. They determine if the weld properties are equal to, greater than, or less than the base metal.

THE BEND TEST

The first test you may encounter is either the free bend or the **guided bend test**. In these tests a properly prepared specimen is bent. The specimen can be flame cut or saw cut across the length of the weld. Controlled bending conditions are used. The conditions are normally described in the code governing the weld service. Coupon specimens for the two methods of bending are shown in Figures 9C-1 and 9C-2.

Three types of bends are used in these tests. They are the face, root, and **side bends**. Figure 9C-3 shows a face bend. The finished weld face is on the outside. It is bent the most in order to stress it the most. The **root bend** puts the most stress on the root pass. It is probably the most important of all the tests. The side bend places most of the stress on one weld side or the other. It is used mainly

Figure 9C-1 Guided bend.

Figure 9C-2 Free bend.

when testing heavy plate, such as 1 in. or more in thickness. However, the side bend test can be used on thinner metals. Normally, the bend test is used to see if a welder can produce sound welds. This is the reason it is one of the first tests the welder encounters. The weld is acceptable when:

1. Single cracks in the stretched surface are less than one eighth of an inch long

   or

2. No group of cracks has a combined length of one eighth of an inch.

This is a rigid test, but the welder who does a good job and prepares the samples properly, will pass the test. Figure 9C-4 shows the correct method to prepare a bend test specimen.

In addition to grinding or milling the weld reinforcement to the plate surface, be sure you put a radius on the four edges along the coupon. Without rounded edges the weld may fail because a crack starts in one of the sharp edges. Here is an important precaution regarding the preparation method. Be sure all grinding, machining, or file marks are in line with

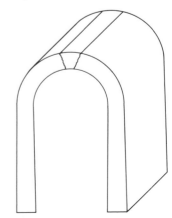

Figure 9C-3 Face bend.

DESTRUCTIVE TESTING METHODS

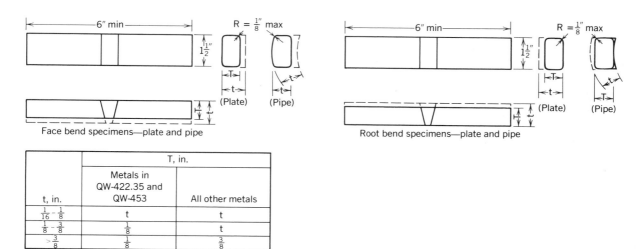

Figure 9C-4 Face and root bends-transverse. (Courtesy American Society of Mechanical Engineers)

Figure 9C-5 Grinding marks should run across the weld

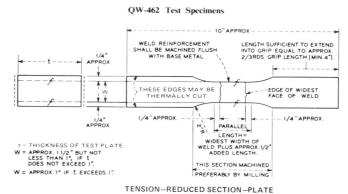

Figure 9C-6 Tension-reduced section-plate (Courtesy American Society of Mechanical Engineers)

the length of the coupon. (See Figure 9C-5.) Marks across the weld may act to start a crack. Even fine marks can cause an otherwise sound weld to crack and fail.

The bend test is simple to perform, and the equipment is not very expensive. It should be available in your school. It is good practice to test your own welds as often as possible while you are training. It will build your confidence. It will also help to develop your ability to produce quality welds. With confidence in your ability, you will not be apprehensive about the test when you seek employment. These tests will be discussed in more detail in the next lesson.

THE TENSILE OR PULL TO DESTRUCTION TEST

When you make a test plate, especially if it is to qualify a procedure, there is a possibility it will be tensile tested. This means that a sample from the welded plate will be prepared as shown in Figure 9C-6. Notice that the center portion of the coupon has been narrowed. This causes it to break somewhere within that section. Take all measurements with care.

Measurements must be accurate to be able to calculate the actual pounds per square inch required to pull the coupon apart.

The correctly prepared specimen is placed in the jaws of a tensile testing machine as shown in Figure 9C-7. Most of these machines use hydraulic power. They exert the force needed to pull the specimen apart.

As the force increases the yield point is eventually reached. The yield point is where the metal begins to neck down. It acts in much the same way that a piece of taffy does before it breaks. After the specimen breaks,

Figure 9C-7 Speciman secured in jaws of tensile tester.

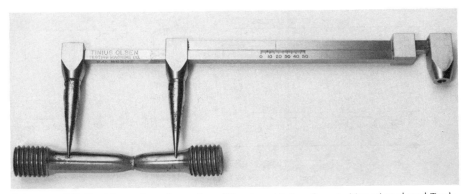

Figure 9C-8 Measuring elongation. (Courtesy Union County Vocational and Technical School)

the operator calculates the yield strength of the metal. The computations depend on the equipment used and the measured cross section of the coupon.

ELONGATION

Elongation is another factor that can be determined by a tensile tester. Elongation is the increase in length of the specimen measured in percentage when it breaks. Elongation is determined from two center-punch-marks on the specimen. The marks are placed at least 1 in. apart. After the specimen has broken, the pieces are placed together. Then, the distance between the two marks is measured. (See Figure 9C-8.) The increase in distance is used to calculate the elongation. This is shown by the following equation:

(Increase in length) (original length) × 100 = (percent elongation)

FOR EXAMPLE

If the center-punch marks were 1 in. apart before the specimen was pulled:
And it stretched 1/16 of an inch apart after breaking:

(1/16 ÷ 1) × 100 = 6.25 percent elongation.
Or 1 1/8 in. apart after breaking:
(1/8 ÷ 1) × 100 = 12.5 percent elongation.
Or 1 3/16 in. apart after breaking:
(3/16 ÷ 1) × 100 = 18.75 percent elongation.

The American Welding Society electrode classification system requires the elongation percentage to be specified by the manufacturer. This is in addition to the tensile strength, yield strength, reduction in area impact value and, in some cases, the hardness of the filler metal.

HARDNESS

Hardness is the ability of a metal to withstand penetration. A harder metal or a diamond penetration is used. The penetrator is pressed into the test sample using a predetermined force. Two common testing machines are the Brinell and the Rockwell hardness testers. These are shown in Figures 9C-9 and 9C-10.

The equipment is different but the results are similar. The Rockwell tester uses a hardened steel ball, 1/16 in. diameter, for softer metals. It uses a conical-shaped diamond for harder metals. (See Figure 9C-11.) The Brinell hardness tester uses a 1/8 in. diameter hardened steel ball for all metals.

IMPACT VALUES

The Charpy test and Izod procedure are used to determine impact strength of metal. For these tests a notch or groove is cut into the specimen. Also, the shape and position of the grooves that are cut into the specimen are different. The Charpy method uses a square groove, with a round hole at the bottom. The Izod method uses a V-notch.

The Charpy method holds the specimen at both ends. The Izod method holds one end only.

As shown in Figures 9C-12 and 9C-13, both methods use the energy stored in a pendulum. It is raised a measured distance and then released, to swing down and strike the specimen. If the specimen fails to break on the first swing, the pendulum is raised to a greater height and released. This is done until the specimen breaks. The pendulum energy absorbed by breaking the specimen determines the impact strength.

MACROSCOPIC AND MICROSCOPIC TESTING

Both of these visual testing methods are very similar. They use instruments to magnify the samples so porosity, slag inclusions, lack of fusion, the grain structure of the metal, as well as other discontinuities may be seen.

Figure 9C-9 Air-O-Brinell air-operated metal hardness tester. (Tinus Olsen Testing Machine Company, Inc.) (Reprinted by permission of John Wiley & Sons, Inc.)

Figure 9C-10 Rockwell Hardness Tester. (Wilson Instrument Division of Acco) (Reprinted by permission of John Wiley & Sons, Inc.)

Figure 9C-11 Brale and ball. These two penetrators are the basic types on the Rockwell Hardness Tester. (*Note:* Brale is a registered trademark of American Chain & Cable Company, Inc., for sphero-conical diamond penetrators.) (Reprinted by permission of John Wiley & Sons, Inc.)

WELDING INSPECTION AND QUALITY CONTROL

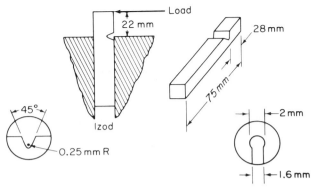

Figure 9C-12 Izod method. (Courtesy Union County Vocational and Technical School)

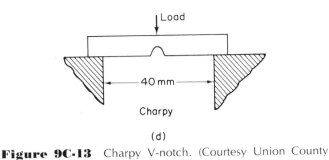

Figure 9C-13 Charpy V-notch. (Courtesy Union County Vocational and Technical School)

The main difference between the two methods is in the equipment used. The macroscopic equipment only magnifies the samples from 10 to 30 times. It views a larger area than when the microscopic method is used. However, it has limitations. Although a large area is visible, the magnification is not as great as with the Microscopic method. The Microscopic equipment provides magnifications of from fifty to as many as five thousand times.

LESSON 9D
ROOT, FACE, AND SIDE BENDS

OBJECTIVES
Upon completion of this lesson you should be able to:
1. Explain the correct methods for preparation of specimens for root, face, and side bend tests.
2. Safely and correctly prepare a specimen for testing.
3. List the important operations and thoroughly explain the reasons for each.

DEFINITION OF ROOT, FACE, AND SIDE BENDS
Root, face, and side bends are all part of the destructive testing procedure. The name of test is determined by which surface of the specimen is stretched the most.

For example, if the root is to be tested, the root side of the specimen will be bent to stretch it the most. This is shown in Figure 9D-1. A face bend stretches the face pass the most. This is shown in Figure 9D-2.

The side bend test applies force to stretch one of the side surfaces.

All three tests are simple to perform. Welding students should be able to prepare and test their own specimens, as well as those of others. The more welds you make that pass the tests, the more confident you will become. Confidence will help you pass the required qualification tests.

PREPARATION OF TEST SPECIMENS
In industry you probably will not prepare your own test specimens. However, in small shops and in your school shop you will have to prepare the specimens. Therefore, it is important that you learn and follow procedures that give you the best chance to pass the tests.

Many a good weld has failed a test

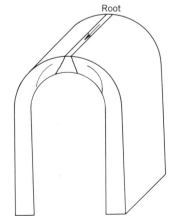

Figure 9D-1 Root bend.

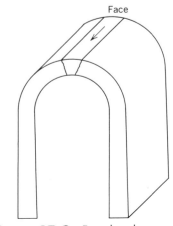

Figure 9D-2 Face bend.

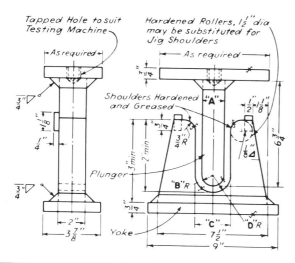

Thickness of Specimens, In.	A In.	B In.	C In.	D In.	Material	Refer To
3/8	1 1/2	3/4	2 3/8	1 3/16	All	
t	4t	2t	6t + 1/8	3t + 1/16	Others	
1/8	2 1/16	1 1/32	2 3/8	1 3/16	P-23 and SB-171, Alloy 628	QW-422.23 QW-422.35
3/8	2 1/2	1 1/4	3 3/8	1 11/16	P-11 and P-25	QW-422.11 QW-422.25
t	6 2/3 t	3 1/3 t	8 2/3 t + 1/8	4 1/2 t + 1/16		
1/16 – 3/8 in. incl	8t	4t	10t + 1/8	5t + 1/16	P-51	QW-422.51
1/16 – 3/8 in. incl	10t	5t	12t + 1/8	6t + 1/16	P-52	QW-422.52

GUIDED-BEND JIG

Figure 9D-3 Guided-bend jig. (Courtesy American Society of Mechanical Engineers)

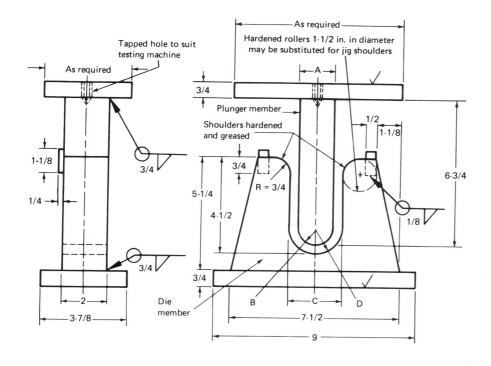

Figure 9D-4 Bending jig for U bends as per AWS specifications. (Courtesy American Welding Society)

due to poor specimen preparation. The organizations that set up the testing procedures provide exact information about the size and shape of the specimen. Testing equipment and methods of specimen preparation are also recommended.

Specimen thickness is determined by the welded plate or pipe. Width and length are specified by the codes.

Section IX of the ASME Boiler and Pressure Vessel Code, and of the AWS "D 1.1 Structural Welding Code-Steel," requires test specimens with the same measurements, configurations, and preparation. In addition, both codes use bending equipment with similiarly shaped anvils. The specimen is bent over the anvil. Sometimes the anvil is called a die. The equipment must be constructed in the same manner and subjected to the same forces during the testing procedure.

Figure 9D-3 and Figure 9D-4 have been taken from these AWS and ASME codes. Note that the shape and size of the dies are the same. At first glance some of the information may seem to be different, but it is not. It is just expressed in a slightly different manner.

METHODS OF PREPARING SPECIMENS

All test specimens or coupons must be prepared in the same manner, regardless of whether they are to be subjected to the root, face, or side bend test. The code will specify how to remove the specimen from the test piece. It will also tell which surfaces must be ground or machined. The codes are exact; follow them.

Radius preparation is of great importance to the welder or student preparing the specimens. Both codes call for specific shapes on all test

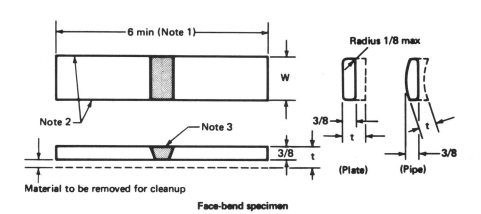

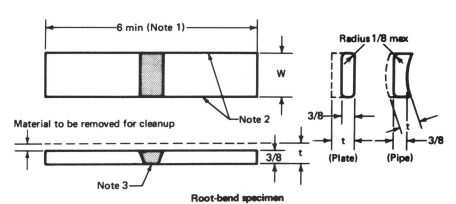

Dimensions	
Test weldment	Test specimen width, in. (W)
Plate	1-1/2
Test pipe 2 in. and 3 in. in diameter	1
Test pipe 6 in. and 8 in. in diameter	1-1/2

Figure 9D-5 Face and root-bend specimens. (Courtesy American Welding Society)

Figure 9D-6 shows a specimen reduced to three-eighths (3/8) of an in. across the weld, and to between three-quarters (3/4) of an in. and one and one-half (1½) in. wide. The symbol for thickness is "t."

THE ROOT BEND TEST

The root bend test determines the welder's ability to produce sound root passes. The specimen is placed on the die with the **root** facing downward, into the gap. The plunger is brought into contact with the specimen. It is used to bend the specimen into a U shape. It is important to center the weld, because this places the maximum stress at that point.

The method used to evaluate a root bend test as well as a face or a side bend follows. It is a direct quote from "D 1.1-79 structural Welding Code-Steel" of the American Welding Society. This code covers structural

1. a. For Procedure qualification of materials other than P-1 in QW-422. If the surfaces of the side bend test specimens are gas cut, removal by machining or grinding of not less than 1/8 inch from the surface shall be required.

1. b. Such removal is not required for P-1 Materials but any resulting roughness shall be dressed by machining or grinding.

2. For performance qualification of all materials in QW-422 if the surfaces of side bend tests are gas cut, any resulting roughness shall be dressed by machining or grinding.

specimens. Examine the end views of the specimen shown in Figure 9D-5.

Notice the lengthwise edges of the specimen are shown as rounded. The previous lesson told you of the detrimental effects of notches and scratches. Sharp edges on weld specimens can cause cracks to start. The cracks can and will lengthen, and cause failure of the specimen.

Side bends are often made on specimens from heavy plate. Therefore, the specimen must be cut to a size that will fit in the die of the bending jig.

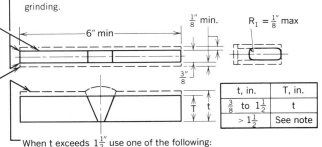

1. Cut along line indicated by arrow. Edge may be flame cut and may or may not be machined.

2. Specimens may be cut into approximately equal strips between 3/4" and 1½" wide for testing or the specimens may be bent at full width (see requirements on jig width in QW-466.1.)

Figure 9D-6 Side bend. (Courtesy American Society of Mechanical Engineers)

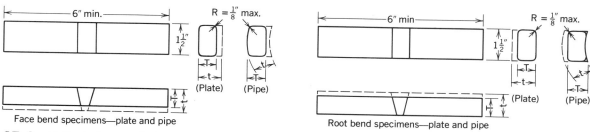

Figure 9D-7 Method of removal of excess metal on face and root bend specimens. (Courtesy American Society of Mechanical Engineers)

welding in general, as well as buildings and bridges.

5.28.1 Root-, Face-, and Side- Bend tests. The convex surface of the specimen shall be examined for the appearance of cracks or other open discontinuities. Any specimen in which a crack or other open discontinuity exceeding ⅛ in. (3.2 mm) measured in any direction is present after the bending shall be considered as having failed. Cracks occurring on the corners of the specimen during testing shall not be considered.

FACE BEND TEST

The face bend specimen is prepared in about the same manner as the root bend specimen. Metal removed, to reduce the specimen to the required three-eighths (⅜) of an inch thickness, must be removed from the root side. This is shown in Figure 9D-7.

In this case, the specimen face must be placed downward on the die. The plunger touches the root side of the joint.

The results are evaluated just as for the root bend test.

SIDE BEND TEST

Side bend tests are used mainly for tests on heavy plate. Side bends are prepared differently than root or face bends. They must be reduced to three-eighths (⅜) of an inch in width so they will fit in the die testing equipment. In order to accomplish this, material must be removed from both sides. If the test plate is thicker than one-half (½) inch, all excess metal must be removed from the weld face surface of the specimen.

The specimen is placed on the die with one side facing down and the other up. The specimen is evaluated with the same criteria as the root bend test.

MISCELLANEOUS

It is important that you follow the test procedures without deviation. The procedures have been well thought out. They are the result of the long-term experience of a large group of knowledgeable and highly qualified individuals. The codes that govern the procedures are under constant revision. Always follow the current version.

As a student welder, you usually cool the workpiece after welding. When you weld a plate or pipe for test purposes, never cool the weldment rapidly. It should be allowed to cool slowly to room temperature. Rapid cooling may produce a brittle structure. A specimen that is not ductile will crack during bending.

LESSON 9E
NONDESTRUCTIVE TESTING METHODS

OBJECTIVES
Upon completion of this lesson you should be able to:
1. List the various types of nondestructive testing methods in use.
2. Describe the principles of each nondestructive testing method.
3. Explain the use for which each method is best suited.

DESCRIPTION OF NONDESTRUCTIVE TESTING
Nondestructive testing is used to detect discontinuities in material without destroying it or impairing its usefulness. Because these methods are **nondestructive**, inspection and quality control costs are reduced. Destructive test methods increase the cost because the article must be replaced or repaired after the testing is over.

Nondestructive welding inspection methods include visual, magnetic particle, liquid penetrant, radiographic, ultrasonic, Eddy current, proof, and leak test techniques.

The method chosen depends upon

equipment availability, defect position, weld shape, and test cost. Only well-qualified people should perform the procedure and interpret the test results.

The American Welding Society has developed material, courses, standards, and tests to insure welding inspectors have the knowledge and ability to make and evaluate the tests. The American Welding Society also has a certification program for welding inspectors. The certification program and tests are offered periodically throughout the country. The program goal is to insure that welding inspectors who perform weld tests are well trained and qualified.

MAGNETIC PARTICLE INSPECTION

As the name implies, magnetic particle inspection uses magnetism. The piece to be inspected is magnetized by placing it in a magnetic field or by passing electrical current through it. Then the surface to be inspected is sprinkled with fine magnetic particles. The particles are attracted to places where the magnetism is strongest, such as at the edges of a crack or other discontinuity. Only metals that can be magnetized may be tested with this method. Both large and small weldments may be tested.

Magnetic particle inspection can be used to inspect the surface and edges of metal plates before welding.

Typical defects that can be detected include surface cracks, undercut, incomplete fusion, and inadequate joint penetration. Laminations and other defects may be detected on the edges of the base metal. Also, cracks slightly below the surface of the piece being tested may be detected. Detection of discontinuities depends on their shape and depth below the surface. For example, cracks running perpendicular to the surface can be easily detected, but round or spherical flaws are difficult to detect.

LIQUID PENETRANT INSPECTION

The liquid penetrant testing method uses special types of liquids. They can enter cracks and crevices by capillary action. Some will remain even when the excess is cleaned away. It is a reliable method of detecting discontinuities that are open to the surface.

RADIOGRAPHIC INSPECTION

Radiographic inspection methods make use of both X-rays and gamma rays. They can detect discontinuities inside of solid materials, such as weld metals. X-rays come from an X-ray machine; gamma rays come from radioactive isotopes.

Both rays have the ability to penetrate solids. The penetration is dependent upon the type and thickness of the material. Special film, placed behind the surface to be radiographed, can make a permanent inspection record. Radiation passing through the part causes the film to "photograph" internal differences in density. These differences show up as shades of light and dark. The ability to interpret the meaning of these contrasting shades is what makes radiography an important inspection tool

When X-rays or gamma rays strike an object, some pass through, some are absorbed, and others are scattered. Those that pass straight through are the ones that expose the film. With training an inspector can interpret the dark and light areas on the film and identify internal flaws.

Two radiographs taken from slightly different positions are needed to calculate the depth of the discontinuity.

The secret of success in effective use of radiography is a competent inspector. The inspector should be trained in the safe usage of the equipment and have the ability to interpret the radiographs. The AWS (Structural Welding Code-Steel), ASME (Boiler and Pressure Vessel Code), various governmental groups, and private organizations have developed radiography acceptance standards. Some states license individuals who operate the radiographic equipment.

A crack is a flaw where solid metal has separated. Cracks are considered the most serious of the discontinuities, and they are likely to elongate with time. It is possible for the crack to cause the weldment to fail while in service. Figure 9E-1 shows an example of a transverse crack in a weld. Transverse cracks are those that run across the weld from side-to-side. This radiograph has excellent contrast, making it easy to read.

Figure 9E-2 shows three different types of cracks in one radiograph.

Figure 9E-3 shows a longitudinal

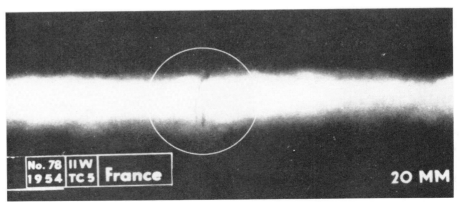

Figure 9E-1 Transverse crack. (Courtesy American Welding Society)

NONDESTRUCTIVE TESTING METHODS

Figure 9E-2 Three different types of cracks in one specimen. (Courtesy American Welding Society)

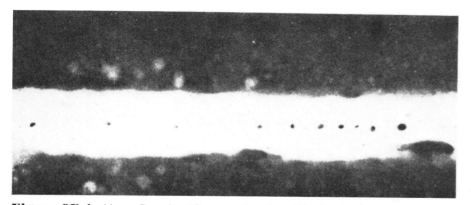

Figure 9E-4 Linear Porosity. (Courtesy American Welding Society)

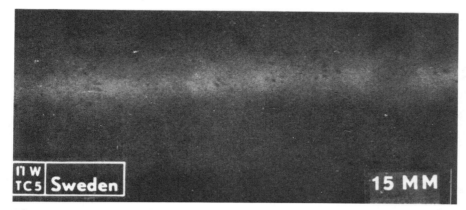

Figure 9E-5 Scattered porosity with clusters. (Courtesy of IIW used by permission of American Welding Society)

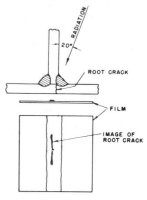

Figure 9E-3 Base metal crack emanating from root of fillet weld (shown in diagram) is indicated on radiograph. (Courtesy American Welding Society)

crack in the root of a tee joint. The line drawing indicates the position of the crack and the angle of radiography.

POROSITY

Empty spaces in the cooled weld metal, caused by trapped gases, are called porosity. These spaces can be shaped like spheres or holes. Elongated holes are somtimes called "worm holes" or "pipes." Porosity can be located along the weld in a line or scattered about. Figures 9E-4 through 9E-8 illustrate various types of porosity.

Depending on the standard, some porosity may be allowed in the weld. The allowable amount depends on where the weld is to be used.

SLAG INCLUSIONS

Slag inclusions are spaces in the cooled weld metal that are filled with foreign material. The slag comes from the welding process. It can be trapped in the weld because of improper welding or by failure to clean the joint properly. Two common types of slag inclusion are shown in Figures 9E-9 and 9E-10.

INCOMPLETE FUSION

The failure of the molten weld puddle to bond to the base metal is called

Figure 9E-6 Small scattered porosity. (Courtesy of IIW used by permission of American Welding Society)

Figure 9E-7 Worm-hole (pipe) porosity. (Courtesy of IIW used by permission of American Welding Society)

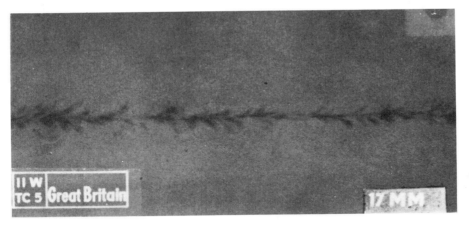

Figure 9E-8 Bad case of worm-hole (pipe) porosity. (Courtesy of IIW used by permission of American Welding Society)

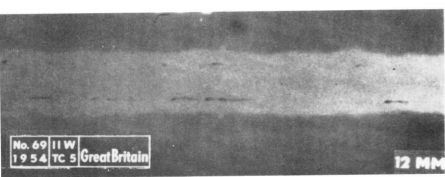

Figure 9E-9 Broken stringers of slag inclusions. (Courtesy of IIW used by permission of American Welding Society)

NONDESTRUCTIVE TESTING METHODS

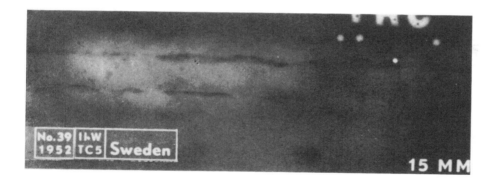

Figure 9E-10 Slag entrapment, both sides of a butt weld. (Courtesy of IIW used by permission of American Welding Society)

incomplete fusion, or lack of fusion. Radiographs of incomplete fusion and slag inclusions are similar in appearance; sometimes they cannot be distinguished. Figure 9E-11 illustrates incomplete fusion.

INADEQUATE JOINT PENETRATION

When the welding arc does not penetrate properly, the result is **inadequate joint penetration**. This condition occurs in both butt and fillet welds. Fine examples of inadequate penetration are shown in Figures 9E-12 and 9E-13.

UNDERCUT

When the groove melted by the arc is left unfilled with filler metal, the result is called undercut. Undercut leaves a groove or crevice along the toes of a weld. The dark lines along the weld edges in Figure 9E-14 are an example of undercut.

TUNGSTEN INCLUSIONS

During Gas Tungsten Arc Welding (GTAW), tungsten particles can drop off the electrode when welding conditions are wrong. The small particles imbed themselves in the weld. Because tungsten has a high density, radiographs of the inclusions appear lighter than the weld metal. See Figure 9E-15 for an example.

In typical radiographs the following is true:

ULTRASONIC INSPECTION

Ultrasonic testing can be used to detect weld discontinuities. This in-

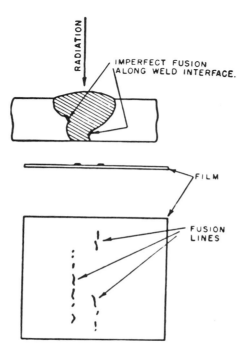

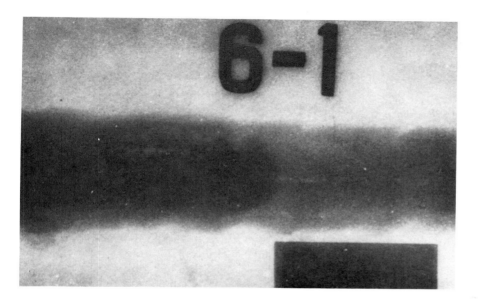

Figure 9E-11 Diagram and radiograph of incomplete fusion at the interface of a weld. (Courtesy American Welding Society)

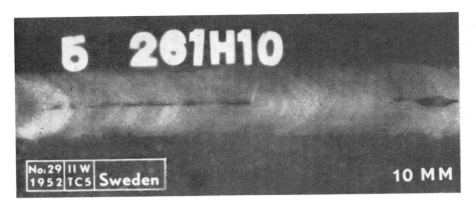

Figure 9E-12 Moderate degree of inadequate joint penetration. (Courtesy of IIW used by permission of American Welding Society)

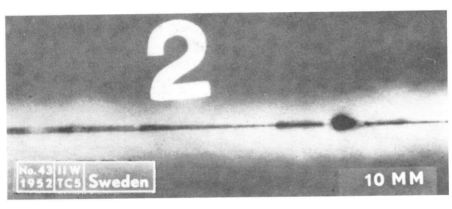

Figure 9E-13 Large degree of inadequate joint penetration. (Courtesy of IIW used by permission of American Welding Society)

spection method uses a beam of high-frequency sound waves that are projected into the test part.

The sound waves are reflected by discontinuities. The reflections are received by a transducer in the testing equipment. A trained operator can determine the size, type, and position of the defect. If there are no defects, the sound waves will be reflected by the opposite surface of the part being tested.

The most important characteristic of the ultrasonic testing method is its ability to determine the depth and position of the defect. Because of this it is a valuable tool in planning the repair of the defect.

EDDY CURRENT (ELECTROMAGNETIC) TESTING

Eddy current or electromagnetic testing uses an alternating electromagnetic field. The alternating field induces Eddy currents in the part. If the part is ferromagnetic, a magnetic field will also be induced. These effects are used to detect discontinuities such as cracks, seams, voids, and so on.

It is difficult to separate the two effects in magnetic materials. However, by use of the correct magnetizing frequency and proper design of the test circuit it is possible.

PROOF TESTS

Proof testing is used to see if the weldment can withstand its service load, without permanent deformation or failure. This test uses stresses higher than those expected in service, but below the elastic strength of the material.

Closed containers such as cylinders, tanks, vessels, and closed piping systems can be tested by either the pneumatic or the hydrostatic method.

The pneumatic method uses air to pressurize the vessel. This method is normally used where test pressures do not exceed 50 pounds per square inch (psi).

The hydrostatic method uses water at very high pressures. First, the vessel is filled to capacity with water, then the pressure is increased. Precautions such as the installation of a pressure relief valve should be followed. A pressure gauge is also needed. It aids in determining the water-tightness of the vessel. After the vessel is filled and pressurized, it should be allowed to sit for a day. If the pressure does not change, the vessel is watertight.

Open containers may be tested for leaks by filling them with water. Another simple leak test for open containers is the kerosene test. A small amount of kerosene is brushed on the inside of the weld seams. If there is a leak, the kerosene will bleed through in a few minutes. Because kerosene is flammable, care must be taken that no hot work is occurring in the vicinity. The remaining kerosene should be covered and removed from the hot work area immediately after the testing is completed. Do not use this test with-

NONDESTRUCTIVE TESTING METHODS

Figure 9E-14 Radiograph showing surface undercut. (Courtesy of IIW used by permission of American Welding Society)

out careful attention to fire prevention.

LEAK TESTS

There are many types of leak tests, including the kerosene test mentioned above. A simple test is to slightly pressurize the vessel or piping system and cover the welds with a soap bubble solution. Bubbles will show up at any point where air leaks through the weld.

Water is the most popular liquid used for leak tests. It has a low cost and is readily available.

Chemicals can be added to reduce the surface tension of water, which makes it easier to detect leaks. Water-soluble fluorescent dyes can also be used. They can be seen with black light where they leak through a flaw.

Gases are sometimes used to detect leaks. Extremely small leaks can be detected by use of gas detectors. Tracer gas may also be added to the air or test gas. A sensor probe, passed over the outside of the vessel, can detect the presence of tracer gas that has leaked through a flaw.

When any gas is used in the testing procedure, including inert gases, all proper precautions must be taken in order to carry out the operation in a safe manner.

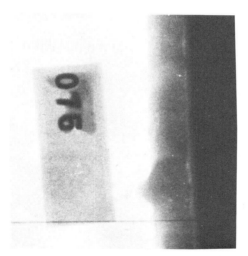
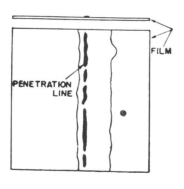
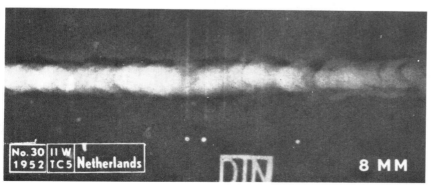

Figure 9E-15 Radiographs showing tungsten inclusions. (Courtesy of American Welding Society)

CHAPTER 10
OXY-FUEL FLAME CUTTING (OC)

Every welder should understand the flame cutting process and know about the safe use and handling of its equipment and gases. There can be no shortcuts if safety is to be achieved.

In most welding schools oxy-fuel welding and cutting are taught before shielded metal arc welding. The basic theory and practice of *oxy-fuel cutting* is reviewed in this chapter.

The first lesson in this chapter reviews the safe practices you should follow.

As a welder, you should be able to sever and cut with accuracy and neatness. Many times you will have to weld joints made from metal you have cut. You will discover a great deal of time and effort go into welding poorly cut and prepared joints. Proper cutting and joint preparation will make your job easier.

Take pride in your work; cutting is part of that work. It is essential that your cutting ability be on a par with your welding ability.

LESSON 10A
SAFE USAGE OF THE EQUIPMENT

OBJECTIVES
Upon completion of this lesson you should be able to:
1. List the precautions to follow to insure safe operation of **oxy-fuel gas cutting** equipment.
2. List in their proper order the assembly steps of an oxy-fuel gas cutting outfit.
3. List in their proper order the steps to pressurize and depressurize the oxy-fuel outfit.
4. Physically demonstrate the proper pressurizing and depressurizing of an oxy-fuel cutting outfit.

TANK
HANDLING AND SECURING
Oxygen and the various types of fuel gases (acetylene, propane, etc.) must be stored and handled properly. Do not store oxygen and fuel gases close to each other. Always follow the codes and regulations. Space should be provided for between them as well as a masonry wall as prescribed by the Fire Underwriters Code.

Gas cylinders should be stored in an upright position. Secure them with a chain or other suitable device so they cannot be knocked over. The caps must be in place while the cylinders are in storage or being moved. Never transport a cylinder with the regulator attached; the regulator connections can be broken by a sharp blow. The regulator could provide the leverage to cause the cylinder valve to break off, and a broken cylinder valve could cause a cylinder containing high pressure to take off like a rocket. Such accidents can kill or cause serious injury. In one instance, the cylinder traveled through two 8-in. thick masonry walls before coming to rest.

A full oxygen cylinder is pressurized to 2200 psi. With enough oxygen, the thrust from the gas could propel the cylinder around the world in 45 minutes. This is illustrated in Figure 10A-1.

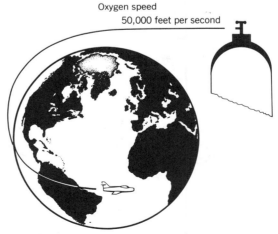

Comparison:
Jet plane = 1000 miles plus per hr
Oxygen speed = around world
25,000 miles in 45 minutes.

Figure 10A-1

SAFE USAGE OF THE EQUIPMENT

Fuel gas cylinder leaks can cause fire or explosion. Always move cylinders carefully by hand, tilting the cylinder so that it is leaning toward your body, or transport them in a cart designed for that purpose. Be sure to secure the cylinders before moving the cart. (Figure 10A-2).

Keep cylinders from where they can be struck by falling objects, moving machinery, vehicles, swinging loads, or other hazards. Be sure the sparks from the cutting or welding operation are not directed toward the cylinders. Never hang either your cutting torch or your electrode holder on them. The heat from the torch or an accidental arc strike could cause the cylinder to rupture.

Keep cylinders from contact with any metal structure that you, or anyone else, is arc welding. There is a possibility of causing an arc between the cylinder wall and the structure. This could cause a cylinder explosion or possibly a fire.

Cylinders should never be placed where their contents can be heated to over 120 degrees Fahrenheit.

Do not drop cylinders or strike them violently. Never use pry bars under the valves or safety caps to pry frozen cylinders loose from the ground. Never use them as rollers or supports. Never lift cylinders by slings or electromagnets. Do not refill cylinders. They should only be filled by a gas supplier. No one, other than a gas supplier, may mix gases in a cylinder. **NEVER USE A HAMMER OR A WRENCH ON A CYLINDER VALVE. IF A CYLINDER LEAKS, REMOVE IT TO AN OPEN AREA.** Report the leak to your instructor, who will call the proper authorities.

ASSEMBLING THE CUTTING OUTFIT

When assembling the oxy-fuel outfit, it is important that you follow the correct procedure. If you form safe work habits, there is little likelihood of making errors. Remember, no oil or grease need ever be used. When oil comes into contact with oxygen, it can burn without having come into contact with open flame. Oxygen makes all fires burn faster. Threads on fittings made of brass are easy to strip. Do not treat them roughly. Use only the special wrench provided with the outfit. The following is one of the correct procedures for assembling an oxy-fuel outfit: (Always follow the manufacturer's instructions.)

1. Secure the cylinders in an upright position, in a cart or tied to an immovable object.
2. Wipe off the seats of the cylinder valves and the regulator and hose connections.
3. "Crack" the valve. This means you should quickly open the valve slightly, then close it. (This is to blow out any dust or foreign particles that may cause leaks or damage the regulators.)
4. Attach the regulators to the cylinders. (Oxygen connections have left-hand threads.)
5. Attach the hoses to the regulators.
6. Make sure the regulator adjusting screws are loosened.
7. Open the oxygen and fuel gas cylinders valves very slowly. (Open the oxygen valve first "all the way" and in the case of acetylene open the valve no more than one and one-quarter turn.)
8. Turn the regulator adjusting screws in one at a time. Then loosen them again to stop the flow of gases. (The purpose of this is to clean out the regulator and the hoses.) Do not let the gas flow uncontrolled. Be sure the area is well-ventilated.
9. Attach the torch to the hoses.

When making connections, use only the wrench recommended for oxy-fuel connectors. Never use an adjustable wrench. Do not use too much leverage on the connector. Be careful not to strip the threads; a snug fit is adequate.

PRESSURIZING THE OXY-FUEL CUTTING OUTFIT

Once you have set up the cutting outfit, and after it is pressurized, check it for leaks.

TESTING FOR LEAKS

This is done by applying special leak detecting solution that you can buy. Some people use soap and water to save money. If soap and water are chosen, only a nonpetroleum-base soap should be used. (Remember, petroleum products and oxygen can cause fires.)

Stir the leak test solution until it is foamy; then brush it on each connection. Bubbles will appear at the leak point (Figure 10A-3). Tighten the connection and retest. Be certain the entire system is leak-free before you start cutting. Checking for leaks is a simple but important process. It should be done each time a connec-

Figure 10A-2 Proper method of storing or removing cylinders in a welding shop.

Figure 10A-3 Soap testing for leaks on an oxygen connection. (Notice the bubble at the bottom?)

tion is made. It is good practice to check for leaks at the beginning of each work day or class session.

The following steps must be performed in sequence to pressurize and leak test an oxy-fuel cutting outfit safely:

1. Make sure that the regulator adjusting screws are loosened.
2. Open the oxygen cylinder valve slowly, "all the way." When the valve is opened "all the way," oxygen cannot leak around the valve stem. (See Figure 10A-4.)
3. Open the oxygen torch valve.
4. Adjust the oxygen regulator until the pressure recommended by the torch manufacturer is reached on the low gauge. Press the cutting oxygen lever and observe the pressure gauge. Readjust the regulator screw to return the gauge to the recommended pressure. (If you do not do this, when you start to cut, the pressure will drop slightly, and you will not obtain a quality cut.)
5. Close the oxygen torch valve.
6. Open the fuel gas cylinder valve, one and one-quarter turn or less.
7. Open the fuel gas torch valve one turn.
8. Adjust the fuel gas regulator until the pressure recommended by the torch manufacturer shows on the gauge.
9. Close the fuel gas torch valve.

The outfit is now pressurized. Also, the gas that flowed through the system put a supply of pure gases up to the torch mixing chamber. In addition, "air" was removed from the system and the chance of oxygen in the fuel gas line or fuel gas in the oxygen line was eliminated. (Either condition could cause an explosion or torch fire or a fire in the regulator or hoses.)

This is not the only method of safely pressurizing a cutting outfit; some manufacturers may have another method. But this method works if it is carried out as prescribed.

DEPRESSURIZING THE OXY-FUEL CUTTING OUTFIT

When you leave the cutting outfit for any length of time, depressurize it. Some people also remove the regulator adjusting screws. By doing so untrained or unauthorized individuals are prevented from using the equipment.

An oxy-fuel outfit should always be purged of all gases after use. This step releases the pressure in the system. The adjusting screws should never be left turned in.* If the screws are left turned in and the cylinder valves are opened rapidly, there is a possibility of an accident. With oxygen, the heat of compression caused when the pressure (2200 psi in a full cylinder) strikes the diaphragm, can cause a regulator explosion.

To safely shut down a cutting outfit, the following steps should be taken to depressurize the system:

1. Close the fuel gas cylinder valve.
2. Open the fuel gas torch valve.
3. Release the fuel gas regulator adjusting screw.
4. Close the fuel gas torch valve. (No pressure should show on either fuel gas regulator gauge.)
5. Close the oxygen cylinder valve.
6. Open the oxygen torch valve.
7. Release the oxygen regulator adjusting screw.
8. Close the oxygen torch valve. (No pressure should show on the oxygen regulator gauges.)

*When the system will not be used for several weeks, turn in the pressure adjusting screws. This is a good practice that lengthens the valve seat life. Repeat the pressurizing procedure when you start work again.

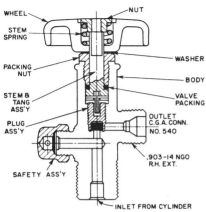

Figure 10A-4 Cut-away drawing of double-seated oxygen cylinder valve. (Courtesy Airco)

SAFE USAGE OF THE EQUIPMENT

These steps relieve all pressure downstream, making the outfit safe for the next person to use.

BACKFIRES AND FLASHBACKS

Before starting to cut you should be aware that mishandling of the torch can cause either **flashbacks** or **backfires.**

A backfire is a momentary flame-out accompanied by a loud snapping or popping noise. Backfires are caused by:

1. Inadequate fuel gas flow (less than that recommended by the manufacturer).
2. Touching, or holding the torch tip too close to the work.
3. Insufficient oxygen pressure.
4. Cutting tip is not screwed firmly into the torch.
5. Dirt on the seat.
6. Overheating the tip.

To remedy the situation do not allow the torch tip to touch the workpiece. Hold the proper flame length, and use the recommended gas pressure. Make sure the joint between the tip and head is clean. Screw the head and tip together securely.

With a flashback the flame burns back inside the torch; it is accompanied by a hissing or squealing sound. Sparks may come out of the torch tip. There can be heavy black smoke coming from the tip, overheating of the blowpipe handle, and even fire bursting through the fuel gas hose. A flashback is much more dangerous than a backfire, because the fire burning inside the torch can progress into the hoses. Flashbacks may be caused by:

1. A small speck of stray hot metal inside the torch tip. (The hot speck can ignite the gases before they leave the torch tip.)
2. Overheating of the tip or the head of the torch. (The torch tip becomes overheated by holding the tip too close to the workpiece.

Working in a tight place, such as in a cavity, can force the flame backward and heat the tip.)

When you hear a whistling or squealing from inside the torch, shut it off immediately. Close the oxygen torch valve first, then shut off the fuel gas torch valve. If you shut the fuel gas valve first, or just shut off the fuel gas valve, the flashback can continue. The reason for first stopping the oxygen is that flowing oxygen continues to support combustion. The metal speck and the metal of the torch itself become the fuel.

Clean the tip when it is cool, then relight the torch. Do not under any circumstances cool the tip or torch by immersing them in water. If the flashback happens again, stop work immediately. Remove the torch from service and return it for repair.

THE IMPORTANCE OF CLEAN CUTTING TIPS

You cannot produce quality cuts with a dirty cutting tip. Lazy welders who don't bother to clean their cutting tips pay for their negligence. Cuts will be poor and weld joints will fit badly, making welding difficult. Poor cuts can be fixed with a chipping hammer and a grinder. But all the extra work is expensive.

Clean the cutting tip often; it will help obtain good cuts and make welding a joint simple. Before cleaning the tip remove it from the torch. Position the torch to let any loose particles fall out of the end. It is bad practice to clean the tip while it is attached to the torch, because dirt particles may be forced into the torch head. The particles could be blown back into the tip when the torch is used. Use a flat file to keep the tip bottom even. Be careful to use the file sparingly or the tip will be worn down rapidly.

Round tip cleaners should be selected to fit very loosely into the orifices of the tin. Figure 10A-5 shows tips with a large orifice in the center.

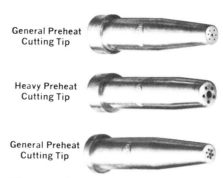

Figure 10A-5 Oxy-acetylene cutting tips showing both the preheat and the high pressure oxygen orifice. (Courtesy Harris Calorific)

This is the high pressure orifice. It is surrounded with a number of preheat holes.

Hold the cleaner very close to its end, between your thumb and forefinger, as shown in Figure 10A-6. Slip it gently into one of the holes. Feed it gradually into the hole until an obstruction is reached or it moves freely. If the tip cleaner cannot penetrate any farther because of an obstruction, do not force it. Remove the tip cleaner and try one with a smaller diameter.

If it moves freely, choose a larger diameter cleaner and reenter the same orifice. Repeat the cleaning steps. Increase the size gradually until the maximum size cleaner is reached. Continue until all orifices are cleaned. Hold the cleaner carefully. Do not let it bend or break off in the torch tip and ruin it.

Reassemble and light the torch, then check the size of the flames. Each of the preheat flames should be the same length and diameter, and they should come straight out from the tip. When the cutting oxygen lever is depressed, the oxygen stream should be straight and in the center of the preheat flame (Figure 10A-7).

ADJUSTING FOR A NEUTRAL FLAME

After pressurizing the oxy-fuel cutting outfit according to the suggested

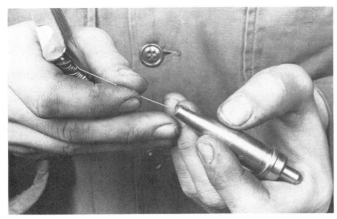

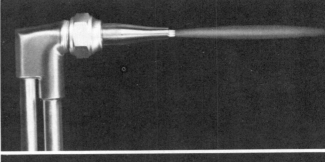

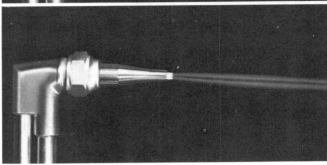

Figure 10A-6 Correct method for holding a tip cleaner when cleaning a torch tip.

Figure 10A-7 *Top:* Properly adjusted, neutral cutting flame (Preheat stage). *Bottom:* Neutral cutting flame with high pressure cutting oxygen steam. (cutting stage). (Courtesy Victor Equipment Co.)

procedure and testing for leaks, the torch may be ignited. To properly ignite the torch follow the manufacturer's instructions. Generally, follow this simple procedure:

1. Open the oxygen torch valve very slightly.
2. Open the fuel gas valve one-quarter to one-half turn.
3. Use a friction striker to ignite the gas mixture. Never use a match.
4. As soon as the flame ignites open the fuel gas valve until the flame is clear of black smoke.
5. Open the oxygen torch valve until the flame shortens and the inner cones of the flame are bright blue.
6. Depress the cutting oxygen lever momentarily. See if the flame jumps away from the torch tip or if the preheat cones become longer. If they do, open the oxygen torch valve a little more; this increases the amount of oxygen to the flame. Momentarily depress the high pressure oxygen lever again. Repeat until the preheat cones remain the same size. When they do, you have not only a neutral flame, but a properly set cutting flame.

LESSON 10B
MAKING STRAIGHT CUTS ON STEEL PLATE

OBJECTIVES
Upon completion of this lesson you should be able to:
1. Safely and correctly adjust the oxy-fuel cutting outfit to a neutral flame.
2. Safely and correctly make straight cuts on steel plate, while following soapstone lines.
3. Safely and correctly assemble, pressurize, depressurize, test for leaks, and operate the oxy-fuel cutting outfit.

NOTE
Being able to make straight or beveled cuts in steel plate is as important as being able to make quality welds. The welder who can make precision cuts will always have weld joints with good fits. Good fitting joints are much easier to weld. A poor fit makes welding more difficult, because you will have to work longer and harder. Poor fit joints also use more filler metal than good fit joints. Such excuses as "Well, I'm in a hurry," or "It's going to get filled in anyway," are poor ones to justify poor cuts.

The truth is the welder either cannot make a good cut or is too lazy to do so. Most good welders also cut well; it is purely a matter of pride. A good welder will also be skilled with the cutting torch.

The key to good cutting is to relax. Position yourself so you are as comfortable as possible. Do not hurry a cut, and do not attempt to make cuts

MAKING STRAIGHT CUTS ON STEEL PLATE

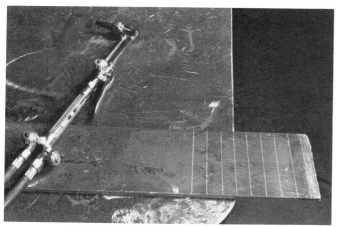

Figure 10B-1 Plate with lines in position for cutting

Figure 10B-2 Correct hand and torch positions for cutting plate

longer than 4 to 5 in. without changing your position.

Before you start, consider the cutting torch you are going to use. It is a precision tool and finely made. It was designed to flame-cut metal. It was not designed to be used as a hammer to knock off pieces of poorly cut metal. If you cannot cut well enough for the scrap portion of the metal to drop off, after the cut is completed, try again or use a chipping hammer to remove it. Never use a cutting torch for any purpose except cutting.

MATERIAL AND EQUIPMENT
1. Proper clothing
2. Welding goggles with either a shade No. 5 or 6 lens
3. Leather gloves
4. Oxy-fuel cutting outfit
5. Chipping hammer
6. Tongs or pliers
7. Soapstone
8. Straight edge
9. Tip cleaner
10. One piece of scrap plate ¼ to ½ in. thick and about 4 in. wide.

GENERAL INSTRUCTIONS
1. Assemble the oxy-fuel cutting outfit according to the instructions of the manufacturer, or your teacher.
2. Pressurize the cutting outfit, following the suggested procedure.
3. Make sure your goggles fit well, are clean, and in good condition. There should not be any openings for sparks to enter.
4. Using the straight edge and the soapstone, mark off a number of lines about ½ in. apart across the width of the piece of scrap.
5. Wear your gloves while practicing and making cuts.

PROCEDURE FOR RIGHT-HANDED WELDERS*
1. Lay the piece of scrap metal on the cutting table as shown in Figure 10B-1. Arrange it so the lines run from left to right as you face them.
2. As shown in Figure 10B-2, pick up the unlit torch. Hold the tubes of the blowpipe in front of the handle between the thumb and first three fingers of your left hand. Don't grab it tightly; just hold it gently between the tips of your fingers. The back of your hand should rest on the plate, with the tips of the fingers pointing upward.

*Reverse the instructions for left-handed welders.

3. Spread your feet apart and lean over the plate as shown in Figure 10B-3. Place yourself to look straight down at both the torch tip and the soapstone lines. Put your right hand on the handle of the torch.
4. Practice moving the torch across the plate from right to left. Follow a line as shown in Figure 10B-4. Do not slide your left hand along. Keep it in one position and allow it to pivot in a rolling motion while your right hand moves the torch slowly to the right. By moving your hands this way the torch tip moves smoothly across the metal. If you allowed the left hand to slide across the metal, the torch movement would be jerky.
5. Practice the movement a few times to acquaint yourself with the feel of the torch and your body position. Now you can proceed to make the first cut. Light the torch and set a neutral cutting flame. Hold the torch as you just learned. Position the tip about ¼ in. in from the right side of the plate. Place the preheat cones approximately ¹⁄₁₆ in. above the plate as shown in Figure 10B-5.
6. Move the tip (flame) in a small

Figure 10B-3 Head over the "kerf"

Figure 10B-5 Flame cone distance from the plate (1/16 to 1/8 in. above plate).

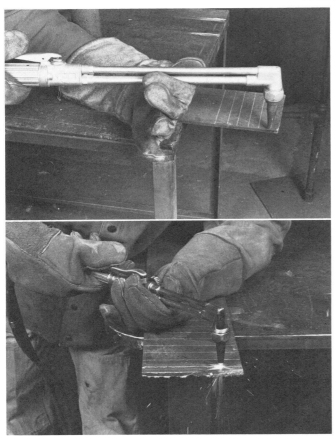

Figure 10B-4 Position of hands for right-handed welder.

circle a little larger than the size of the tip. When the metal is heated cherry red, move the tip to the right-hand edge of the plate. Next, depress the cutting oxygen lever and immediately begin to move the torch tip slowly and steadily toward the left. (See Figure 10B-6.)

7. If you move too rapidly, you will lose the cut. When this happens, release the cutting oxygen lever and go back to where the cut was lost. Preheat that spot, depress the cutting oxygen lever, and move along the line again. Do not move too slowly. The metal will fuse together as you move along and will not separate when you reach the end of the plate. (See Figure 10B-6.)

After you have made a cut of about 4 in., you will reach a point where it is no longer comfortable to handle the torch. At that point move the torch tip at a right angle, and cut away from the line. Move into the scrap side of the cut for a distance of no more than 3/8 in. At that point move the torch tip in a small circle to form a hole approximately 1/4 in. in diameter. Then release the cutting oxygen lever. (Figure 10B-7.) This hole will give you a place to restart the cut. It is located away from the slag that may be on the opposite side of the plate. Slag can deflect the oxygen stream and cause unacceptable gouges in the kerf* and plate surface.

8. Practice on the remaining lines. Become familiar with the feel of the torch.

*The space left by the stream of cutting oxygen is called the **kerf.**

MAKING STRAIGHT CUTS ON STEEL PLATE

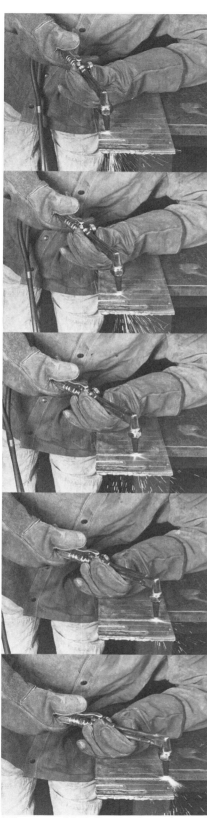

Figure 10B-6 Hand and torch movement.

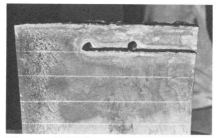

Figure 10B-7 Ending cut away from the cut line

9. To obtain the most accurate cuts it is important to watch the side of the **kerf** nearest you. The edge of the kerf should touch the soapstone line. Do not try to follow the center of the line. You will find the cut is not straight and the piece being cut may be the wrong length, depending on how well you followed the line and how thick it was drawn.
10. Practice until you can produce a straight cut with an acceptable edge. See the examples of good and bad cuts in Figure 10B-8.
11. When you are finished, shut off the torch as you have been taught. Cool off any specimens that you wish to show to the instructor. Always use the tongs or pliers to pick up any hot metal.

POINTS TO REMEMBER
1. Follow the procedure when pressurizing and depressurizing the cutting outfit.
2. Pivot the left hand—do not slide it across the metal.
3. Get into a comfortable position before starting to cut.
4. Keep the preheat flame cones about 1/16 in. above the metal being cut.
5. When you depress the cutting oxygen lever, press it all the way.
6. Make sure the kerf edge nearest you just touches the soapstone line.
7. Learn to judge the travel speed. Never take your eyes off the torch tip or the kerf, even for a moment.
8. Grasp the torch gently. You will not tire as quickly and you will be able to hold it more steadily.

HELPFUL SAFETY REMINDERS
1. Always wear the prescribed clothing.
2. Be sure your gloves are in good condition.
3. Be sure you wear goggles that fit and use the proper shade lens for the job.
4. Prevent fires. Make sure there are no combustibles in the cutting area. Do not let the cutting sparks hit the gas cylinders.
5. Pressurize and depressurize the cutting outfit as prescribed. Be sure to test for leaks each time you use the equipment.
6. Do not pick up hot metal with your gloves. Use the tongs or pliers for this task.

OXY-FUEL FLAME CUTTING (OC)

Figure 10B-8 Examples of good and bad cuts. (Courtesy Harris Calorific)

PRODUCTION CUT has moderately sloping drag lines and reasonably smooth surface. This type represents best combination of quality and economy.

PERFECT CUT shows regular surface with slightly sloping drag lines. Surface can be used for many purposes without machining.

DIRTY TIP or scale in the tip will deflect oxygen stream and cause excess slag, pitting and undercutting.

SLIGHTLY TOO FAST makes drag lines incline backwards, but a "drop cut" is still attained. Quality is satisfactory for production work.

EXTREMELY FAST. Not enough time is allowed for slag to blow out of the kerf. Cut face is often slightly concave.

SLIGHTLY TOO SLOW produces high quality cut, although there is some surface roughness caused by vertical drag lines.

EXTREMELY SLOW produces pressure marks which indicate too much oxygen for cutting conditions.

LESSON 10C
BEVELING PLATE

OBJECTIVES
Upon completion of this lesson you should be able to:
1. Safely and correctly bevel carbon steel plate.
2. Safely and correctly assemble, pressurize, depressurize, test for leaks, and operate the oxy-fuel cutting outfit.

NOTE
Many of the joints you weld must first be beveled. Beveling will allow you to fill in the entire joint and, in most cases, penetrate through to the other side. Precision is more important in beveling than in making straight cuts. The accuracy of the bevel determines the type of root opening you will weld.

The root, or first pass is the most important pass in welding a joint. The root opening or gap must be as even as possible. This minimizes the amount of filler metal that must be deposited and reduces the difficulty of welding.

It is important to bevel the plate on as straight a line as possible and hold the same bevel angle for the entire length of the joint. When the bevel angle changes, the root face of the plate will vary. It will become either shorter or longer depending on how the bevel angle changes.

Beveling is not a difficult operation to learn, but it does require concentration and effort on the part of the welder. Your efforts will pay off in dividends, especially when you weld a set of test plates or weld a joint in a difficult position.

MATERIAL AND EQUIPMENT
1. Proper clothing
2. Welding goggles with either a shade No. 5 or 6 lens
3. Leather gloves
4. Oxy-fuel cutting outfit
5. Chipping hammer
6. Tongs or pliers
7. Soapstone
8. Straight edge
9. Tip cleaner
10. One piece of scrap plate, 3/8 to 1/2 in. × 6 in. × random length.

GENERAL INSTRUCTIONS
1. Assemble and pressurize the oxy-fuel cutting outfit according to the instructions of the manufacturer, or your instructor.
2. Make sure your goggles fit well, are clean, and in good condition.
3. Use the straight edge and the soapstone and make a line across the width of the plate. The line should be thin, so be sure the soapstone has been properly sharpened.
4. Wear your gloves while practicing and making bevel cuts.

PROCEDURE
1. The best method for obtaining a good bevel with a minimum of slag is to cut the bevel at the same time. Some welders first make a square cut, and bevel along this cut. This requires more work than beveling and cutting in a single operation. (This two-step operation is inefficient and unacceptable). The two passes with the torch mean you have to chip the slag off after the cut is made, then chip it off again after the bevel is made.
2. Assume the same practice position as you did in the lesson on cutting plate. However, as shown in Figure 10C-1, line your shoulders up at a 45° angle to the line of the plate.

Figure 10C-1

3. Hold the torch in the same manner as shown in the previous lesson, but tilt it 30 to 37½° off the perpendicular. This is shown in Figure 10C-2.
4. The side of the torch tip nearest to you should almost touch the plate as shown in Figure 10C-3. The torch should be held gently in the left hand. It should slide

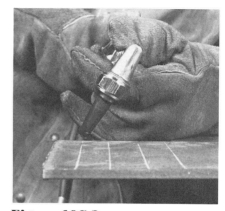

Figure 10C-2

Figure 10C-3 Torch movement sequence.

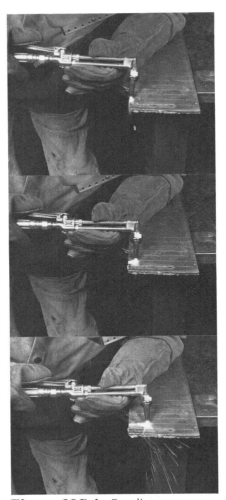

Figure 10C-4 Beveling sequence.

Figure 10C-5 Incomplete cut.

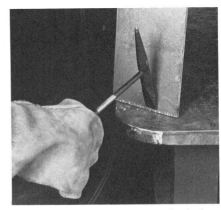

Figure 10C-6 Correct method of removing slag with a chipping hammer.

through the fingers in the same manner as a pool cue. Practice moving the torch over the line, across the plate. When you feel you are ready, light the torch. Make sure the tip is clean and you have the correct cutting flame.

5. Start heating the plate as described in the previous lesson. When the spot becomes cherry red, depress the cutting oxygen lever and begin to cut toward the left side of the plate. Advance the torch as shown in Figure 10C-4, pivoting your left hand and sliding the torch gently through its fingers. Do not grip the torch too tightly with the left hand. A tight grip may cause your movements to be jerky.

6. As you cut look directly into the kerf behind the advancing torch tip. You will be able to see if you are making a clean cut. If the kerf seems to be filling with slag behind the tip, increase the rate of travel.

7. Do not stop when you reach the end of the plate. Continue on past the end of the plate for about ¼ in. Otherwise a small section at the bottom left side of the plate will not be completely severed. This is shown in Figure 10C-5.

8. When you are finished, shut off the torch as you have been taught. Cool the plate. Chip off slag by striking it at a right angle as shown in Figure 10C-6.

GOOD QUALITY is shown by excellent top edge and extremely smooth cut face. The cut part is dimensionally accurate.

POOR QUALITY results in gouging is the most common fault, and is caused by either excess speed or too mild preheat flame.

Figure 10C-7 Examples of good and bad bevels. (Courtesy Harris Calorific)

vel angle. Any change in the cutting angle will change the finished dimensions of the plate.
5. Cut and bevel in a single operation. It will save time, material, and effort, as well as produce a better bevel.
6. Always work with a clean torch tip.

HELPFUL SAFETY REMINDERS
1. Clear the area of combustibles before starting to cut or bevel.
2. Follow the recommended pressurizing and depressurizing procedures. Test for leaks at the start of each work session.
3. Do not pick up hot metal with your gloves. Use pliers or tongs.
4. Place the oxy-fuel cutting outfit out of the path of the flame and flying sparks.
  Have you:
  (a) Cleaned the area so it is ready for the next student?
  (b) Cooled off all hot metal?

9. Compare your bevel with the samples in Figure 10C-7. Practice until you are able to produce good quality bevels.

POINTS TO REMEMBER
1. Never use any oxy-fuel outfit unless you have tested for and repaired all for leaks.
2. Always follow the correct procedure when you pressurize and depressurize the system.
3. Do not grip the torch tightly with the left hand. Hold it gently in your fingers. If you do this, the torch will move freely in your left hand as you cut the bevel.
4. Pay very close attention to the be-

LESSON 10D
FLAME CUTTING AND BEVELING PIPE

OBJECTIVES
Upon completion of this lesson you should be able to:
1. Safely and correctly flame cut and bevel pipe as you rotate it to keep the cut near the top of the pipe.
2. Safely and correctly assemble, pressurize, depressurize, test for leaks, and operate the oxy-fuel cutting outfit.

NOTE
Cutting pipe requires more care than almost any cutting operation. Usually the welder cuts and bevels the pipe in a single operation. This saves time and produces superior cuts, relatively free of slag.

A great deal of skill is needed to control the straightness of the cut and the angle of the bevel, while constantly following the cut line around the pipe.

Care is also required if you wish to obtain quality cuts. As with any welding or cutting operation, it is important to place yourself in a comfortable position. If you are not fairly comfortable, your straining muscles can begin to shake. It is difficult to follow a line accurately when you are not relaxed.

MATERIAL AND EQUIPMENT
1. Proper clothing
2. Welding goggles with shade No. 5 or 6 lens
3. Leather gloves
4. Oxy-fuel cutting outfit
5. Chipping hammer
6. Ballpeen hammer
7. Tongs or pliers
8. Soapstone
9. Wrap-around straight edge
10. Tip cleaner
11. One piece of 4 or 6 in. standard pipe

GENERAL INSTRUCTIONS

1. Assemble and pressurize the oxy-fuel cutting outfit according to the manufacturer's instructions or as you have been taught.
2. Make sure that your goggles fit well and are clean.
3. The area should be free of combustibles before you start to cut.
4. Make sure the cutting tip is clean before you start to cut.
5. Provide a piece of channel, a rest made from angle bar, or some other device to hold the pipe while you make the cut. (See Figure 10D-1.)

PROCEDURE

1. Use the torch flame on the pipe surface. Burn away any coating along the path where the cut is to be made. Thoroughly wire-brush the path after burning.
2. Use the wrap-around and a properly sharpened soapstone to scribe a straight line around the pipe. Keep the line as narrow as possible.
3. Stand to the right of the pipe. Left-handed welders should stand to the left of the pipe. Place yourself to face toward the end that will become scrap. Practice the cutting motions until you feel comfortable, as shown in Figure 10D-2.
4. Rest your left forearm on the pipe, as shown in Figure 10D-2. Hold the torch tip approximately 30 to 37½ degrees away from the perpendicular, pointed toward the scrap piece, as shown in Figure 10D-3. Light the torch, making sure you have a neutral flame.
5. Start at a point near the top center of the pipe and begin to preheat the metal. As shown in Figure 10D-4, the distance from the top should be no more than 3 or 4 in. You can comfortably cut that much pipe without moving your position.

Figure 10D-1 Pipe secured in channel on table.

Figure 10D-2 Welder in position on right side of pipe with left forearm on pipe.

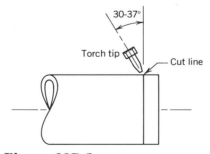

Figure 10D-3

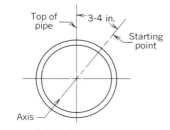

Figure 10D-4

6. Heat the pipe until the zone along the line is cherry red. Then move the torch tip away from the line, about ¼ in., and pierce a hole in the scrap end of the pipe. As soon as you pierce the pipe, rotate the torch to form a small hole. Then, move the torch toward the line to be cut, and start. Cut the bevel from right to left as shown in Figure 10D-5.
7. Cut along the line until your movements feel uncomfortable. Then cut away from the line into the scrap zone. Cut into the scrap zone the same distance you started with and finish the cut with a small circle or keyhole. By starting and finishing the cut sections in the scrap side of the pipe, you can establish the correct bevel angle before you reach the actual cut line. You eliminate the blow back and gouging problems caused by slag. (See Figure 10D-6.)
8. Rotate the pipe after each cut. Start again and make cuts until the next cut will be the last. (See Figure 10D-7.) Before you begin the last section, strike the uncut area with a ballpeen hammer. The hammer blows should remove most of the slag on the remaining section. Look inside the pipe to see if there is any slag left in the cut area. Remove any slag you find or it can spoil the end of your cut.
9. When you have assured yourself that the inside of the cut area is relatively free from slag, continue to cut the last section. Clean the cut face and remove the remaining slag.
10. Tap the bevel lightly with the ballpeen hammer to remove any slag that may adhere to the feather edge. The hammer may also be used to even up the cut, if needed. (See Figure 10D-8.)
11. Inspect the cut for straightness and evenness of the bevel. After

FLAME CUTTING AND BEVELING PIPE

Figure 10D-5

Figure 10D-6 Kerf with keyholes at start and end.

Figure 10D-7 Last ½ in. of cut left until slag is removed.

Figure 10D-8 Hammer being used to remove slag and even up the cut.

cooling it, show it to your instructor.

POINTS TO REMEMBER

1. Never use an oxy-fuel cutting outfit unless you first test for leaks.
2. Always pressurize and depressurize using the proper sequence.
3. Do not grip the torch too tightly.
4. Make sure that you maintain the correct bevel angle.
5. Always work with a clean torch tip. Dirty tips cause poor cuts.

HELPFUL SAFETY REMINDERS

1. Clear the area of combustibles before starting to cut.
2. Follow the procedure you have been taught for pressurizing and depressurizing the system.
3. Do not pick up hot metal with your gloves.
4. Position the oxy-fuel outfit so it will not be in the path of the cutting flame or flying sparks.

CHAPTER 11
THE AIR CARBON ARC CUTTING PROCESS (AAC)

There are a number of methods used to remove or sever metal. The **air carbon arc cutting and gouging process** is one of the most popular. One reason for its popularity is equipment availability. Most of the required equipment is normally on hand in the average welding shop. Usually the only other equipment needed is a torch and suitable carbon arc electrodes.

Another reason for its widespread use is its versatility. It can be used on most metals, unlike the oxy-fuel gas process, which is limited to steel. The air carbon arc process can be used to sever and gouge carbon and low alloy steels, as well as cast iron, stainless steel, copper, aluminum, and magnesium.

Before the air carbon arc process came into use, welding preparation of metals (other than carbon steel) was a tedious process. Good cuts made in a machine shop were expensive, whereas cuts made in the welding shop were of poor quality. They needed a lot of grinding to prepare a joint properly.

You should master the use of the air carbon process. It will serve you well in your chosen field. There are companies that pay as much for proficient air carbon arc cutters as they do for a first-class welder. Many companies expect their welders to be proficient with the air carbon arc equipment.

LESSON 11A
EQUIPMENT AND FUNDAMENTALS OF THE PROCESS

OBJECTIVES
Upon completion of this lesson you should be able to:
1. List the equipment needed for cutting and gouging with the air carbon arc process.
2. Explain the principles of the air carbon arc cutting process.
3. List the safety precautions to be used with the air carbon arc cutting process.

THEORY OF OPERATION
The air carbon arc process uses the heat of an arc and a jet of compressed air. The arc melts the metal and the air jet blows it away. The arc electrodes are made of carbon and graphite.

The process can be used to cut or sever metal, or to gouge grooves. It is especially useful for preparing joints for welding or for removing welds. Figure 11A-1 shows a typical use of the process.

MATERIAL AND EQUIPMENT
The materials needed for air carbon arc (AAC) include a power supply, an electrode holder, a supply of compressed air, and electrodes.

The typical AAC electrode holder or torch differs from those used for welding. The jaws are designed to hold flat or round electrodes. The jaws can pivot so the holes for the high pressure air jet may be aimed along the electrode, into the molten puddle. Figure 11A-2 shows a typical AAC torch. Some torches are designed for

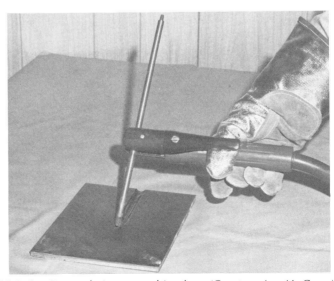

Figure 11A-1 Groove being gauged in plate. (Courtesy Arc Air Corp.)

EQUIPMENT AND FUNDAMENTALS OF THE PROCESS

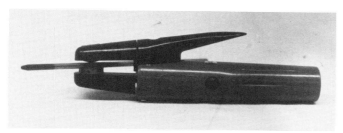

Figure 11A-2 Typical Air Carbon Arc Torch. (Courtesy Arc Air Corp.)

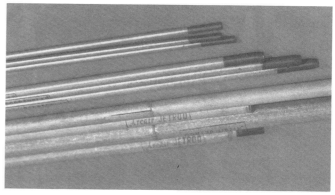

Figure 11A-3 Various size air carbon arc electrodes. (Courtesy Arc Air Corp.)

semiautomatic and automatic operations.

The electrodes are made of carbon and graphite and may be bare or copper coated. A variety of electrode shapes are available as shown in Figure 11A-3.

The most common is the round electrode. But flat and half-round electrodes are often used. For example, when a flat groove is required or when a wide surface needs to be removed, flat electrodes are used. Half-round electrodes are used when the job requires a groove of that shape.

Copper coated electrodes are noted for their resistance to breakage, long life, stable arc characteristics, high current carrying capability, and cutting performance. Bare electrodes (which cost less) can be used for many of the same jobs.

Electrodes are available from 5/32 in. (4.0 mm) to 1 in. (25.4 mm) in diameter.

AIR SUPPLY

The compressed air supply for the air carbon arc process must be about 80 to 100 pounds per square inch. A reliable air supply is needed to blow the molten metal and slag away from the cut.

Low air pressure will lower the cut quality. Leaks can reduce cut quality; therefore, use proper size hoses and fittings. Hoses and fittings with a diameter of at least 3/8 in. (9.5 mm) will be sufficient for torches with a capacity up to 3/8 in. diameter electrodes. Larger hoses are recommended for larger diameter electrodes.

POWER SUPPLY

Almost any welding machine which can deliver the current required can be used for normal applications. However, make sure the manufacturer of the power supply rates it for AAC operation. Some units can be damaged by use with the AAC process.

Automatic operations that use high current and large diameter electrodes use constant current power sources.

The chart in Figure 11A-4 shows electrode recommendations for specific alloys as well as the type of current needed for the best results. It also includes some helpful remarks on the use of the electrodes. The electrode manufacturers also provide recommendations for obtaining quality cuts. The wise welder will always contact the equipment and materials manufacturers for information on the best use of their product.

Figure 11A-4 Electrode and current recommendations for air carbon arc cutting of several alloys. (Courtesy Arcair Corp.)

Alloy	Electrode Type	Current Type	Remarks
Carbon, low alloy, and stainless steels	DC	DCRP	
	AC	AC	Only 50% as efficient as DCRP
Cast irons	AC	DCSP	At middle of electrode current range
	AC	AC	
	DC	DCRP	At maximum current only
Copper alloys:			
Copper 60% or less	DC	DCRP	At maximum current
Copper over 60%	AC	AC	
Nickel alloys	AC	AC	
	AC	DCSP	
Magnesium alloys	DC	DCRP	Before welding, surface must be cleaned.
Aluminum alloys	DC	DCRP	Electrode extension should not exceed 100 mm (4 in.). Before welding, surface must be cleaned.

THE AIR CARBON ARC CUTTING PROCESS (AAC)

Figure 11A-5 gives the recommended current ranges for various sizes of electrodes. The ranges provide a starting point for the welder when setting up the equipment. As a general rule you should use the most current you can work with comfortably. High current cuts more effectively than low current.

AIR CARBON ARC SAFETY REMINDERS

You must be very careful in making preparation for air carbon arc cutting or gouging. Make sure there are no combustibles nearby. Remember that the sparks given off by this process travel quite a long distance. Because of this, greater care is necessary than for welding or cutting with the oxyfuel process.

You must also protect others in the area from the arc glare, the spatter, and noise. Never use the process without proper ventilation. When working in an enclosed area, forced ventilation must be provided. Sometimes it is necessary for a welder to use a respirator. Noise from the cutting operation can readily exceed acceptable levels, especially when working in a small enclosed area. A welder should always use proper ear protection such as ear plugs or over-the-ear safety devices. The protection also prevent sparks and spatter from getting into your ears, as well as protecting against the noise.

Remember to follow all safety precautions you have been taught. A very useful reference is "Recommended Practices for Air Carbon-Arc Gouging and Cutting," Document C5.3, published by the American Welding Society.

Figure 11A-5 Suggested current ranges for the commonly used AAC electrode types and sizes. (Courtesy American Welding Society)

Electrode Diameter		DC electrode with DCRP Amperage		AC electrode with AC Amperage		AC electrode with DCSP Amperage	
(mm)	(in.)	(min)	(max)	(min)	(max)	(min)	(max)
4.0	5/32	90	150	—	—	—	—
4.8	3/16	150	200	150	200	150	180
6.4	1/4	200	400	200	300	200	250
7.9	5/16	250	450	—	—	—	—
9.5	3/8	350	600	300	500	300	400
12.7	1/2	600	1000	400	600	400	500
15.9	5/8	800	1200	—	—	—	—
19.1	3/4	1200	1600	—	—	—	—
25.4	1	1800	2200	—	—	—	—

LESSON 11B
PRECAUTIONS, TROUBLESHOOTING, AND MAINTENANCE

OBJECTIVES
Upon completion of this lesson you should be able to:
1. List the precautions to be taken when using the air carbon arc process.
2. List the problems, causes and solutions involved in troubleshooting with the air carbon arc process.
3. List the important maintenance items pertaining to the air carbon arc process.

GENERAL PRECAUTIONS
As with any process it is always necessary to take precautions. Safe operation of the equipment depends on correct operating conditions. Properly operated equipment produces quality cuts, the end product that it was designed for.

Follow the precautions below. Work in a safe manner and use the equipment according to the conditions recommended by the manufacturer.

Fumes. Never use air carbon arc equipment unless there is adequate ventilation. Where needed, use a respirator for protection against particles in the air.

Combustibles. Always make sure the work area is free of any combustible materials.

Shields. Use protective shields such as sheet metal, to protect other workers from flying sparks.

Safety equipment. Use safety glasses, face shields, leathers, ear protection, and a respirator if required.

Note: for eye comfort and safety a shade 14 lens should be used, especially when using large diameter carbon electrodes. A good rule is to wear a shade that is too dark to see the cutting action, then use one shade lighter.

When in doubt consult the manufacturer for recommendations on the proper safety equipment.

Compressed air. Be sure that you

PRECAUTIONS, TROUBLESHOOTING, AND MAINTENANCE

have at least 80 psi. The air supply lines should be at least 3/8 in. (9.5 mm) I.D. (Inside Diameter) and unrestricted. Blow out the air line before attaching it to the torch. This gets rid of moisture in the lines.

IMPORTANT. Be sure the air is on, and the torch valve open, before striking an arc.

AMPERAGE. Always start with the highest recommended amperage. If the **amperage** is too low, the torch will not operate efficiently.

ELECTRODES. DC electrodes may *only* be used with:

DC. **Reverse Polarity,** electrode positive (+).

AC. Electrodes may be used with either AC or DC **straight polarity,** electrode negative (−).

ARC LENGTH. Maintain a short length arc but *do not* touch the electrode to the work except to start the arc.

ELECTRODE ANGLE. Be sure the electrode angle is correct for the type and depth of groove desired.

TRAVEL SPEED. Adjust the speed of travel to produce a continuous hissing sound and a clean surface.

TROUBLESHOOTING

As with any welding process you may run into trouble of one kind or another. The troubleshooting information in Figure 11B-1 will help you find the cause of the problem.

Study the chart in Figure 11B-1 fully. Memorize as much of the information as possible. Remember where you can find the chart and use it as a reference whenever it is needed.

MAINTENANCE

One of the main causes of trouble is the lack of maintenance. Many welders do not take the time to check out their equipment. Poor equipment maintenance causes lost time and makes it difficult to cut easily and efficiently.

It is far better to do a little preventive maintenance than have a piece of equipment ruined, or have a job

Figure 11B-1 Trouble Shooting Chart (Courtesy Arcair Corp.)

Trouble	Cause	Solution
Hard, irregular start	Air not on before striking arc	Be sure valve is open and air on before striking arc.
Sputtering arc, with electrode slow in heating	Low amperage	Increase amperage and check circuit for poor ground or loose connections.
Sputtering arc, with electrode heating rapidly	Wrong polarity	Change polarity (sometimes polarity switch on welding machine is incorrect.)
Intermittent gouging action	Travel speed too slow	Increase travel speed.
Carbon deposit	Touching electrode to work	Hold and maintain short arc.
Irregular groove	Unsteadiness of operator	Relax, loosen grip on torch. Use steady rest if possible.
Slag adhering to edges	Low air pressure	Increase air pressure, check lines. If pressure cannot be raised reduce travel speed, make lighter cuts.
Groove gets deeper	Electrode angle too steep, travel speed too slow	Reduce electrode angle. Increase travel speed.
Groove gets shallower	Electrode angle too flat, travel speed too fast	Increase electrode angle. Reduce travel speed.

held up because the equipment is not in good condition.

Some maintenance check points to ensure that the equipment is in good condition and is being used properly are listed below:

1. Be sure all threaded electrical connections are tight. This keeps the threads from heating up and reducing current due to increased resistance.
2. Remove moisture from the air lines. Drain the moisture from the bottom of the compressor. Blow down the air lines for a few moments before using them.
3. Make sure the electrode is held firmly in the jaws of the holder.
4. Wire brush the torch head occasionally. Maintain a bright contact surface. This avoids poor contact and arcing.
5. Do not allow the hot end of an electrode to contact the torch.
6. Renew damaged insulators before damage occurs to the head, body, and upper arm of the holder. Figure 11B-2 shows an exploded view of a typical AAC torch.
7. When used in the overhead position, hold the torch at an angle to prevent molten metal from dropping on it.
8. Perform routine maintenance as recommended by the manufacturer of the torch you are using.
9. Treat the torch as a precision tool. Do not use it for a hammer, toss it on the floor, or mistreat it in any manner.

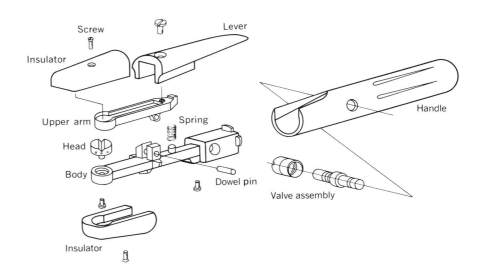

Figure 11B-2 Exploded view of a popular torch in use in industry. (Courtesy Arcair Corp.)

CHAPTER 12
AIR CARBON ARC PROCEDURES

The Air Carbon Arc (AAC) cutting and gouging process gives you the ability to cut and gouge metals that were once difficult, if not impossible, to work with.

A welder without the ability to use this process will have limited employment opportunities. The ability to use the AAC process is a fine addition to a welder's skills. This extra skill can help in finding employment as well as make the welder a more valuable employee.

LESSON 12A
ASSEMBLING THE EQUIPMENT

OBJECTIVE
Upon completion of this lesson you should be able to safely and correctly assemble the air carbon arc equipment in preparation for cutting and gouging.

MATERIAL AND EQUIPMENT
1. AAC torch
2. Power cable and air hose
3. Rubber "boot"
4. Air supply hose
5. AC or DC power supply

GENERAL INSTRUCTIONS
1. Check all fittings for proper size and freedom from foreign matter or nicks.
2. Check for an adequate air supply. The pressure should be 80 to 100 pounds per square inch (psi).
3. Make sure that the power supply is adequate for the job. Do not use whatever power supply is around without checking to see if it can be used.
4. Know the current range for the electrode used.
   Example: Electrode Diameter
   (4.0 mm) 5/32 in. = 90 to 150 amperes.
   (4.8 mm) 3/16 in. = 150 to 200 amperes.
   (6.4 mm) 1/4 in. = 200 to 400 amperes.

CAUTION: **MAKE SURE THE POWER SUPPLY (WELDING MACHINE) IS *OFF* BEFORE STARTING THE FOLLOWING PROCEDURE**

PROCEDURE
1. "Peel back" the insulation boot as shown in Figure 12A-1. This exposes both the air and power connections.
2. When using a standard electrode holder to provide power, clamp the holder on the power cable of the air and cable assembly. Place the jaws of the electrode holder over the connection as in Figure 12A-2. You may also make a permanent connection by using a standard cable connector as shown in Figure 12A-1.
3. Blow out the air hose. Be sure there is no moisture in the line.

Figure 12A-1 (Courtesy Arcair Corp.)

Figure 12A-2

4. Connect the air supply hose to the air and power cable assembly as in Figure 12A-1.
5. Slide the rubber "boot" over the power and air assembly as in Figure 12A-3.
6. Make sure that all connections are tight. If the torch is not fastened tightly, there is a possibility that the air hose will loosen while you are working. If it does, the air hose can whip around and may cause considerable damage or injury.

POINTS TO REMEMBER
1. Make sure the hose connections are clean and undamaged.
2. Check the air and power supply.
3. Tighten all connections.
4. Replace the rubber insulator before turning the power on.

HELPFUL SAFETY REMINDERS
1. A loose connection at the electrode holder is dangerous.
2. An exposed cable or fitting can cause shock or burns.

Figure 12A-3 Rubber boot covering connection.

LESSON 12B
GOUGING PLATE

OBJECTIVE
Upon completion of this lesson you should be able to safely and correctly adjust the amperage and make acceptable grooves in carbon or stainless steel plate.

NOTE
The ability to **gouge** a groove of a desired depth is important to a welder. By being able to determine the groove depth, by control of the electrode angle and travel speed, you can remove the desired amount of metal.

Removal of weld metal is a very important function of the air carbon arc process. Weld defects, such as porosity, slag-inclusions, gas pockets, or other discontinuities must be repaired. Before repairs can be made, the bad spots must be removed.

Remember that any metal that is removed will usually have to be replaced or rewelded. It is important that you remove only enough metal to effect the repair.

Skillful control of the cut depth reduces the amount of welding filler metal needed to weld the groove. This helps to reduce welding costs. Such performance will help make you a valued employee.

MATERIAL AND EQUIPMENT
1. Welding shield (start with No. 14 Lens. Go to No. 12 if you cannot see well.)
2. Safety glasses
3. Proper clothes
4. Protective leather and gloves, full jacket and cap
5. Ear protection
6. Chipping hammer
7. 3/16 in. (4.8 mm) diameter copper coated electrodes
8. 1 pc. 1/2 in. thick scrap carbon steel or stainless steel plate

GENERAL INSTRUCTIONS
1. Position the plate as shown in Figure 12B-1.
2. Make sure the power cable and the work piece are secure.
3. Direct Current Reverse Polarity (DCRP)—electrode positive (+) amperage range: 150 to 200.
4. Adjust the current control to approximately 200 amperes. If you are not sure how to adjust the current, ask the instructor.

PROCEDURE
1. Place an electrode in the holder. The pointed end should not stick out more than 6 in. from the jaws as shown in Figure 12B-2.
2. Make sure the air jet holes are on the side of the electrode opposite the direction of travel. (Refer to Figure 12B-3 and Figure 12B-6).
3. Start up the air jets.* (See Figure 12B-2.) Then strike an arc at the right side of the plate. Hold the electrode perpendicular to the plate as in Figure 12B-4. This is called the **work angle.** *Note:* Left-handed welder should start at the left side of the plate.
4. Tilt the electrode holder so that the electrode is at a 35 to 45° angle at the surface of the plate as in Figure 12B-5. This is the **push angle.**

*The method varies, depending on the manufacturer of the torch. One version is shown in Figure 12B-6.

Figure 12B-1 Plate position.

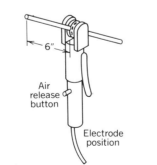

Figure 12B-2 Electrode position.

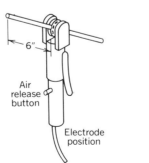

Figure 12B-3 Position of air jets.

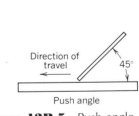

Figure 12B-4 Work angle.

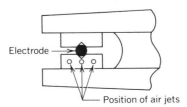

Figure 12B-5 Push angle.

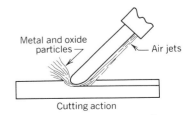

Figure 12B-6 Cutting action.

CUTTING LIGHT PLATE

5. After the arc ignites move steadily toward the left side of the plate. If you cannot maintain the arc, gradually reduce the amperage setting on the power supply until you can. Then move steadily across the plate. If you still have difficulty, you may not have enough current. (Hold a close arc.) The groove should look similar to Figure 12B-7.

6. Make another groove with the push angle reduced to about 22½°. Notice the groove is not as deep as the first one. By adjusting the push angle you can vary the depth of the groove. With practice you will be able to remove

Figure 12B-7 Sample gouged groove. (Courtesy Arcair Corp.)

as little as ¹⁄₁₆ in. (1.6 mm) of metal at a time.

7. If slag piles up in front of the electrode, stop. Chip it off with your scaling hammer and start again.

Slag is difficult to remove with the arc.

8. Practice gouging with various push angles. Present the plate to your instructor for inspection.

POINTS TO REMEMBER
1. Set the power supply at the highest recommended amperage. Adjust the current as necessary.
2. The greater the electrode push angle the deeper the groove.
3. Start the air jets before striking an arc.
4. Make sure the air jets are behind the electrode. This makes sure they blow the metal in the direction of travel as in Figure 12B-6.

LESSON 12C
CUTTING LIGHT PLATE

OBJECTIVE
Upon completion of this lesson you should be able to safely and correctly cut plate, ⅜ in. or less in thickness.

In the previous lesson you learned to gouge grooves of various depths in plate. There are also times you will be called upon to sever, or cut, plate. It is important for you to know the easiest and best method to perform this task.

Three-eighths inch thick plate is the most you should try to cut with a single pass. This lesson covers cutting the lighter thicknesses. Heavy plate is covered in the next lesson.

MATERIAL AND EQUIPMENT
1. Welding shield (Start with No. 14 Lens. Go to No. 12 if you cannot see.)
2. Safety glasses
3. Proper clothes
4. Protective leather and gloves, full jacket and cap.
5. Ear protection
6. Chipping hammer
7. ³⁄₁₆ in. (4.8 mm) diameter copper coated carbon electrodes.
8. 1 pc. carbon steel plate ¼ in. or ⅜ in. × 4 in. × 12 in. lg.

GENERAL INSTRUCTIONS
1. Position the plate as in Figure 12C-1.
2. Make sure the power cable and the work piece are secure.
3. Use Direct Current Reverse Polarity (DCRP-electrode positive)
4. Adjust the current control to approximately 150 to 200 amperes. Start out at the high end of the range. If you are not sure how to adjust the current, ask your instructor.

PROCEDURE
1. Place an electrode in the holder. The pointed end should not stick out more than 6 in. from the jaws as shown in Figure 12C-2. Make

Figure 12C-1 Plate position.

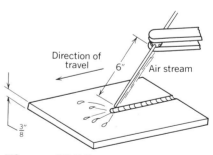

Figure 12C-2

sure the air jets are on the side of the electrode opposite the direction of travel. (Refer to Figure 12C-2).

2. Hold the same work angle as in the last lesson (Figure 12C-3), but use a push angle about 10 to 15° off the perpendicular. (Refer to Figure 12C-4.)

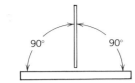

Figure 12C-3 Work angle.

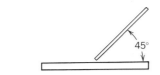

Figure 12C-5 Gouging angle.

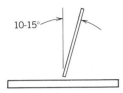

Figure 12C-4 Push angle.

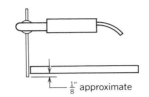

Figure 12C-6

3. Start up the air jets and strike the arc on the right side of the plate.
4. After ignition takes place, allow the electrode to penetrate through the entire thickness of the plate, then steadily move toward the left.

If you have the proper heat setting and travel speed, the plate will cut easily.

5. If you have trouble making the cut in one pass, try again. First lower the lead angle as in Figure 12C-5 and proceed as in gouging. Next, after you gouge out one pass reenter the groove at the right side and cut out the remaining metal.
6. If the finished cut is not straight and smooth you can use an arc off the side of the electrode to even the metal out. (Refer to Figure 12C-6).
7. Experiment with your electrode angles, vary them slightly. Attempt to produce as smooth a cut as possible.

POINTS TO REMEMBER
1. Set your power supply at the highest recommended amperage before you start.
2. Always start the air jets before striking an arc.
3. Make sure the air jets are behind the electrode so they blow the metal in the direction of travel.

LESSON 12D
CUTTING HEAVY PLATE

OBJECTIVE
Upon completion of this lesson you should be able to safely and correctly cut heavy plate, over 3/8 in. in thickness.

NOTE
Cutting heavy plate with AAC is no more difficult than cutting lighter plate. Slightly different procedures are used and when they are followed, excellent cuts are produced.

The single most important difference to remember when cutting heavy plate is the *air jets must be on the side of the electrode facing the direction of travel*. This is opposite to the side you have used to the present.

MATERIAL AND EQUIPMENT
1. Welding shield (Start with No. 14 Lens. Go to No. 12 if you cannot see.
2. Safety glasses
3. Proper clothes
4. Protective leather and gloves
5. Ear protection
6. Chipping hammer
7. 3/16 in. (4.8 mm) diameter copper-coated electrodes
8. 1 pc. scrap carbon or stainless steel plate 1/2 in. or more in thickness.

GENERAL INSTRUCTIONS
1. Position the plate as in Figure 12D-1.
2. Make sure the workpiece clamp and the workpiece are secure.
3. Adjust the welding current: Direct Current Reverse Polarity (DCRP) electrode positive (+) ampere range: 150 to 200.

Figure 12D-1 Plate position.

PROCEDURE
1. Turn the electrode holder and the jaws so the air jet holes will be on the left side of the electrode, as you travel from right to left as shown in Figure 12D-2.

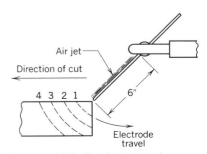

Figure 12D-2 Cutting action.

2. Place the electrode in the holder. The pointed end should not stick out more than 6 in. as in Figure 12D-2.
3. Hold the electrode at a 90° work angle as in Figure 12D-3. Start the cut with a lead angle of 45° as in Figure 12D-4. When the electrode angle is in the direction of travel it may be called the "push" angle.
4. Strike the arc on the right side of the plate. Then rotate the electrode down as it cuts into the plate, until the electrode angle is about 60° from the plate. Refer to Figure 12D-4. This first move will not take you all the way through the plate. Notice the first dashed line in Figure 12D-2.
5. Swing the electrode back to the top of the plate. Do not break the arc. When you reach the top, push the electrode deeper into the puddle. Then swing the electrode down and to the right again, as with the first move. Refer to dashed line No. 2 in Figure 12D-2.
6. Repeat this swinging and cutting motion as you travel toward the left. As you move to the left you may notice the cut is not uniform and the kerf may be uneven. If this happens, you can use the side of the electrode to even out the cut. This is done by moving the electrode back and forth during the cutting process. (See Figure 12D-5.)
7. Practice a few cuts. Then try to produce an acceptable accurate cut as follows:
   (a) Draw a line across a plate using a straight edge and a well-sharpened soapstone.
   (b) Place center punch marks 3/8 in. apart along the line.
   (c) Using the procedure above, cut along the line. Make sure the edge of the kerf removes one-half of the center punch marks; leave the half closest to you intact. (See Figure 12D-6). Compare your cut with the sample in Figure 12D-6.

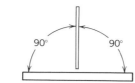

Figure 12D-3 Work angle.

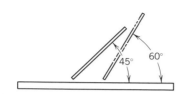

Figure 12D-4 Push angle.

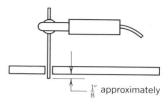

Figure 12D-5

Figure 12D-6 Partially completed cut in ¾ in. plate showing half of the center punch marks left on the plate surface.

CHAPTER 13
FLAT OR DOWNHAND WELDING

The most commonly used welding position is the **flat,** or **downhand,** position. This means the metal to be welded will be placed in front of you just like a book on your desk or food on the table. (See Figure 13-1). Employers like you to use this welding position as much as possible. It is very efficient because specially designed electrodes with high rates of deposition can be used. This simply means they deposit weld metal faster than electrodes used in all positions. These high deposition electrodes use iron powder in their coatings. It melts rapidly, and adds up to 50 percent more filler metal to the puddle than the out-of-position electrodes. Electrodes in this category are the E-7024, E-6027, and E-7020. These electrodes are known as fast fill electrodes. Another reason for the popularity of this welding position is that it is less tiring for the welder. One more reason for using the flat position is that it requires the least welding skill. Welders with a little training can be hired to do downhand welding.

Employers often purchase expensive fixtures to position heavy parts for welding in the flat position; this is illustrated in Figure 13-2.

The **flat position** is the first one you will be taught in welding class. It is important that you learn your tasks well. Everything you learn will be helpful in the other phases of your welder training. The good habits you develop will stay with you, but so will the bad ones. Do the best you can from the beginning. Bad habits are very difficult to break.

Welding in this position is fairly simple. The molten metal deposited by the electrode transfers through the arc into the weld. Gravity does most of the work for you. In out-of-position welding, such as **horizontal, vertical,** and **overhead,** you have to combat the force of gravity. It makes learning to weld in these positions more challenging. In the flat position, the electrode is normally held perpendicular to the plate with a 10 to 15° tilt away from the direction of travel. This is shown in Figure 13-3.

After the lessons in this chapter you will be able to deposit smooth, even weld beads of good quality in the flat position. Although you will not be considered a welder in the true sense of the word, you may be able to obtain a job with the skill you develop in this portion of your training.

In the previous lessons on the air carbon arc process you made some power supply adjustments to obtain a reasonable current with which to work. In this, and in all future lessons, and throughout your welding career, you will have to make many critical adjustments if you wish to produce quality welds.

It is important for a welder to be able to set the arc current for weldments with different size, thickness, material, or position. All of these factors are important to

Figure 13-1 Plate in position for downhand (flat) welding.

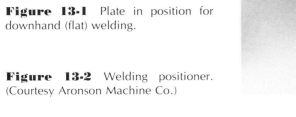

Figure 13-2 Welding positioner. (Courtesy Aronson Machine Co.)

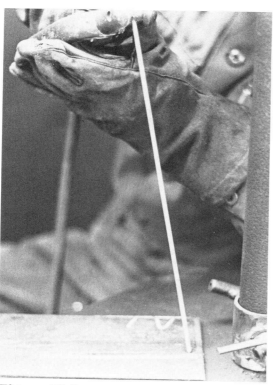

Figure 13-3

know when setting the current. For this reason the welder must learn to "set his heat," not only by setting the welding current adjustment control, but by sight and sound. Through practice a welder can recognize the correct current by the arc sound and the way the molten pool reacts.

Use the current adjustment indicator on the welding machine as a guide to obtain the approximate current setting. Also, use it as a reminder of where to reset the control when starting out on a similar weldment.

The chart in Figure 13-4 lists the **ampere range** for carbon steel electrodes as recommended by the American Welding Society. This chart can be used to determine the approximate amperage setting for a particular size and type electrode, but you must make the final choice.

The amperage range for the electrodes used in each lesson will be shown in the materials and equipment section.

HELPFUL SAFETY REMINDERS

Although welding in the downhand position is not particularly dangerous, it is important to follow safe work practices. Always wear the proper clothing and protective equipment. This chapter deals with welding in the flat position so it is not necessary to wear a full leather jacket. However, some sort of arm protection should be used to protect them from sparks.

Figure 13-4 Typical amperage ranges. (Courtesy American Welding Society)

Electrode diameter in.	mm	E6010 and E6011	E6012	E6013	E6020	E6022	E6027 and E7027	E7014	E7015 and E7016	E7018 and E7018-1	E7024 and E7028	E7048
1/16	1.6	...	20 to 40	20 to 40	...	...	...	...	...	...	...	...
5/64	2.0	...	25 to 60	25 to 60	...	...	...	...	...	...	...	...
3/32[a]	2.4[a]	40 to 80	35 to 85	45 to 90	...	...	...	80 to 125	65 to 110	70 to 100	100 to 145	...
1/8	3.2	75 to 125	80 to 140	80 to 130	100 to 150	110 to 160	125 to 185	110 to 160	100 to 150	115 to 165	140 to 190	80 to 140
5/32	4.0	110 to 170	110 to 190	105 to 180	130 to 190	140 to 190	160 to 240	150 to 210	140 to 200	150 to 220	180 to 250	1.50 to 220
3/16	4.8	140 to 215	140 to 240	150 to 230	175 to 250	170 to 400	210 to 300	200 to 275	180 to 255	200 to 275	230 to 305	210 to 270
7/32	5.6	170 to 250	200 to 320	210 to 300	225 to 310	370 to 520	250 to 350	260 to 340	240 to 320	260 to 340	275 to 365	...
1/4	6.4	210 to 320	250 to 400	250 to 350	275 to 375	...	300 to 420	330 to 415	300 to 390	315 to 400	335 to 430	...
5/16[a]	8.0[a]	275 to 425	300 to 500	320 to 430	340 to 450	...	375 to 475	390 to 500	375 to 475	375 to 470	400 to 525	...

a. These diameters are not manufactured in the E7028 classification.

In addition to the helpful safety reminders contained in the lessons you should always work in a safe manner and remember to:

1. Pick up hot metal with tongs, pliers, clamps. **NEVER PICK UP HOT METAL WITH YOUR GLOVES OR BARE HANDS.** Gloves used to pick up hot metal will quickly become stiff, brittle, and unusable.
2. Cool all metal after welding.
3. Remove all welding electrode stubs from the area.
4. Leave the area clean for the next student.
5. Store the welding cables in their proper place at the end of the lesson.

GENERAL MATERIAL AND EQUIPMENT

Item one of the Material and Equipment section of each lesson will be personal equipment welding. For all the lessons in this chapter this means:

1. Welding shield
2. Safety glasses
3. Protective leathers and gloves
4. Chipping hammer and wire brush

Throughout the lessons in this chapter you may use either an AC or DC welder, depending on their availability. When a DC welder is used, the polarity will be indicated in the Material and Equipment section, along with any other items necessary to properly complete the job.

One final note before you start your "hands on" training in shielded metal arc welding: The lessons in this book have been written for the right-handed student. However, left-handed students should not have any trouble. They should start at the opposite end of a plate and reverse the right-hand electrode drag and push angles.

Everyone should learn to weld both ways, from right to left and from left to right. In addition, they should also learn to weld toward and away from themselves. The proficient welder can weld in any direction.

This chapter begins your "hands on" welding training. But there is something very important to remember before you start. **NAMELY, YOUR INSTRUCTOR IS THE EXPERT.** Go to your instructor whenever you have a problem. Don't move on to a new lesson before the instructor has lectured and demonstrated that lesson—unless of course, you have permission to do so.

LESSON 13A
STRIKING AND MAINTAINING AN ARC

OBJECTIVE
Upon completion of this lesson you should be able to safely and correctly strike, maintain, and restart an arc.

NOTE
Learning to strike and maintain the arc is very important. Each time you weld with a new electrode, or a partially used one, you must strike an arc.

If you cannot master this task, you will have difficulty learning to weld. The instructor will waste time while waiting for you to strike an arc successfully. Also, productive welding time is lost as you attempt to start an arc. Finally, misstarts leave scars on the workpiece.

Starting and maintaining an arc is not very difficult to learn if you follow directions and practice correctly. Do not attempt to see how long you can keep the arc going. That is not the purpose of this lesson.

The purpose of this lesson is to learn to strike an arc, where you want to strike it, accurately and consistently.

MATERIAL AND EQUIPMENT
1. Personal equipment (as listed under General Material and Equipment)
2. E-6010 or E-6011 5/32 in. (4.0 mm) diameter electrodes
3. One piece of scrap plate 1/4 in. (6.35 mm) or 3/8 in. (9.52 mm) × 4 in. × 8 in. lg.

GENERAL INSTRUCTIONS
1. Wire brush the welding surface of the plate. The axis of the weld will be as shown in Figure 13A-1.
2. Position the practice plate at the right-hand corner of the work table. See Figure 13A-2.
3. Secure the workpiece clamp and adjust the welding current. Use Direct Current Reverse Polarity (DCRP) electrode position (+), ampere range: 75 to 125.

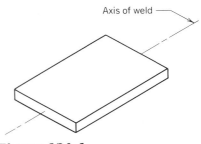

Figure 13A-1

Figure 13A-2 Position of piece on table.

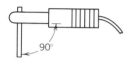

Figure 13A-3 Position of electrode in holder.

CAUTION: When AC and DC currents are recommended, *only* **DC current can be used with E-6010 electrode.** Either AC or DC can be used with E-6011 electrode. This statement holds true throughout this book.

PROCEDURE

1. Place the electrode in the holder at a 90° angle. (See Figure 13A-3.)
2. Allow your right hand and the electrode holder to rest in the left hand. (See Figure 13A-4.)
3. Hold the electrode at approximately a 30° angle off the perpendicular about 1 in. above the left-hand edge of the plate. (See Figure 13A-5.)
4. Lower your shield, then bring the electrode down and to the right using a swinging motion (like striking a match). (See Figure 13A-5.)
5. When the electrode momentarily touches the plate, it will strike an arc. (When this happens, raise the electrode slightly, about 3/16 in. (See Figure 13A-5.)
6. Immediately lower the electrode to reduce the **arc gap**. The distance between the tip of the electrode and the **workpiece** should be approximately the thickness of the electrode core wire (5/32 in.). (See Figure 13A-5.)
7. *Do not lift your shield,* but break the arc by snapping the electrode up to the left. The power supply cannot provide the correct arc current to a long arc. As you increase the arc length the current is decreased. There will be some arc length that causes the arc to suddenly extinguish. When this happens, restart the arc exactly as you did the first time. (Continue

Figure 13A-4 Hand positions.

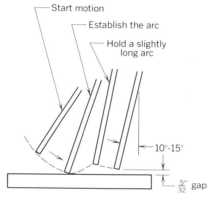

Figure 13A-5 Scratch start method.

Figure 13A-6 Plate with starts and restarts.

to strike and strike the arc until the electrode burns down to approximately 2 in.) (See Figure 13A-6.)

8. The electrode melts with each start. It gets shorter and you must constantly feed the electrode toward the puddle. (If you do not lower the electrode, the arc gap will increase, and the arc will go out.)
9. Don't strike the arc just anywhere. Pick a point before you start, and strike the arc as close to that point as you can. Attempt to have your completed plate resemble Figure 13A-6.

POINTS TO REMEMBER:

1. If you start out with too much of an angle with the plate, the electrode will not skip off the plate. It will stick to the plate or "stub out."
2. There is another method used with a tapping motion that you will learn later. For the present always use a swinging or scratching motion.
3. Sometimes an electrode can be accidentally spot welded to the plate during an arc start. This is called "stubbing out." If your electrode sticks to the plate, snap it loose by rocking the electrode holder rapidly from side to side.

Remember the current is still flowing while you are doing this. **DON'T TRY TO PULL THE ELECTRODE OFF OR RELEASE THE JAWS OF THE HOLDER.** The current will jump across the gap you create and damage the holder. It may also startle you.

4. The purpose of this operation is to teach you to strike and restrike the arc. **DO NOT TRY TO SEE HOW LONG YOU CAN MAINTAIN THE ARC.** Each time you start a new electrode you must strike an arc. If you do it well, you save your time and the instructor's time.

5. Constantly feed the electrode into the puddle to maintain a proper arc gap.

HELPFUL SAFETY REMINDERS
DO NOT PICK UP ANY HOT METAL WITH YOUR BARE HANDS OR YOUR GLOVES. Use a pair of tongs or pliers to grip the metal. (If you use your gloves, they will become stiff and brittle and be unusable in a very short time.)

LESSON 13B
WELDING STRINGER BEADS

OBJECTIVES
Upon completion of this lesson you should be able to:
1. Safely and correctly weld **stringer beads** in the flat or downhand position.
2. Understand the effects of different arc lengths, welding currents, and travel speeds.

NOTE
A good welder should be able to weld stringer beads correctly. Although there are other methods used, almost all types of joints can be welded successfully with the stringer method.

A stringer bead is a single weld bead. It varies in width from 2 to 2½ times the diameter of the metal electrode. For example, if you use a ⅛ in. (3.2 mm) electrode, the bead should be between ¼ in. (6.4 mm) to ⁵⁄₁₆ in. (8.0 mm) wide. (See Figure 13B-1).

The metal electrodes have coatings of varying thicknesses; therefore, it is important that you only consider the diameter of the metal core wire when judging the width of the bead.

You should become proficient in the use of the stringer bead. It will make **out-of-position welding** (in the following chapters) easier.

In this chapter you will learn to control the width, height, and bead shape by changes in travel speed, electrode angle, welding current, and arc length.

Do not start to weld until you have heard your instructor's lecture, seen a demonstration, and have received all the necessary information. Read this entire lesson before starting. If you have any questions, ask the instructor before you begin.

MATERIAL AND EQUIPMENT
1. Personal equipment

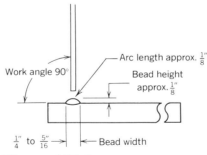

Figure 13B-1

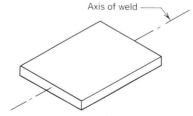

Figure 13B-2 Plate position.

2. E-6010 or E-6011 ⅛ in. (3.2 mm) electrodes
3. One piece of carbon steel plate approximately ⅜ in. (9.5 mm) × 4 in. × 8 in. lg.

GENERAL INSTRUCTIONS
1. Wire brush the welding surface of the plate.
2. The plate must be welded with the axis as shown in Figure 13B-2.
3. Position the practice plate at the right-hand corner of the work table just as you did in the previous lesson. (See Figure 13B-3.)
4. Attach the workpiece clamp and adjust the power supply as di-

Figure 13B-3 Position of piece on table.

rected. Use alternating current (AC) for E-6011 electrodes or Direct Current Reverse Polarity (DCRP), electrode positive (+) in the range of 75 to 125 amperes for either electrode.

5. Use a clamp, if provided, to hold your **workpiece** in place. If you do not have a clamp, tack weld the plate to the table. Use a single tack weld.

6. Make sure the workpiece cable is securely attached.

CAUTION: When AC and DC currents are recommended, *only* **DC current can be used with E-6010 electrode.** Either AC or DC can be used with E-6011 electrode. This statement holds true throughout this book.

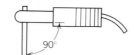

Figure 13B-4

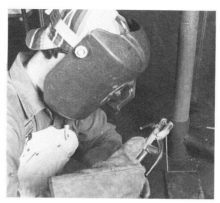

Figure 13B-5 Welder—elbow and hand positions.

PROCEDURE

1. Place the electrode in the holder at a 90° angle. (See Figure 13B-4.)

2. Assume a comfortable position, with your left elbow resting on the table. Place your right hand, the one with the electrode holder, in the palm of your left hand. (Reverse the position if you are left-handed.) (See Figure 13B-5.) Do not grip the holder tightly. If you do, you will tire quickly and your hand will become unsteady.

3. Strike the arc on the left side of the plate. Tilt the electrode holder slightly to the right. The electrode should point at about a 10 to 15° angle from the vertical into the puddle. This is commonly called the *lead* or drag *angle*. (See Figure 13B-6.)

4. Maintain an arc length of about ⅛ in. (3.2 mm) and keep the width of the puddle to 2 to 2½ times the diameter of the core wire and approximately ³⁄₃₂ in. in height to ⁵⁄₁₆ in. wide. (See Figure 13B-1.)

5. Run the stringer bead for about 3 in. **STOP**—examine the stringer for evenness of the ripples, width, and height. (Chip the slag away and wire brush thoroughly before continuing.)

6. Strike another arc about ¾ in. to the right of the weld puddle **crater** at the end of the stringer. Then move the arc directly over the crater. It should be a slightly long arc. Hold it in position for a moment. Then close the arc gap to normal and proceed as before.

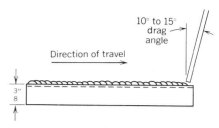

Figure 13B-6

7. Continue in the same manner, but start the next stringer ½ in. below the first. When the plate is completely covered, **COOL IT** and present it to the instructor for his or her comments or a grade.

8. Weld the opposite side of the plate as follows:
   1st **Stringer**—Correct arc length, too short an arc, too long an arc. Note the results.
   2nd **Stringer**—Correct travel speed, too slow a travel speed, too fast a travel speed.
   3rd **Stringer**—Correct current setting, too low a setting, too high a setting.

POINTS TO REMEMBER:

1. Use your left hand to brace your right hand and the electrode holder, unless your instructor prefers an alternate method. **DON'T GRIP THE ELECTRODE HOLDER TIGHTLY—BE RELAXED.**

SOME SAFETY REMINDERS

1. DO NOT PICK UP ANY HOT METAL WITH YOUR BARE HANDS OR YOUR GLOVES.

2. Have you cooled off all hot metal?

LESSON 13C
THE ESSENTIALS OF PROPER WELDING PROCEDURES

OBJECTIVES
Upon completion of this lesson you should be able to:

1. List the essential variables that influence weld quality.

2. List the causes of common welding problems.

3. List the methods used to solve common welding problems.
4. Identify common welding problems.

NOTE

By now you have had some "hands on" welding practice and you probably have encountered some of the most common welding problems. This lesson describes the essential variables and how they affect the finished weld. There are five essential variables. If you vary or change them beyond permissible limits, they will affect the appearance and quality of the weld. To obtain a proper weld you must consider these essential variables:

1. Electrode size
2. Current
3. Arc length
4. Travel speed
5. Electrode angle

Electrode Size

When choosing an electrode, you must consider the size, shape, and position of the weld joint.

Usually, you should choose the largest diameter electrode that can be used for the particular weldment and position. Large diameter electrodes have a higher deposition rate than small ones. With large electrodes the job will be finished in a shorter time, as well as at lower cost. In addition, larger electrodes use more current than small ones. Therefore, the arc has more energy and you can weld faster.

Current

The correct polarity of current is important as the amount of current. Some electrodes require Alternating Current (AC), whereas others require Direct Current (DC). Some electrodes can use either AC or DC. Always read the information on the box of electrodes. It will suggest the amount of current to be used and its polarity. (Lesson 5A discusses polarity and direction of the welding current.)

Arc Length

The arc length you should use is approximately the same as the diameter of the electrode you use. Correct arc length is important; the wrong arc length can cause poor welds. Most of the time you will know the correct arc length by its sound. An arc of the correct length sounds like the crackling of bacon and eggs frying. An arc that is too long has a loud humming sound. One that is too short has a sputtering sound.

Travel Speed

Travel speed must be matched with the amount of arc current. High currents require a high travel speed. If the current is low, your rate of travel should be slow.

Electrode Angle

The angle at which you hold the electrode is very important. The molten metal leaving the electrode transfers quite forcefully. In most cases it tends to transfer where the electrode is aimed. The metal transfer can be compared with water from the nozzle of a water hose. If you have ever pointed a water hose at the ground, you were able to observe the water flows in the direction of the hose nozzle. (See Figure 13C-1.) The molten metal transfers in the direction of the electrode. The arc force on the weld puddle can force it to one side in a direction due to electrode angle. (See Figure 13C-2.)

The partial list of conditions that follows describes some of the results they cause:

1. Current too low (low arc energy and low arc force)
   (a) Poor penetration
   (b) Small molten puddle, narrow bead
   (c) Excessive piling up of the weld metal
   (d) Slow progress
   (e) Electrode sticks to the base metal
2. Current too high (high arc energy and high arc force)
   (a) Electrode melts too fast, too much metal for travel speed.
   (b) Excessive **spatter**
   (c) Undercut along the edges
   (d) Irregular deposit
3. Arc length too long
   (a) Irregular bead
   (b) Poor **penetration**
   (c) Wavering arc
   (d) Poor fusion
4. Arc length too short: (low arc energy and high arc force)
   (a) Not enough heat to melt the base metal
   (b) High, uneven bead
   (c) Poor fusion
   (d) Slag holes, **gas pockets,** and **porosity**
   (e) Possibility of the electrode stubbing out
5. Travel speed too fast (low heat input)
   (a) Puddle cools too rapidly
   (b) Impurities and gases are locked in

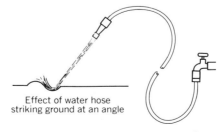

Figure 13C-1 Effect of water hose striking ground at an angle.

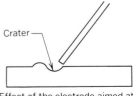

Figure 13C-2 Effect of the electrode aimed at a low angle to the plate.

(c) Narrow irregular bead
(d) Ripples in the shape of a half moon

Assignment: An experiment in the essential variables.

MATERIALS AND EQUIPMENT
1. Personal equipment
2. E-6010 or E-6011 electrodes ⅛ in. (3.2 mm) diameter
3. 1 pc. carbon steel plate approximately ⅜ in. × 4 in. × 8 in. lg.

GENERAL INSTRUCTIONS
1. The plate must be welded in the flat position as shown in Figure 13C-3.
2. Position the plate so the narrow side is facing you. (See Figure 13C-4.)
3. Adjust the welding current: AC or DCRP electrode positive (+), ampere range: 75 to 125.

CAUTION: When AC and DC currents are recommended, **only DC current can be used with E-6010 electrode.** Either AC or DC can be used with E-6011 electrode. This statement holds true throughout this book.

PROCEDURE
1. Start ½ in. from the top of the plate. Listen to the sound of the arc as you run a stringer from left to right. Keep the electrode perpendicular to the plate. Use a 10 to 15° drag angle. (See Figures 13C-5 and 13C-6.) Familiarize yourself with the sound you hear.
2. Hold a very close arc. Repeat the procedure of step No. 1 and weld another stringer. Place it ½ in. below the first. Familiarize yourself with the new sound you hear. Compare the sounds in your mind.
3. Hold a long arc and weld another stringer. Use the same technique as in steps Nos. 1 and 2. Note the difference in the sound?
4. Inspect the appearance of the three stringers and make note of the dif-

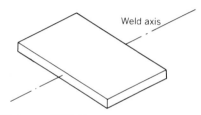

Figure 13C-3

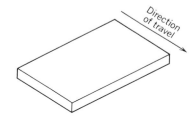

Figure 13C-4 Position of plate.

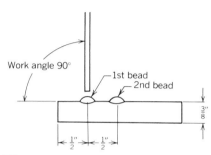

Figure 13C-5

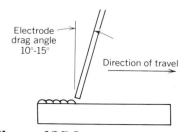

Figure 13C-6

ference. Note the appearance of the weld after you run a bead holding:
(a) Correct arc gap—After you run a bead holding, note the appearance of the weld.
(b) Too long an arc
(c) Too short an arc
5. Run three stringers at the travel speeds indicated. Use the correct arc length and follow Procedure 1.
(a) Correct **travel speed**
(b) Too slow a travel speed
(c) Too fast a travel speed
Note the appearance of the weld.
6. Use the same welding conditions as before. Run three stringers using:
(a) The correct electrode angle
(b) Angle the electrode away from you
(c) Angle the electrode towards you
Note the appearance of the welds.
Cool and clean your workpiece. Then present it, along with your written findings, to the instructor for evaluation.

POINTS TO REMEMBER
1. Use the largest electrode that you can comfortably control. Size may vary because of the joint configuration and welding position.
2. Maintain an arc length approximately equal to the diameter of the electrode core wire.
3. Travel speed must be steady or the width of the bead will vary.
4. Electrode angles are important. The molten metal will flow in the direction the electrode is pointing.
5. You must listen for the sound the electrode makes as well as watching the arc length and puddle shape.

SOME SAFETY REMINDERS
1. Do not pick up hot metal with your gloves—use tongs or pliers.
2. Are welding cables hung in their proper place?
3. Are stubs and material removed from the booth?
4. Is booth clean and ready for the next student?
5. Is all hot metal cooled off?

LESSON 13D
WELDING A PAD OF STRINGERS
(E-6010 OR E-6011 ELECTRODES)

OBJECTIVE
Upon completion of this lesson you should be able to safely and correctly weld a pad of stringers in the flat or downhand position using E-6010 or E-6011 electrodes.

NOTE
Most welds can be made with stringer beads. But to properly complete a weld with stringer beads, you must be careful.

A series of overlapping stringers is called a "weld pad." When finished, you should be able to see one-half to two-thirds of each stringer deposited, as well as all of the last stringer.

There are many industrial applications where "padding" is used. Remember to practice your welding skills when padding the plate. The practice will help you to progress in your training.

MATERIAL AND EQUIPMENT
1. Personal welding equipment
2. E-6010 or E-6011 electrodes 1/8 in. (3.2 mm) diameter
3. One piece of carbon steel plate approximately 3/8 in. × 4 in. × 10 in. long

GENERAL INSTRUCTIONS
1. Wire brush the welding surface of the plate.
2. The plate must be welded with the axis as shown in Figure 13D-1.
3. Position the practice plate at the right-hand corner of the work table just as you did in the previous lesson. (See Figure 13D-1.)
4. Secure the workpiece and the workpiece clamp as in the previous lesson.
5. Adjust the welding current: AC or DCRP, electrode positive (+), ampere range: 75 to 125 amps.

CAUTION: When AC and DC currents are recommended, **only DC current can be used with E-6010 electrode.** Either AC or DC can be used with E-6011 electrode. This statement holds true throughout this book.

PROCEDURE
1. Make sure that the electrode is held firm in the electrode holder, at a 90° angle.
2. Start at the left side of the plate, about 1/2 in. from the top edge. (See Figure 13D-2.) Use about a 10 to 15° drag angle. (See Figure 13D-3.)

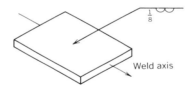

Figure 13D-1

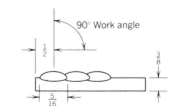

Figure 13D-2

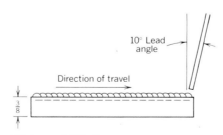

Figure 13D-3

3. Strike and hold a slightly long arc until a puddle begins to form. Then shorten the arc length and move steadily to the right. Use a "C" motion. (See Figure 13D-4.) With a "C" motion the bottom swing of the electrode moves further to the right than the top portion. Pause slightly at the top of the "C." Then move down, out and back somewhat faster. **DO NOT ATTEMPT TO USE THE "C" MOTION UNTIL THE INSTRUCTOR DEMONSTRATES IT.**
4. Clean the stringer thoroughly using the chipping hammer and the wire brush.
5. The next bead should overlap about one-third to one-half of the first bead. Hold the electrode at a work angle of 10 to 15° (into the puddle) from the perpendicular. Use a drag angle of 10 to 15°.
6. Use the angles mentioned above for this bead and all the remaining beads in this assignment. Strike the arc at the left side of the plate. Hold the electrode so that the edge of the coating that is farthest from you is pointed at the nearest edge of the first bead. (See Figure 13D-5.)
7. In Figure 13D-5, notice the angle of the electrode with relation to the first bead and the surface of the plate.
8. Strike and hold a long arc. Then proceed along the closest edge

Figure 13D-4 "C" motion

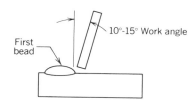

Figure 13D-5

of the first bead. Make sure that you limit the motion of the electrode. The stringer should not be wider than 2 to 2½ times the diameter of the core wire.

9. The second stringer must cover about ⅓ to ½ of the first one, as shown in Figure 13D-2. Notice there are no valleys or low spots between the beads on the sample weld shown in Figure 13D-6.
10. Continue to weld stringer beads in this manner. Cover the entire plate. Then cool the plate, **CLEAN IT, AND PRESENT IT TO THE INSTRUCTOR.**

POINTS TO REMEMBER
1. Clean the surface of the plate before welding.
2. Make sure the workpiece and the lead are securely fastened.

Figure 13D-6 Sample pad of stringers welded from right to left.

3. Make your heat adjustments on a piece of scrap metal.
4. Be sure to use the correct electrode angle. Different electrode angles create different bead shapes.
5. Hold a long arc after each arc start.

SOME SAFETY REMINDERS
1. Use tongs or pliers to pick up hot metal.
2. Are your welding cables hung in their proper place?
3. Have you removed all the welding stubs from the booth?
4. Have you cooled off all hot metal?
5. Is the booth clean and ready for the next student?

LESSON 13E
T-JOINT FILLET, 1-F POSITION, STRINGER BEADS (E-6010 OR E-6011 ELECTRODES)

OBJECTIVES
Upon completion of this lesson you should be able to:
1. The student will safely and correctly weld a T-joint fillet in the 1-F portion in accordance with QW461.4 ASME Section IX Boiler and Pressure Vessel Code and Figure 5.8.1.3 (A) AWS D1.1-80 Structural Welding Code—steel, using E-6010 or E-6011 electrodes.

2. Produce a finished weld with stringers of proper width and height, and with uniform ripples and without undercut.

NOTE
The T-joint fillet is one of the most commonly used joint configurations due to the minimum amount of preparation required to prepare the metal for welding.

It is used in structural iron fabrication, ship building, storage tank construction, bridges, buildings, and general fabrication.

Quite often, the T-joint is the first test piece that is required of a new welder in the average shop.

Mastering the T-joint in the flat position, using stringer beads, will make it a lot simpler to weld them in the horizontal and overhead position.

FLAT OR DOWNHAND WELDING

MATERIAL AND EQUIPMENT
1. Personal welding equipment (same as previous lesson)
2. E-6010 or E-6011 electrodes 1/8 in. (3.2 mm) diameter.
3. 1 pc. carbon steel plate 3/8 in. × 4 in. × 12 in. long, 2 pcs. carbon steel plate 3/8 in. × 1 3/4 in. × 12 in. long

GENERAL INSTRUCTIONS
1. Remove all slag, rust, or mill scale from the weld surface of the plates.
2. Tack the three plates as shown in Figure 13E-1. Be sure that there is little or no space between surfaces. Make sure that the plates are perpendicular to each other.
3. Position the practice piece as shown in Figure 13E-2.
4. Be sure that the workpiece and the work lead are secure.
5. Adjust the welding current: AC or DCRP, electrode position (+), ampere range: 75 to 125.

PROCEDURE
1. With the electrode perpendicular to the joint run a stringer from left to right. Maintain an electrode drag angle of 10 to 15°. (Refer to Figure 13E-3.) Use either a whipping motion or the "C" motion as in Figure 13E-4.
2. Chip and wire brush the bead thoroughly. Cool the plate. Then inspect it for undercut and unevenness of height or width. Undercut may be caused by too long an arc, too high a current, wrong electrode angle or manipulation, or too fast a rate of travel.
3. Overlap a second stringer bead on the first one. Run half on the first bead and half on the side of the workpiece farthest from you. Use a 10 to 15° drag angle and a 15 to 20° work angle. This will cause the metal to flow up the side of the workpiece. (See Figure 13E-3.) One-half of the bead should be on the plate and the other half covering about one-half of the first bead. (See Figure 13E-3.) (Using this angle you will find that the molten metal flows or washes up the side of the plate that the electrode is pointed toward.)
4. Clean the weld thoroughly. Run another stringer with the electrode pointed toward you, at about a 15 to 20° angle. This bead should cover about one-half of the last stringer and flow up the side of the plate. The complete weld should be flat or slightly convex, and have legs of equal size.
5. **STOP** at this point. Have your instructor evaluate your work.
6. After the instructor has commented on your work, repeat the task. This time use the joint opposite to the one you just welded.
7. When you have completed two of the joints, fill them in as shown in Figure 13E-5. This task will improve your ability to run proper stringers without undercut.

SOME SAFETY REMINDERS
Have your cooled off all hot metal?

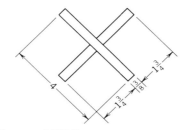

Figure 13E-1

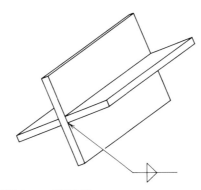

Figure 13E-2

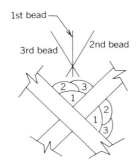

Figure 13E-3 Electrode angle and bead progression.

Figure 13E-4 "C" motion.

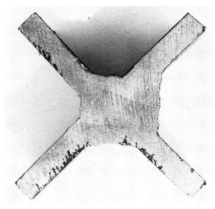

Figure 13E-5 Cross completely welded.

LESSON 13F
T-JOINT FILLET, 1-F POSITION, STRINGER BEADS (E-7014 AND E-7024 ELECTRODES)

OBJECTIVES
Upon completion of this lesson you should be able to:
1. Safely and correctly weld a T-joint fillet in the 1-F position in accordance with QW 461.4 ASME Section IX Boiler and Pressure Vessel Code and Figure 5.8.1.3 (A) of AWS D1.1-80 Structural Welding Code-steel. Fillet in the flat or downhand position using E-7024 and E-7014 iron powder electrodes.
2. Explain the characteristics of the iron powder electrode and its uses.

NOTE
In this lesson you will be introduced to the E-7014 and E-7024 Iron Powder Electrode.

The coating of typical E-7014 electrodes contain approximately 30 percent iron powder and coatings of the E-7024 electrodes contain about 50 percent iron powder. The iron powder enables these electrodes to deposit 30 to 50 percent more filler metal than the E-6010 or E-6011 electrodes.

Because of their increased deposition these electrodes are extremely popular.

If you become proficient in the use of these electrodes, and the welding of the T-joint fillet, you can get a job in some shops, at a level just below a first class, or "A" welder.

MATERIAL AND EQUIPMENT
1. Personal welding equipment
2. E-7014 electrodes 1/8 in. (3.2 mm) or 5/32 in. (4.0 mm) diameter E-7024 electrodes 1/8 in. (3.2 mm) or 5/32 in. (4.0 mm) diameter
3. Use the two joints that you left unwelded from your last assignment

GENERAL INSTRUCTIONS
1. Wire brush the surfaces to be welded.
2. Fit, tack, and position the practice piece as shown in Figure 13F-1.
3. Make sure the workpiece clamp and the workpiece are secure.
4. Adjust the welding current: AC or DCSP electrode negative (−) with E-7014 or DCRP electrode positive (+) with E-7024.
Ampere range—
E-7014—1/8 in.(3.2 mm)110–160
        5/32 in.(4.0 mm)150–210
E-7024—1/8 in.(3.2 mm)140–190
        5/32 in.(4.0 mm)180–250

PROCEDURE
1. Weld the first joint with E-7014 electrodes. Use a 10 to 15° drag angle as you did with the E-6010 and E-6011 electrodes. **BUT YOU DO NOT HAVE TO USE THE "C" MOTION**—just move the electrode along the joint, without any movement other than the forward travel.
2. After each pass chip and wire brush thoroughly. Although slag from the E-7014 electrode comes off quite easily it often sticks to tiny crevices, especially in the corners or toes of the weld.
3. Pay close attention to the edges or toes of the weld. Make sure you are not leaving any undercut. (If you are, slow down. Use a lower rate of travel. Make sure you are not holding too long an arc.)
4. After cooling and cleaning, have the weld evaluated by the instructor.
5. Place the second workpiece in the same position as the first. Use the E-7024 electrodes.
6. The E-7024 electrode is a heavily coated iron powder electrode. Because of this, it can be dragged along the joint with the coating touching the metal being welded. This prevents it from stubbing out.
7. Drag the first pass along the root of the joint. Keep the electrode perpendicular to workpiece. Use a drag angle of approximately 30°. (Strike the arc by tapping the tip of the electrode in the groove of the workpiece to break the coating off the tip.) (See Figure 13F-2.)

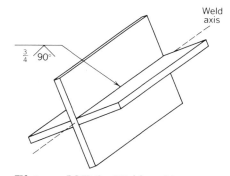

Figure 13F-1 Weld position.

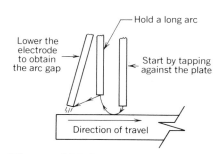

Figure 13F-2 Tapping method.

8. When the arc starts, hold a long arc for a moment. Then shorten the arc length until the coating on the right-hand side of the electrode touches the metal, at a 30° lead angle. (See Figure 13F-3.) Weld at a travel speed that gives the proper bead width for the electrode diameter used.

9. Be sure you maintain the correct drag angle. If you reduce the drag angle too much, the electrode can stub out.

10. Continue, as with the E-7014 electrode, using the same side angles.

11. After you have welded the first three stringers, cover over the weld with a weave bead. (Increase the arc current setting and with the same electrode angles strike the arc on the left. Then move to the right, weaving the electrode from side to side.) (See Figure 13F-4.) Pause slightly at the sides to minimize undercut.

12. Part of the next weave bead should cover slightly more than

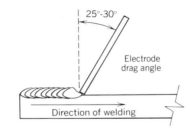

Figure 13F-3 Drag technique.

Figure 13F-4 Weave method.

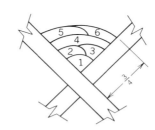

Figure 13F-5 Bead progression

half of the previous bead, while the rest of the bead washes up the side of the joint. (Figure 13F-5.)

13. Part of the next weave bead should cover about one-third of the previous bead, while the rest of the bead washes up the opposite side of the joint. (See Figure 13F-5.) (Continue to run beads in this manner. Make sure they are laced together, the same as you did with stringer beads.)

POINTS TO REMEMBER

1. The heavy coating that covers the tip of iron powder electrodes requires the tapping method be used to strike the arc.

2. Increase the arc current when you use the weave method of arc welding. Weaving covers a wider area, so more heat is needed to melt the base metal.

3. The E-7014 electrode can be used in all positions (horizontal, vertical, and overhead). It can be dragged along the work or a whipping motion may be used.

4. The E-7024 is an iron powder electrode that may be used in the flat and horizontal positions *only*.

LESSON 13G
FILLING CRATERS AT THE END OF WELDS

OBJECTIVES
Upon completion of this lesson you should be able to:

1. Safely and correctly fill **craters** at the end of welds.
2. Pass a test on the theory involving the proper procedures for filling craters and the detrimental effects of unfilled craters.

NOTE
Every time you stop during welding, come to the end of a weld joint, or finish an electrode, you have to extinguish the arc. Each stop can cause a potential weak spot in the weld, if not performed correctly.

Whenever the arc heat is removed from the puddle, a weld crater will form. The crater is a hollow spot in the weld bead. If left unfilled, the crater can become a weak spot in the weld. Unfilled craters can crack and the **crater cracks** can cause the weld to fail.

This lesson teaches you the methods used to eliminate this problem.

MATERIAL AND EQUIPMENT
1. Personal welding equipment
2. E-6010 or E-6011 electrodes ⅛ in. (3.2 mm) diameter
3. 1 pc. hot rolled steel plate ¼ in to ⅜ in. × 4 in. × 12 in. long

GENERAL INSTRUCTIONS
1. Wire brush the surface of the plate.
2. Position the plate as shown in Figure 13G-1.

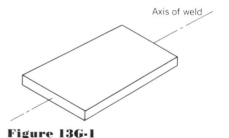

Figure 13G-1

3. Secure the plate and attach the workpiece clamp.
4. Adjust the welding current AC or DCRP, electrode position (+), ampere range: 75 to 125.

PROCEDURE

1. Start at the left. Run a short stringer, about 2 or 3 in. Stop. Then clean the weld.
2. Strike another arc about 1 in. ahead of the crater. Then move the arc on top of the crater. (See Figure 13G-2.) Whenever you strike an arc, it leaves a hard spot in the metal. When you begin ahead of the crater, the hard spot is covered over by the bead, and is eliminated.
3. Hold a normal arc on the crater until it is filled to the same height as the bead. Then continue the stringer. Stop every few inches and repeat the process on the new crater. This will give you practice with this method. **DO NOT GO TO THE END OF THE PLATE UNTIL YOU READ STEP NUMBER FOUR.**
4. When you reach the end of the plate, reverse the direction of travel. Raise the electrode slowly to keep the same arc length as you weld up and over the finished bead. Weld back over the bead for ½ in. Hold a close arc, then stop welding. Use a 10 to 15° push angle. (See Figure 13G-3.) If you burn a hole through the weld, or the weld metal spills over and runs off the end of the plate (see Figure 13G-4), you either moved too slowly or carried too long an arc. Practice this method until you can do it well.
5. Another method of crater filling is to break the arc about 2 in. from the end of the plate. Clean the bead. Then start at the right side of the plate and weld back toward the crater with the electrode pointing into the puddle at a drag angle of about 15 to 30°. (See Figure 13G-5.) Weld over the crater of the bead and break the arc slowly or with a slight circular motion (Figure 13G-5).
6. Practice all methods until your instructor ok's your work.

POINTS TO REMEMBER

1. Keep the height of the weld uniform when starting, restarting, or welding over a crater.
2. Always strike the arc in the weld groove or on an area to be welded. This way all arc starts will be welded over as the bead progresses.
3. Craters are a potential starting point for cracks that can cause weld failure.

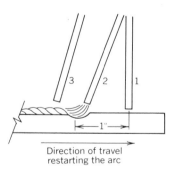

Figure 13G-2 Restarting the arc.

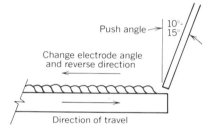

Figure 13G-3 Fill craters at the ends of a weld.

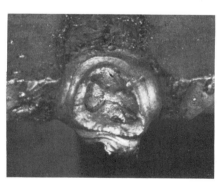

Figure 13G-4 End of plate overheating with metal dripping over edge.

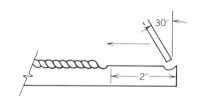

Figure 13G-5 Alternate method.

LESSON 13H
SQUARE GROOVE BUTT JOINT WITH BACK-UP BAR, STRINGER BEADS (E-7024 IRON POWDER ELECTRODES)

OBJECTIVES
Upon completion of this lesson you should be able to:
1. Safely and correctly weld a square groove butt joint in the 1-F test position, in conformation with "Figure 5.22.2—Fillet-weld root-bend test platewelder qualification—option 2," American Welding Society Structural Welding Code D1.1-80 using E-7024 Iron Powder Electrodes.
2. Prepare and test the weld as indicated in Figure 5.22.2 of the AWS Structural Welding Code D1.1-80.
3. Pass the test indicated in Figure 5.22.2 AWS D1.1-80.

NOTE
Successful completion of this test qualifies the welder for flat position *fillet* welding of plate, pipe and tubing according to AWS D1.1-80.

MATERIAL AND EQUIPMENT
1. Personal welding equipment
2. E-7024 electrodes, 5/32 in. (4.0 mm) diameter
3. 2 pcs. hot rolled steel 3/8 in. × 3 in. × 8 in. lg.
   1 pc. hot rolled steel 3/8 in. × 2 in. × 8 in. × lg.

GENERAL INSTRUCTIONS
1. Clean all slag, rust, millscale, and other dirt from the parts to be welded.
2. Tack weld the pieces as shown in Figure 13H-1. It is important that you have as good a fit as possible. Be sure that there are no spaces between the parts.
3. Position the test piece as shown in Figure 13H-2.

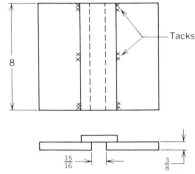

Figure 13H-1 Tacking procedure.

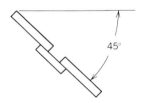

Figure 13H-2 Weldment position.

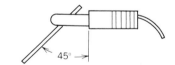

Figure 13H-3

4. Make sure the work clamp and the test piece are securely fastened.
5. Adjust the welding current: AC or DCRP, electrode positive (+) ampere in the range 180 to 250.

PROCEDURE
1. Make sure your tack welds are at least 3/4 in. long and are placed according to Figure 13H-1. If the tacks are not strong enough or in the proper places, the plates may distort and make the test more difficult to pass.
2. Place the electrode in the holder

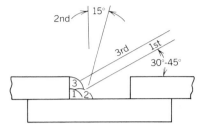

Figure 13H-4 Fillet progression. Electrode angles.

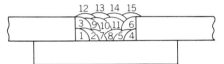

Figure 13H-5 Bead sequence.

    at a 45° angle. (See Figure 13H-3.)
3. Fill the two fillets with three stringer beads. (See Figure 13H-4.)
4. Hold the electrode at approximately 30 to 45° from the surface of the plate with a drag angle of about 30°. Drag it along the bottom leg of the joint. The electrode coating should touch both the back-up bar and the vertical leg. (See Figure 13H-4.)
5. Move along steadily. Make sure the bead equally covers both legs of the joint.
6. Clean and run the second bead with the electrode held about 15° off the perpendicular as in Figure 13H-4.
7. Remove the slag from the weld deposit. Run the third bead with the same electrode angle used with the first bead. (See Figure 13H-4.) Pay close attention to the top, or leading edge of the upper plate. Watch the top edge

of the puddle. Make sure that you do not cut down the top edge. The fillet is not complete if you do not cover the entire vertical leg.

8. Weld the other fillet in the same manner.

9. Fill in the remainder of the joint with either the stringer bead or the weave method as in Figure 13H-5. (See sample weld in Figure 13H-6.)

CAUTION: **DO NOT QUICK QUENCH** the weldment, or cool the plates rapidly. They will be hardened to a certain extent, and will not bend as readily. Then it will be more difficult for you to pass the bend tests.

10. Use an oxy-acetylene cutting torch or any carbon arc cutting torch to remove as much of the backing strip from the test piece as possible. Grind or machine the remainder of the backing strip, and the face of the weld, flush with the base metal.

11. After removal of the excess material, saw or flame cut two strips, as shown in Figure 13H-7. Each should be 1½ in. wide, and taken across the weld from the center section of the test piece. Remove all slag and grind a one-eighth in. (⅛") radius on the four long edges of the strip or "coupon" as they are called.

12. When the coupons are ready, they should be placed in the bending jig. Place the root of the fillets so they face the opening or female member.

13. Bend the coupons as instructed. Then present them to the instructor for examination.

POINTS TO REMEMBER

1. Don't quench the metal used in a test at any time.
2. This is a test of your ability to produce a sound weld with excellent penetration and fusion.
3. Poor fit up, or tacking, will make it difficult to pass the test.

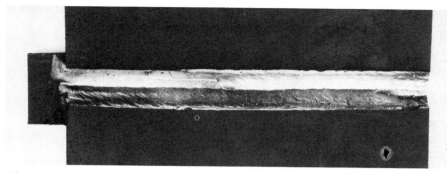
Figure 13H-6 Partially completed weld.

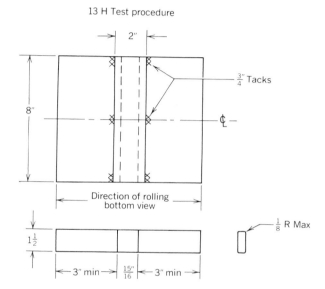

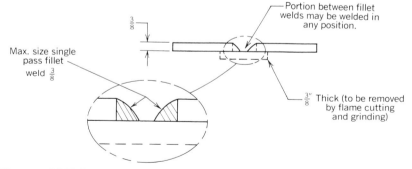
Figure 13H-7 Test procedure as per Figure 5.22.2 of D1.1-79, AWS structural code-steel.

4. Prepare the coupons properly for the bend test. Don't forget to put the radius on the four long edges of the coupons. This will decrease the probability of a weld failure due to cracks along the edge.

SOME SAFETY REMINDERS

1. Wear your safety glasses when bending the coupons.
2. Always wear goggles or a shield when grinding the coupons.

LESSON 131
OUTSIDE CORNER JOINT 1-G POSITION, STRINGER AND WEAVE BEADS, (E-6010 OR E-6011 ELECTRODES)

OBJECTIVE
Upon completion of this lesson you should be able to safely and correctly weld an outside corner joint in the flat, or downhand, position using E-6010 or E-6011 electrodes.

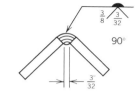

Figure 131-1 Bead sequence.

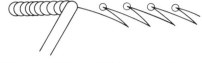

Figure 131-4 Whipping motion.

MATERIAL AND EQUIPMENT
1. Personal welding equipment
2. Two spacers (metal strip 3/32 in. × 3/4 in. × 5 in. long)
3. E-6010 or E-6011 electrodes 1/8 in. (3.2 mm) diameter
4. 2 pcs. hot rolled steel, 3/8 in. × 2 in. × 12 in. long

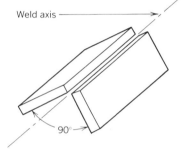

Figure 131-2

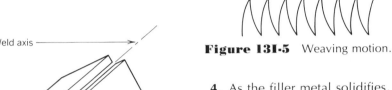

Figure 131-5 Weaving motion.

GENERAL INSTRUCTIONS
1. Clean all the parts to be welded.
2. Use the spacers to position the pieces. Then tack weld the two pieces together to form a 90° angle as in Figure 131-1. Be sure the space along the entire joint is even.
3. Remove the spacers and place the weldment in position as in Figure 131-2.
4. Make sure the workpiece clamp and workpiece are securely fastened. Adjust the welding current: AC or DCRP, electrode positive (+), ampere range 175 to 125.

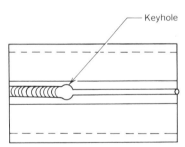

Figure 131-3

PROCEDURE
1. The tacks must be at least 1 in. long or the root opening may close up as you weld the first pass.
2. Hold the electrode perpendicular and use a drag angle of 10 to 15°. Weld over the tack at the left side of the joint. When you reach the end of the tack, pause until a keyhole appears. (See Figure 131-3.)
3. When the keyhole appears, whip the electrode forward and slightly up one side of the joint. (See Figure 131-4.) This motion of whipping out and away from the puddle, allows the molten metal to cool and **FREEZE**. Otherwise it would fall through the root and cause excessive melt through, or even an **icicle**.

4. As the filler metal solidifies, return the electrode to the forward edge of the keyhole. Hold it there until it fills the keyhole and another keyhole opens. Continue whipping the arc across the entire joint, adding a dab of metal each time.
5. The second pass should be a stringer bead. Make sure you fuse both sides of the joint at the toes into the edges of the weld.
6. The third pass is a weave bead. Pause slightly at the sides for proper fusion. (See Figure 131-5.)
7. The final pass is also a weave bead. Use the motion shown in Figure 131-5.
8. Watch out for excessive **overlap**. Avoid extra weld metal hanging over the edge of the joint.
9. The completed weld should have a slightly convex (outwardly curved) appearance as in Figure 131-6.
10. Cool, clean, and present the completed weld to the instructor.

POINTS TO REMEMBER
1. Use the spacers. Otherwise, it will

Figure 131-6 Completed corner joint.

be very difficult to obtain a uniform spacing across the entire joint length.

2. You need a **keyhole** along the root pass to have penetration.
3. When using the whipping motion, be sure to carry the arc a short distance up the side of the groove. If you carry the arc along the root, there is a possibility of some filler metal lodging or sticking in the root opening. It is difficult to obtain a keyhole when you reach these deposits.
4. Your first pass should penetrate about $1/16$ to $3/32$ in. below the surface of the metal at the bottom of the joint. It should protrude the same amount as **reinforcement** on the finish pass. (See Figure 131-2).

CHAPTER 14
HORIZONTAL WELDING OF PLATE

The horizontal welding of this lesson will be your first attempt at out-of-position welding. Horizontal welding requires more skill than downhand welding. However, it is not very difficult if you apply yourself to the task.

Horizontal welding is done on vertical plates. The weld bead is placed on the vertical surface, running from either left to right or right to left. This is illustrated in Figure 14-1.

Some fillet joints are horizontal welds. In this case one leg of the fillet joint is vertical and the other is on a flat plane. Two examples are shown in Figures 14-2 and 14-3.

There are two differences that you will notice in horizontal welding versus downhand welding. First, the weld bead has a tendency to sag, and second, there is a tendency for undercut to appear at the top edge of the bead. It all depends on whether you manipulate the electrode correctly, have an improper rate of travel, or too high an amperage setting.

You can make these differences disappear and produce satisfactory welds by paying close attention to details with the proper heat (adjusting the welding current), electrode angle, manipulation of the electrode, and correct travel speed.

A word of advice before you start this chapter. The methods, electrode angles, and so on, are those that the author found to work best for him as a welder. Your own instructor may want to change some of the procedures. Therefore, never start a new lesson without the approval of your instructor, or without being present for the lecture and demonstration on that lesson.

REMEMBER ALTHOUGH THIS BOOK IS A GUIDE FOR YOU—YOU SHOULD ALSO FOLLOW YOUR TEACHER'S INSTRUCTIONS.

Figure 14-1 Pad of stringers. Horizontal position.

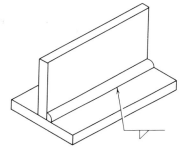

Figure 14-2 T-joint fillet. Horizontal position 2F.

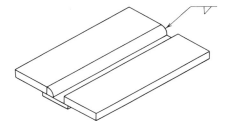

Figure 14-3 Lap joint fillet. Horizontal position 2F.

LESSON 14A
PAD OF STRINGERS (E-6010 or E-6011 ELECTRODES)

OBJECTIVE
Upon completion of this lesson you should be able to safely and correctly use either E-6010 or E-6011 electrodes to weld a pad of stringers in the horizontal position.

NOTE
Plate padding is widely used to build up worn parts. It is used on things such as the teeth on earth digging equipment, worn shafts, and so on.

Stringer beads are also used for welding butt joints in the horizontal position. Therefore, the practice you get from padding plate will aid you in the butt welding.

MATERIAL AND EQUIPMENT
1. Personal welding equipment (as in previous lessons)
2. E-6010 or E-6011 electrodes, ⅛ in. (3.2 mm) diameter
3. 1 pc. carbon steel plate, ⅜ in. × 6 in. × 6 in. long
4. 1 pc. soapstone
5. One straight edge

GENERAL INSTRUCTIONS
1. Be sure the edges of the plate are clean, then wire brush the surface of the plate.

LESSON 14A

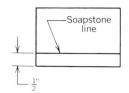

Figure 14A-1

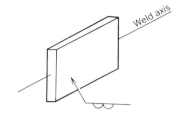

Figure 14A-2 Plate position.

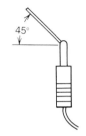

Figure 14A-3

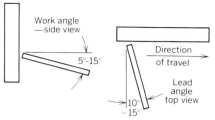

Figure 14A-4

Figure 14A-5 Whipping motion.

Figure 14A-6 "C" motion.

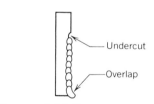

Figure 14A-7

2. Draw a line across the plate with the soapstone. Place the line ½ in. from the bottom edge of the plate as in Figure 14A-1.
3. Position the plate as in Figure 14A-2, with the line at the bottom.
4. Adjust the welding current: AC or DC reverse polarity, electrode positive (+), ampere range: 75 to 125.

PROCEDURE

1. Place the electrode in the holder. Clamp it at a 45° angle as shown in Figure 14A-3.
2. Start at the left side of the plate. Hold the electrode pointing upward at about a 5 to 15° angle. Use a 10 to 15° drag angle as shown in Figure 14A-4.
3. Move to the right with a slight whipping motion as in Figure 14A-

5. You can also use the "C" motion shown in Figure 14A-6. Each of these motions gives the molten puddle a chance to cool slightly. This helps it solidify and keeps it from sagging. When using the "C" motion, do not forget to pause at the upper left part of the C.
4. The second stringer, and all that follows, should cover the top third of the bead below it. This keeps the weld surface even in height, without valleys or low spots.
5. After each stringer inspect carefully for undercut and overlap as shown in Figure 14A-7.
6. Compare your completed pad with Figure 14A-8.
7. Clean and cool the plate and present to your instructor.

POINTS TO REMEMBER

1. If the plate was flame cut, remove all slag from the edges and wire brush it before welding.
2. Pay close attention to the top edge of the puddle. Be sure the undercut zone is filled in as the weld progresses.
3. Sag is caused by holding the arc in one place on the puddle or working with too much current. Both of these cause too much heat in the puddle.

Figure 14A-8 Partially completed pad of stringers in horizontal position. (This pad was welded from right to left.)

LESSON 14B
PAD OF STRINGERS (E-7014 AND E-7018 LOW HYDROGEN ELECTRODES)

OBJECTIVE
Upon completion of this lesson you should be able to safely and correctly weld a pad of stringers in the horizontal position using E-7018 low hydrogen electrodes.

NOTE
The purpose of this lesson is to introduce you to the low hydrogen electrode. Code welders should be skilled in the use of this group of electrodes.

Low hydrogen electrodes are less crack sensitive than other electrodes. They are recommended for welding high strength low alloy steels, high carbon steels, and high sulphur steels.

They are in general use on bridges, structures, and ships. Low hydrogen electrodes were used on parts of the Alaskan pipeline. They are also used a great deal in the construction of nuclear and fossil fuel power plants. By now you can understand the importance of mastering the use of these electrodes.

If you have the ability to produce sound welds using low hydrogen electrodes, you should be able to pass the welder qualification tests needed for high paying jobs. Your instructor can discuss the job opportunities and pay scales in your particular area.

The weld puddle of the low hydrogen electrode is fairly fluid. You must keep the size of the puddle to a minimum so it is easier to control. On critical applications the puddle should not be more than three electrode diameters wide. For example: When the electrode has a 1/8 in. core wire, the bead width should not be more than 3/8 in.

There are exceptions to the rule. Sometimes a width of five diameters is allowed on less critical work.

Although the weld puddle of the low-hydrogen electrode is fairly fluid, it will trap gases when the weave is excessive. Weaving cools the puddle and it becomes less fluid. When the gases are unable to "bubble to the top," porosity results. The coating of the E-7018 electrode contains about 30 percent iron powder to increase the deposition rate. This makes it a little more difficult to handle in out-of-position welding. When used properly, this electrode can produce welds in excess of 75,000 psi tensile strength.

Remember to hold a close arc when you weld with low hydrogen electrodes. A slight electrode oscillating motion is acceptable within the puddle, but you must never use a whipping motion. It will cool the puddle, trap gases, and cause porosity.

The slag can be removed fairly easily. However, never try to chip it off while it is still red hot. Wait until it has cooled to where the color is gone. If you try to chip it too soon, it will just move around like putty. Remember, hot chips are dangerous to the welder. When it is possible cool the weld completely before chipping. Another purpose of this lesson is to introduce you to the use of the E-7014 electrode in the out-of-position work. Like the E-7018, its coating contains some iron powder. This iron powder increases the deposition rate. Because of this increase, it is very popular in some welding shops. **DO NOT ATTEMPT TO USE THESE ELECTRODES BEFORE YOU READ AND UNDERSTAND THIS LESSON. WAIT UNTIL YOU HAVE HEARD THE INSTRUCTOR'S LECTURE AND WITNESSED THE DEMONSTRATION.**

MATERIAL AND EQUIPMENT
1. Personal welding equipment
2. E-7018 and E-7014 electrodes, 1/8 in. (3.2 mm) diameter
3. 1 pc. hot rolled steel, 3/8 in. × 6 in. × 6 in. long
4. Soapstone and straight edge

GENERAL INSTRUCTIONS
1. Remove all slag and other impurities from the plate. Clean the area to be welded.
2. Place a line at the bottom of the plate as in Lesson 14A.
3. Position the plate as in Lesson 14A. Refer to Figure 14B-1.
4. AC or DC reverse polarity, electrode positive (+). Adjust the welding current. Ampere range: E-7014 110-160 + E-7018 115 to 165.

PROCEDURE
1. Place an E-7018 electrode in the holder at a 45° angle as in the previous lesson. (See Figure 14B-2.)

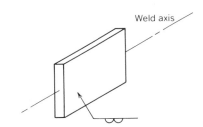

Figure 14B-1 Plate position.

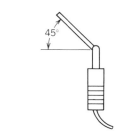

Figure 14B-2

2. Start at the left side of the plate. Use the tapping method to start the arc. See Figure 14B-3.
3. Hold the same electrode angles, 5 to 15° upward with a drag angle of about 10 to 15°, as shown in Figures 14B-4 and 14B-5.
4. Run the first stringer from left to right. Hold a very close arc. A small amount of oscillation in the puddle is permitted but **DO NOT WHIP OUT OF THE PUDDLE.**
5. Allow the bead to cool until the color leaves the slag at the end of the bead. Then remove the slag. Place your gloved hand a short distance from the weld. Strike the slag with a chipping hammer, with your hand placed to deflect the slag as it flies off. Your welding shield and safety glasses protect your eyes, but flying slag has a way of getting into openings in your clothing, gloves, shoe tops, and down your collar. Wear your personal safety equipment.
6. Check for undercut and overlap, then continue to weld stringers

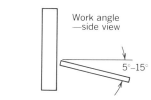

Figure 14B-4

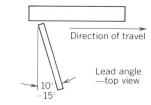

Figure 14B-5

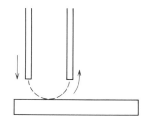

Figure 14B-3 Tapping.

until there is about 1 in. of plate that is not covered. (Compare your pad when you have finished this step to Figure 14B-6.)
7. Set the welding current for E-7014 electrodes. Use straight polarity and try to run a stringer. Use the same technique as you have been using with an E-7018 electrode. Remember to keep all electrode movement to the minimum.
8. Clean and cool your plate and present it to the instructor.

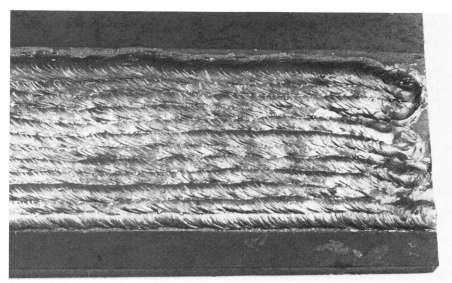

Figure 14B-6 Pad completed to 1 in. from the top in the horizontal position with E-7018.

LESSON 14C
LAP JOINT FILLET 2-F POSITION, MULTIPASS (E-6010 or E-6011 ELECTRODES)

OBJECTIVE
Upon completion of this lesson you should be able to safely and correctly weld a lap joint fillet in the horizontal position using E-6010 or E-6011 electrodes in accordance with Figure 5.8.1.3 B of D1.1 American Welding Society Structural Welding Code and QW-4614 (b), Section IX, Boiler and pressure vessel code of the American Society of Mechanical Engineers.

NOTE

The lap joint is in common use in industry today and has many applications. It is used for welding such things as storage tank roofs and bottoms, shipbuilding, and general fabrication. The lap joint requires very little preparation before welding. It is low in cost which is what prompts manufacturers to use it whenever possible.

Some lap joint applications require a single pass of a specific size. Because of this you must learn to judge the size of the weld bead. Practice judging the size while you are welding so you can consistently weld beads of the size required.

Bead appearance is usually of great importance to your employer. If the weld is the wrong size or of poor quality, rewelding or grinding may be necessary. These steps all add to the cost of the job. Employers are very cost conscious, and they will not continue to employ a worker who consistently produces improperly welded joints. Repairs cause them to lose money.

Even though your first pass is covered over in multipass welding it presents an excellent opportunity to gain skill in welding first passes. Thoroughly examine the first bead of every plate; don't be satisfied with anything but your best effort. With a little practice you will soon be producing acceptable welds.

MATERIAL AND EQUIPMENT

1. Personal welding equipment
2. E-6010 or E-6011 electrodes, 5/32 in. (4.0 mm) diameter
3. 4 pcs. hot rolled steel, 3/8 in. × 2 in. × 12 in. long

GENERAL INSTRUCTIONS

1. Remove all slag and clean the area to be welded.
2. Tack the plates as shown in Figure 14C-1. Make sure that there are no openings where the plates meet.

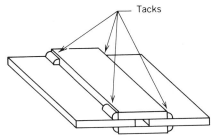

Figure 14C-1 Tacking procedure

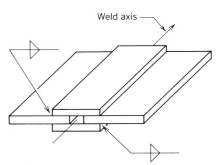

Figure 14C-2 Plate position

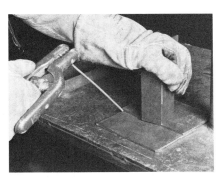

Figure 14C-3

3. The weldment position should be as shown in Figure 14C-2.
4. Use AC or DC reverse polarity electrode positive (+). Adjust the welding current: range 110 to 170

CAUTION: When fitting and tacking do not hold the plates with your gloves. Use a small block of steel or other suitable object to hold the plates together as shown in Figure 14C-3.

PROCEDURE

1. Place the electrode in the holder at a 45° angle as in Figure 14C-4.

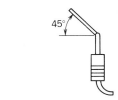

Figure 14C-4

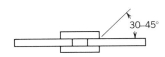

Figure 14C-5 Work angle

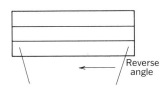

Figure 14C-6 Drag angle

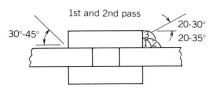

Figure 14C-7 Bead sequence

2. Weld the first stringer from left to right. Hold the electrode at about a 30 to 45° angle from the surface of the bottom plate as in Figure 14C-5. Use a drag angle of about 10 to 15° as in Figure 14C-6.
3. When you reach the end of the joint, reverse the drag angle as in Figure 14C-6. Then to eliminate the crater weld back over the bead for about 3/4 in.
4. Weld the remaining three joints, alternating the direction of travel to obtain practice in welding in other directions.
5. The second bead should cover about three-quarters of the first. (See Figure 14C-7.) Hold the electrode at about the same angle as for the first pass. Keep the edge of the electrode coating closest to

you, almost touching the lower plate. Try to hold it about 1/8 in. from the toe of the first bead.

6. For the third stringer bead decrease the electrode angle slightly. Run the bead so that almost half of the second bead is covered. Pay careful attention to the top edge of your puddle. Make sure you are not undercutting the top plate. Similarly, be sure you are not overlapping it either.
7. Complete the other three welds in the same manner.
8. Clean and cool your plates and present them to your instructor.

POINTS TO REMEMBER
1. Use a block of steel or scrap to hold the metal when tacking.
2. The face of the completed joints should make a 45° line from toe to toe.

LESSON 14D
LAP JOINT FILLET 2-F POSITION, MULTIPASS (E-7014 and E-7024) ELECTRODES

OBJECTIVE
Upon completion of this lesson you should be able to safely and correctly weld a lap joint fillet in the horizontal position using E-7014 and E-7024 electrodes in accordance with Figure 5.8.1.3 B of D1.1 American Welding Society Structural welding code and QW-461.4 (b), Section IX, Boiler and pressure vessel code of the American Society of Mechanical Engineers.

NOTE
Your approach to this lesson will be similar to the previous one, except in this lesson you will use E-7014 and E-7024 electrodes. It is important for you to learn to use as many of the most popular electrodes as you can. This lesson will increase your capabilities for welding different joint configurations.

MATERIAL AND EQUIPMENT
1. Personal welding equipment
2. E-7014 electrodes 1/8 in. (3.20 mm) or 5/32 in. (4.0 mm) diameter
   E-7024 electrodes 1/8 in. (3.20 mm) or 5/32 in. (4.0 mm) diameter
3. 4 pcs. hot rolled steel, 3/8 in. × 2 in. × 12 in. long

GENERAL INSTRUCTIONS
1. Remove all slag and clean the area to be welded.
2. Tack the plates as in the last lesson and as shown in Figure 14D-1. Use a steel block or some other object to hold the plates in position while tacking; this way you will not burn your hands or ruin your gloves. Refer to Figure 14D-2.

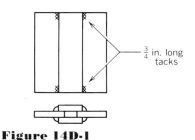

Figure 14D-1

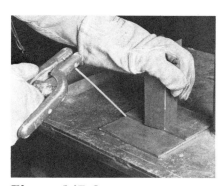

Figure 14D-2

3. Position the weldment as shown in Figure 14D-3.
4. Adjust the welding current: AC or DC straight polarity (−) for the E-7014 electrode; AC or DC reverse polarity (+) for the E-7024 electrode.
   E-7014 1/8 in. (3.2 mm) = 110 to 160 amperes
   E-7014 5/32 in. (4.0 mm) = 150 to 210 amperes
   E-7024 1/8 in. (3.2 mm) = 140 to 190 amperes
   E-7024 5/32 in. (4.0 mm) = 180 to 250 amperes

PROCEDURE
1. Place the E-7014 electrode in the holder at a 45° angle. Be sure to use the correct polarity and current range.

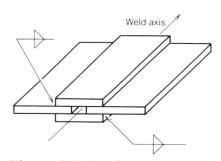

Figure 14D-3 Plate position.

2. Run the first stringer from left to right. Hold the electrode at about a 30 to 45° angle from the bottom surface of the plate. Use a drag angle of approximately 30°. (Refer to Figures 14D-4 and 14D-5.) Remember to eliminate the **crater** at the end of the weld. This can be done by removing the electrode with a slow spiral motion.
3. Use the E-7024 electrode for the next pass. Hold an electrode work angle of about 90°. (Refer to Figure 14D-4.) Use a 30° drag angle (see Figure 14D-5).
4. For finish bead, point the electrode into the joint with the outer coating of the electrode resting on the top of the second bead. Hold it at a 30 to 45° angle as shown in Figure 14D-4. Do not move too slowly or use too small an electrode angle, or you will **overlap** the top of the joint, burn it away, or cause undercut.
5. Compare with sample in Figure 14D-6.
6. Clean and cool the plates. Then present them to your instructor for inspection before welding the three remaining joints.
7. Alternate both the electrodes and the direction of travel when welding the other joints.

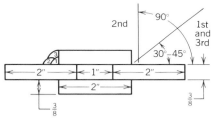

Figure 14D-4   Work angles.

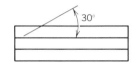

Figure 14D-5   Drag angle.

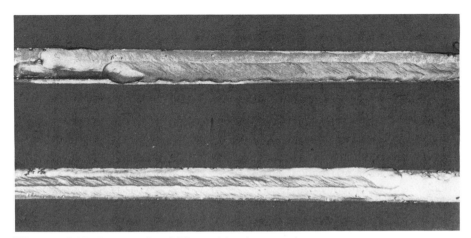

Figure 14D-6   Completed fillet weld in horizontal position.

LESSON 14E
T-JOINT FILLET 2-F POSITION, MULTIPASS (E-6010 OR E-6011) ELECTRODES

OBJECTIVE
Upon completion of this lesson you should be able to safely and correctly weld a T-joint fillet in the horizontal position using E-6010 or E-6011 electrodes in accordance with Figure 5.8.1.3 B of D1.1 AWS Structural Welding Code and QW-461.4 b Section IX, Boiler and pressure vessel code of the ASME.

NOTE
Welding a T-joint fillet is very similar to welding a lap joint. The main difference is in the technique used for welding the last bead.

There is a problem of possible undercut at the top edge of the last stringer. However, the problem can be eliminated with a little effort on your part. It is simply a matter of learning to manipulate the electrode properly and maintain the correct electrode angles.

MATERIAL AND EQUIPMENT
1. Personal welding equipment
2. E-6010 or E-6011 electrodes, ⅛ in. (3.2 mm) diameter
3. 1 pc. hot rolled steel, ¼ in. or ⅜ in. × 4 in. × 12 in. long
4. 2 pcs hot rolled steel, ¼ in. or ⅜ in. × 1½ in. × 12 in. long

GENERAL INSTRUCTIONS
1. Remove all slag and other impurities from the plates. Clean the area to be welded.
2. Tack the plates to form a cross as shown in Figure 14E-1. Do not hold the metal in place with your gloves. Use some scrap metal or a clamp to position the plates.

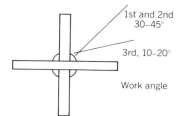

Figure 14E-1 Work angle.

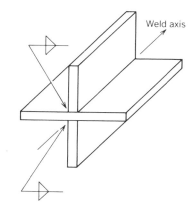

Figure 14E-2

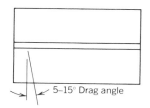

Figure 14E-3

Figure 14E-4 Electrode angle for the second bead.

3. Position the weldment as in Figure 14E-2.
4. Adjust the welding current: AC or DC reverse polarity electrode positive (+), ampere range: 75 to 125.

PROCEDURE

1. Place the electrode in the holder with the same 45° angle you have been using.
2. As in the previous lesson hold the electrode at about a 30 to 45° angle off the bottom plate as in Figure 14E-1. Use a 5 to 15° drag angle for the first pass as shown in Figure 14E-3. Fill the crater at the end of the bead.
3. Hold the same electrode angles with the second bead. Point the tip of the electrode so it is partially on the plate and partially on the first bead as shown in Figure 14E-4. Slightly more than half of the filler metal should be on the stringer bead.
4. For the third stringer lower the electrode angle to between 10 and 20°, as in Figure 14E-1. Use a whip or "C" motion. Undercut can be reduced if you whip properly or pause at the upper left side of the puddle when using the "C" motion.
5. Clean and inspect the last bead for undercut. Weld three more stringers over the first three passes to complete the joint. See the bead sequence in Figure 14E-5.
6. Weld the other three joints as follows:
   (a) One with a single bead
   (b) One with two stringers over the first
   (c) The last with a total of six stringers
7. Clean you weldment and present to your teacher for inspection. Then complete the unfinished welds, so all welds have six stringers.

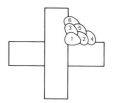

Figure 14E-5 Bead sequence.

LESSON 14F
T-JOINT FILLET 2-F POSITION, SINGLE PASS (E-7024) ELECTRODES

OBJECTIVE
Upon completion of this lesson you should be able to safely and correctly weld a T-joint fillet in the horizontal position using E-7024 iron powder electrodes. You will prepare coupons for test purposes as described in Figure 5.22.1, D1.1 AWS Structural Code. You should test the specimens and they should meet or exceed the minimum requirements of the code.

NOTE
There are many applications in the welding industry where a single pass weld is required. This lesson and the test you will make are designed to give you practice with this type of weld. It also gives you practice preparing and testing weld specimens.

In this lesson, you will also test your ability to weld first passes in T- and lap joints.

To pass the visual examination, fillet welds need to present a reasonably uniform appearance and be free of overlap, cracks, and excessive undercut. Porosity should not be visible on the surface of the weld.

The test results required according to Figure 5.28.2.2 D1.1 AWS Structural Welding Code-Steel are as follows:

The specimen shall pass the test if it bends flat upon itself. If the fillet weld fractures, the fractured surface shall show complete fusion to the root of the joint and shall exhibit no inclusion or porosity larger than 3/32 in. (2.4 mm) in greatest dimension. The sum of the greatest dimensions of all inclusions and porosity shall not exceed 3/8 in. (9.5 mm) in the 6 in. (152 mm) long specimen.

MATERIAL AND EQUIPMENT
1. Personal welding equipment
2. E-7024 electrodes 1/8 in. (3.2 mm) diameter
3. 2 pcs. hot rolled steel plate, 1/2 in. × 4 in. × 8 in. long

GENERAL INSTRUCTIONS
1. Remove all slag and mill scale from the plate. Clean the area to be welded.
2. Tack weld the pieces as in Figure 14F-1. Tilt the vertical plate a few degrees away from the side to be welded. Place the tacks at each end of the joint, on the side to be welded. The tacks should pull the vertical plate so it is perpendicular.
3. Position the tacked pieces in the position shown in Figure 14F-2.
4. Place the E-7024 electrode in the holder at a 45° angle.

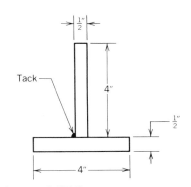

Figure 14F-1

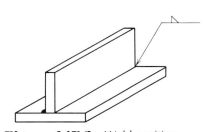

Figure 14F-2 Weld position.

5. Adjust the welding current: AC or DC reverse polarity, electrode positive (+). Ampere range: 140 to 190.
6. Check to insure that the work lead clamp and the workpiece are secure.

PROCEDURE
1. Set the welding current by welding on scrap metal. Then run a stringer bead from left to right. Use a drag technique. Hold the electrode at about 30 to 45° to the bottom plate (see Figure 14F-3) with a drag angle of about 30°, as shown in Figure 14F-4. **STOP WELDING AT THE CENTER OF THE JOINT—CLEAN THE CRATER AREA. THEN COMPLETE THE STRINGER.**
2. Pay very close attention to the electrode angles. Be sure you do not change angles during stopping and starting.
3. Watch your travel speed. It must be steady if the bead width is to remain constant.
4. Be sure the legs of the weld are of equal length.

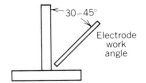

Figure 14F-3 Electrode work angle.

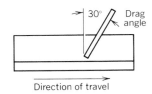

Figure 14F-4 Drag angle.

5. Remove the slag and prepare the specimen by flame cutting or saw cutting according to Figure 14F-5.
6. Clean and polish the ends of the joint and inspect for voids of any sort.
7. Inspect the surface of the weld for porosity and undercut. **SHOW THE TEST COUPON TO YOUR INSTRUCTOR BEFORE YOU DO THE DESTRUCTIVE TEST.**
8. Break the weld, if possible, using the procedure shown in Figure 14F-5. Remember to wear your safety glasses while performing the test.
9. Present the completed piece to your instructor for evaluation and grading.

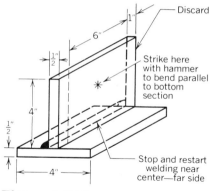

Figure 14F-5

POINTS TO REMEMBER
1. This is a test of your ability to produce a quality weld and follow directions. **PAY CLOSE ATTENTION TO THE DIRECTIONS AND FOLLOW THEM EXACTLY.**
2. Be sure that the edges of the plates are square and clean before welding.
3. Use the drag technique. Be careful to hold the same electrode angles for all welds. Preposition the plates slightly before welding.
4. Be sure to clean the crater thoroughly before starting the weld at the center of the joint.

LESSON 14G
SINGLE BEVEL BUTT JOINT WITH BACK-UP BAR (E-7018 ELECTRODES)

OBJECTIVE
Upon completion of this lesson you should be able to:
1. Safely and correctly weld a single bevel butt joint in the horizontal position, using E-7018 low hydrogen electrodes, in accordance with Figure 5.19-A D1.1 AWS Structural Welding code.
2. Remove, prepare, and test weld specimens.
3. Meet or exceed the minimum requirements of the code.

NOTE
The single bevel butt joint is often used to test welders who work on bridges, buildings, ships, and so on. It is a good test of a welder's ability to make good quality welds in the horizontal position. This joint shape is also popular because it requires the beveling of only one plate. It also uses less filler metal than a full V-joint.

When you can successfully make this weld, you will be qualified to weld flat groove, horizontal groove, flat position fillet, and horizontal position fillet joints up to ¾ in. in thickness on plate, pipe, and tubing.

MATERIAL AND EQUIPMENT
1. Personal welding equipment.
2. E-7018 electrodes, ⅛ in. (3.2 mm) diameter
3. 2 pcs. hot rolled steel plate ⅜ in. × 3 in. × 8 in. long
   1 pc. hot rolled steel plate ⅜ in. × 1½ in. × 8 in. long

GENERAL INSTRUCTIONS
1. Flame cut and grind a 45° bevel, on the long edge of one plate.
2. Remove all slag, mill scale, and so on, from the area to be welded on all three plates.
3. Fit and tack the plates as shown in Figure 14G-1. Tack at the ends of the joint out of the test area.

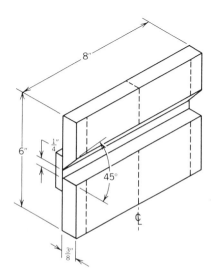

Figure 14G-1 Test plate.

4. Position the test piece as in Figure 14G-2. The weldment *may not* be moved until all welding is completed.
5. Secure the weldment and workpiece clamp.
6. AC or DC reverse polarity electrode positive (+). Adjust the welding current. Ampere range: 115 to 165.

PROCEDURE

1. For the first pass hold an electrode work angle of about 45° as shown in Figure 14G-3. Use a drag angle of about 20 to 30°. Don't let the puddle cover more than one-half of the root spacing. The electrode should penetrate deeply into the bottom plate and wash up on the *back-up bar* as shown in Figure 14G-4.
2. Clean the first bead thoroughly. Then run the second bead over the remaining root area and the beveled plate. Use an electrode work angle of about 10 to 20° upward into the root as shown in Figure 14G-3. For good fusion be sure to burn into the top plate, the remaining root space, and first bead as shown in Figure 14G-4.
3. Clean the second bead thoroughly. Cover it with two stringers, or one wide stringer depending on the width of the first bead. Stringers at the bottom plate must be made with an electrode angle of about 45° if you are to get deep penetration. Refer to Figures 14G-3 and 14G-4.
4. All other stringers, until the last pass should be at about a 10 to 20° upward angle. Follow the bead sequence shown in Figure 14G-4. The exception to this is stringer 5 which requires the same angle as 1 and 3.
5. The next to last pass (beads #5, 6, and 7 in Figure 14G-4) should be about 1/16 in. below the surface of the plate.
6. All the stringers in the finish pass (beads #8 to 11 in Figure 14G-4) may be put in with an electrode angle of about 10 to 15° upward.
7. Remove the back-up bar by flame cutting or air carbon arc cutting and grinding. Any remaining piece of back-up bar should be ground off flush with the surface of the plate. The face of the weld should also be ground flush in the same manner.
8. Saw or flame cut two coupons from the center section of the test piece. Each of them should be 1½ in. wide.
9. Break all edges. "Breaking" means to file or grind a slight radius on all four of the long sides as shown in Figure 14G-5.
10. Use one coupon for a face bend and one coupon for a root bend. Present the bent specimens to your instructor for evaluation and grading.

POINTS TO REMEMBER

1. Prepare and fit the pieces properly.
2. Electrode work angles and drag angles are important. They control the arc stream, the location of the weld puddle, and the point at which the metal is deposited.
3. Don't forget to clean the welding area on the plates thoroughly. A little slag in an undercut area can be the cause of a weld failure, especially in root passes.

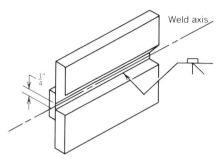

Figure 14G-2 Test position.

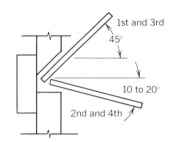

Figure 14G-3 Electrode work angles.

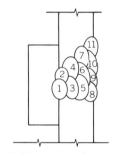

Figure 14G-4 Bead sequence.

Figure 14G-5 Test coupon preparation.

LESSON 14H
V-GROOVE, BUTT JOINT, OPEN ROOT 2-G POSITION (E-6010 OR E-6011 ELECTRODES)

OBJECTIVES
Upon completion of this lesson you should be able to:
1. Safely and correctly weld an open root V-groove butt joint in the **horizontal position,** using E-6010 or E-6011 electrodes, in accordance with Figure 5.20A D1.1 AWS
2. Remove, prepare, and test weld specimens.
3. Meet or exceed the minimum requirements of the code.

NOTE
All position open root butt joints are among the most difficult to weld. They require a great deal of skill to produce acceptable welds. when you successfully complete this lesson, you will be qualified to make flat and horizontal position groove welds on plate. Another benefit you will gain from this lesson is the ability to weld open root butt joints on pipe in the (2-G) position.

MATERIAL AND EQUIPMENT
1. Personal welding equipment
2. Two spacers 3/32 in. × 3/4 in. × 5 in. long
3. E-6010 or E-6011 electrodes, 1/8 in. (3.2 mm) diameter
4. 2 pcs. hot rolled steel 3/8 in. × 4 in. × 8 in. long

GENERAL INSTRUCTIONS
1. Use a hand cutting torch or mechanical flame cutter to bevel one of the long edges of each plate. Cut the bevel at a 30° angle. This makes a 60° included angle as in Figure 14H-1.
2. Remove all slag, mill scale, and foreign matter by grinding or sanding.
3. Grind or file a 3/32 in. (2.4 mm) root face on each beveled edge as shown in Figure 14H-1.
4. Lay the plates, face down, on a flat surface with the two spacers in place as in Figure 14H-20. (top view).

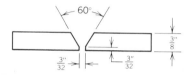

Figure 14H-1 Joint dimensions.

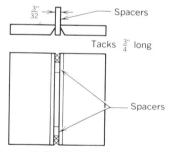

Figure 14H-2 Tacking procedure.

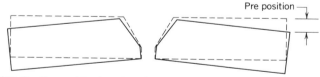

Figure 14H-3 Prepositioning the plates.

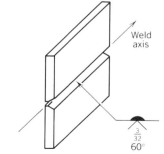

Figure 14H-4

5. Tack the two plates as indicated in Figure 14H-2.
6. Remove the spacers and preposition the plates as in Figure 14H-3.
7. Position the plate as in Figure 14H-4, secure the plate and workpiece clamp.
8. AC or DC reverse polarity, electrode positive (+). Adjust the welding current: ampere range: 75 to 125.

PROCEDURE
1. After setting your heat on scrap metal, start the weld bead on the left. Use an electrode angle of about 5 to 10° upward into the root. (See Figure 14H-5.) Use a 10 to 15° drag angle.
2. Pause at the end of the tack until a keyhole appears.
3. When the keyhole appears, whip the electrode forward and down onto the wall of the lower bevel. Without stopping make a slight "C" motion and return to the key-

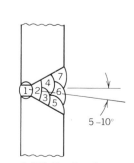

Figure 14H-5 Bead sequence.

hole. As you do this the rear of the puddle solidifies as shown in Figure 14H-6. The whipping allows the puddle to cool. It keeps the keyhold small and stops the puddle from spilling out the back of the joint. (You remember the **keyhole** from the flat welding section, don't you?)

4. With a little practice you will master this technique. If you pause too long in the puddle, you will "blow through" or spill the molten metal out of the rear of the joint. If you move too fast, you will not penetrate the joint properly.
5. The second pass is a stringer bead. Make it with the "C" motion. Take care to pause at the upper left hand portion of the "C" shown in Figure 14H-6.
6. Finish the remainder of the joint with stringers as shown in Figure 14H-5. Don't forget to leave about a 1/16 in. space between the face of beads number 3 and 4 (the next to last pass) and the surface of the plate. (See Figure 14H-7.)
7. Saw or flame cut two coupons. Make them each 1½ in. wide. Take them from the center section of the test piece.
8. Break the edges as in the previous lesson.
9. Use one coupon for a face bend and one for a root bend. Present the bent pieces to your instructor for evaluation and grading.

POINTS TO REMEMBER

1. When whipping out, as in step 3 of the procedure, be sure you do not carry the arc to the surface of the plate. The heat of the arc will cause hardened spots on the surface of the plate. As long as you whip onto the inside of the weld groove there won't be any problems. The spots will be heat treated by welding over them with the next bead.
2. Be sure that the root space is uniform throughout the joint.
3. Pay close attention to the keyhole. Get a rhythm going by counting under your breath. Try it; it works.
4. Remember to break the edges with a file or grinder. If you don't, you will leave a sharp edge where cracks start easily. These small cracks can cause *your* weld to fail when placed in the bender.

Figure 14H-6 "C" motion.

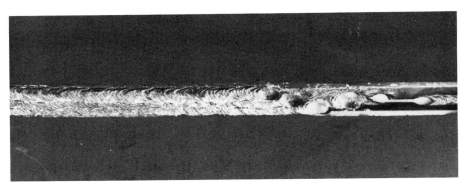

Figure 14H-7 Completed weld.

CHAPTER 15
VERTICAL WELDING OF PLATE

Welding plate in the vertical position is another true test of a welder's ability. The molten puddle is fluid and wants to run downhill. It is difficult to control at first, but after some practice you will be able to control it. You can learn to produce quality welds.

When you heat the **base metal,** it changes to liquid at the melting point, As you know liquids tend to flow downhill because of the pull of gravity. However, something called surface tension tends to keep the weld puddle in the same place. Surface tension force holds the droplet of water on the end of your finger after you dip it into a glass of water. To be a successful welder you must set the welding current properly, use the correct electrode angles and the correct electrode movements.

This chapter shows you how to weld most of the vertical joint configurations, or shapes, that are in common use in the trade today. A few of the joints may not be used in the type of industry near where you live, but it would serve you well to learn to weld them all. You may move, or a new employer may locate in your area. It is always better to know more than is actually required. The extra skills you learn and knowledge you obtain makes you a more valuable welder.

You will be taught to weld the various vertical joints in both the **uphill** and **downhill** modes. When welding uphill, you start at the bottom of the joint and weld toward the top. In downhill welding you start at the top and weld downward toward the bottom. Do not confuse "downhill" and "downhand," which is the flat position.

The downhill method is used extensively by pipeline welders. It was used on some construction of the Alaskan pipeline. It is used mainly with fairly thin wall pipe where a small included angle keeps the weld bead size to a minimum. Downhill welding is not suitable for every job; when it can be used, it is a fast, economical method of welding.

Practice in these positions will increase your ability to weld all thicknesses of metal in the vertical position. It will also make it possible for you to pass the code welding tests. Vertical welding of plate is also excellent training for pipe welding which is discussed later.

At one time the E-6010 electrode was considered the best general purpose electrode. Many people still feel this to be true. Newer electrodes such as the E-7018 have been gaining in popularity because they can join the steels that are difficult to weld, or are not readily weldable, with E-6010 electrodes.

The E-6010 electrodes are still in widespread use and will remain so for quite some time. The E-6010 electrode is still preferred for root passes on open root butt joints, including pipe.

The E-6010 electrode requires a great deal of skill. It is usually found where high quality welds are needed. It has a fast freezing characteristic, which makes it good for welding in the vertical position, both uphill and downhill.

As mentioned in the other lessons, the methods, electrode angles, and so on, are those the author found to work best. Never start a new lesson, such as this, without the approval of your instructor. Be sure to attend the lectures and demonstrations given by your instructor. Some slight variations may be made to the lessons in this chapter.

REMEMBER, THIS BOOK IS A GUIDE FOR YOU; BUT YOUR INSTRUCTOR'S PREFERENCES SHOULD BE FOLLOWED AT ALL TIMES.

LESSON 15A
T-JOINT FILLET, VERTICAL, 3-F POSITION, WEAVE BEADS (UPHILL)
E-6010 OR E-6011 ELECTRODES

OBJECTIVE
Upon completion of this lesson you should be able to safely and correctly weld a T-joint fillet in the 3-F position, using E-6010 or E-6011 electrodes, while conforming to the requirements of QW 461-4 (C) ASME Section IX and AWS Structural Code D1.1 Figure 5.8.1.3 C.

NOTE
T-joints in the vertical position are in common use. This joint is used as a test of a welder's ability to weld out of position. It is even used in shops where the welding does not have to meet the strict requirements of a code.

You will have many opportunities to put the skill you develop in this lesson to good use. Structural members, stiffeners, and wear plates on vessels are a few of the applications that need the skill you will develop.

MATERIAL AND EQUIPMENT
1. Personal welding equipment
2. E-6010 or E-6011 electrodes ⅛ in. (3.2 mm) diameter
3. 2 pcs. ⅜ in. × 4 in. × 8 in. long mild steel plates

GENERAL INSTRUCTION
1. Prepare, clean, and tack the plates to form a T-joint as in previous lessons. Refer to Figure 15A-1.
2. Place the electrode in the holder at a 45° angle as shown in Figure 15A-2. Use this electrode position for all vertical welds.
3. Set your welding current on a piece of scrap metal in the vertical position.
4. Position the joint so the weld axis is vertical as in Figure 15A-3.
5. Adjust the welding current: Direct Current Reverse Polarity (DCRP) electrode positive. Ampere range: 75 to 125.

PROCEDURE
1. Position yourself so the electrode is centered between the two plates as in Figure 15A-2. Point the electrode slightly upward, at a 10 to 15° angle as in Figure 15A-4.
2. First pass: Strike the arc at the bottom of the joint where the two plates meet. Hold the arc until the puddle is approximately ¼ to 5/16 in. wide where it joins the two plates. Whip the electrode up and to the right as in Figure 15A-5. Hold a long arc as you move out of the puddle and as you drop back into it.
3. As the electrode drops back into the puddle, hold a normal length arc. Pause slightly and whip out of the puddle again.
4. Second pass: Start at the left side of the weld groove at point 1.

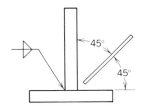

Figure 15A-1 Electrode work angle.

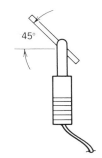

Figure 15A-2

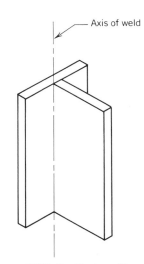

Figure 15A-3 Test position 3F.

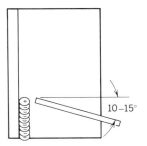

Figure 15A-4 Electrode push angle.

(See Figure 15A-6.) Pause slightly to allow the metal to form a puddle. When the puddle forms, move the electrode slowly to point 2 on the right side. Pause again to allow any undercut to fill in. Then whip the electrode to point 3 slightly above the center of the puddle. As the weld puddle changes color, return the electrode to point 4. Pause again while the undercut fills in, then slowly move across to point 5. Repeat the motions on the left as shown by points 6, 7, and 8. As you continue the sequence, remember the electrode is always returned to the side where

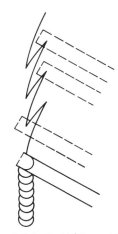

Figure 15A-5 Whip motion.

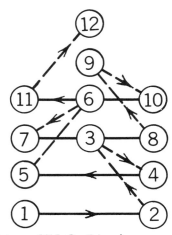

Figure 15A-6 Triangle.

it left the puddle. **ALWAYS HOLD A LONG ARC ON BOTH THE UPSTROKE AND THE DOWNSTROKE SO YOU DO NOT DEPOSIT ANY METAL DURING THE ELECTRODE MOTION.**

5. If the center of the weld is too high or if it sags, you are moving too slowly across the face of the weld. Move faster.
6. If the center of the weld is too low or has holes, you are moving too fast across the face. Move slower.
7. Move with a rhythm. Count to yourself and the weld will be more uniform. Always pause at the sides of the groove.
8. As the electrode moves from side to side, make sure you stop with the electrode coating at the edge of the previous pass. The puddle will flow out past the edge and provide the correct pass width.
9. Finish the second pass in this manner; repeat the sequences for the third pass. The finish of third pass should not exceed five electrode diameters in width.
10. Cool and clean the finished weld. Inspect for unevenness, undercut, holes, and bead contour. Your weld should look like the one in Figure 15A-7.

POINTS TO REMEMBER
1. Pause at the sides. It fills in the undercut.
2. Clean thoroughly between passes.
3. Voltage and amperes are controlled by arc length.
4. Do not whip out excessively.
5. Move with rhythm; your weld will be more uniform.

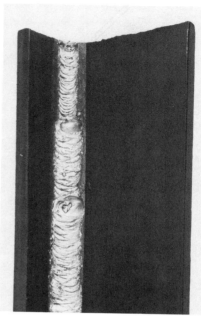

Figure 15A-7 Completed fillet weld.

LESSON 15B
T-JOINT FILLET, VERTICAL, 3-F POSITION, WEAVE BEADS (UPHILL) E-7018 ELECTRODE

OBJECTIVE
Upon completion of this lesson you should be able to safely and correctly weld a T-joint fillet in the 3-F position, using low hydrogen E-7018 electrodes, while conforming to the requirements of QW-461.4 (C) ASME and AWS Structural Code D1.1. Figure 5.8.1.3. C.

NOTE
This lesson will help you develop the ability to produce fillet welds with the E-7018 electrode. The fillet weld and the E-7018 electrode are used in many areas including the construction of bridges, buildings, and ships.

The low hydrogen electrode produces a more fluid puddle than the E-6010 electrode. It also produces a heavier slag. The iron powder in the electrode coating increases the deposition rate. With a little practice you should be able to produce acceptable welds. The welding experience you gain with low hydrogen electrodes will help a great deal when you use stainless steel electrodes.

MATERIAL AND EQUIPMENT
1. Personal welding equipment
2. E-7018 electrodes ⅛ in. (3.2 mm) diameter
3. 2 pcs. ⅜ in. × 3 in. × 12 in. long hot rolled steel

GENERAL INSTRUCTIONS
1. Prepare, clean and tack plates to form a T-joint as shown in Figure 15B-1.
2. Place the electrode in the holder at a 45° angle, as shown in Figure 15B-2.
3. Position the joint so the weld axis is vertical as in Figure 15B-3.
4. Adjust the welding current: Direct Current Reverse Polarity, DCRP electrode positive (+) ampere range: 115 to 165 (slightly lower than for horizontal welding).

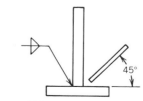

Figure 15B-1

VERTICAL WELDING OF PLATE

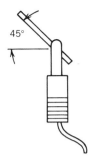

Figure 15B-2

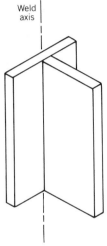

Figure 15B-3 Test position 3F.

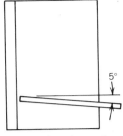

Figure 15B-4 Electrode push angle.

Figure 15B-5 First pass weave.

Figure 15B-6 Alternate weave.

Figure 15B-7 Completed weld.

PROCEDURE

1. Hold the electrode midway between the two plates as in Figure 15B-1. Point the electrode slightly upward, at about a 0 to 5° angle, as in Figure 15B-4.
2. *First pass:* Start at the bottom of the joint. Start the arc with a tapping motion. This breaks the coating covering the end of the electrode.
3. Weave the electrode from side to side, advancing upward slightly to keep the bead progressing. (Refer to Figure 15B-5.) The bead should only be about 5/16 in. wide.
4. Pause at each side and watch for the weld puddle to sag in the center. If sag is present:
   (a) reduce the amperes or
   (b) increase the travel speed or
   (c) hold a closer arc
5. *Second pass:* Clean the first pass; then strike the arc on the the bottom center of the groove. Move quickly to the lower left edge of the first pass.
6. Pause until the puddle forms. Then move the electrode steadily to the right, drawing the puddle along.
7. Pause on the right when the edge of the electrode reaches the edge of the weld. The puddle will flow out another 3/32 in. (2.4 mm) to 1/8 in. (3.2 mm).
8. Use either the basic weave or the alternate weave shown in Figure 15B-6.
9. *Third pass:* Weave another bead over the second pass. Cover slightly more than one-half of the right side of the weld. Watch for undercut on the right side of the weld.
10. Weave the fourth bead on the left side. Cover the second pass and overlap about 1/4 in. of the third pass. This is called a double cap. (Refer to Figure 15B-7.) Be careful of **undercut,** both on the left side and where this bead joins the other bead.

POINTS TO REMEMBER

1. Clean thoroughly between passes.
2. Pause at the sides of each bead to eliminate undercut.
3. Do not weave more than three electrode diameters wide.
4. Maintain a close arc at all times.

LESSON 15C
LAP JOINT FILLET, VERTICAL, 3-F POSITION, STRINGER AND WEAVE BEADS (UPHILL), E-6010 AND E-6011 ELECTRODES

OBJECTIVE
Upon completion of this lesson you should be able to correctly and safely weld a lap joint fillet in the Vertical (3-F) position, with E-6010 or E-6011 electrodes, using the uphill method in accordance with accepted industrial procedures.

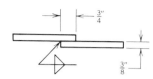

Figure 15C-1 Tacking procedure.

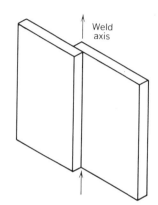

Figure 15C-2 Test position.

Figure 15C-3 Whipping motion.

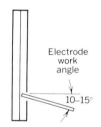

Figure 15C-4 Side view.

Figure 15C-5

NOTE
The lap joint is similar to the T-joint. Both are variations of fillet welds. The major difference is one lap joint weld leg is on the end instead of the surface of the plate.

This smaller surface presents the only problem. The edge has a tendency to "burn away" unless you handle the electrode skillfully. Too much heat can melt down the weld toe and reduce the effective weld throat. This, in turn, means the weld will not be strong enough.

You must take care to keep the legs equal. This can be done by keeping the electrode away from the toe of the joint. Stop the electrode movement when the edge of the puddle just reaches the plate edge. If this is done, the joint will be filled properly.

MATERIAL AND EQUIPMENT
1. Personal welding equipment
2. E-6010 or E-6011 electrodes ⅛ in. (3.2 mm) diameter
3. 2 pcs. ⅜ in. × 3 in. × 12 in. long hot rolled steel plate

GENERAL INSTRUCTIONS
1. Thoroughly wire brush the intended joint area.
2. Overlap the plates ¾ in. as shown in Figure 15C-1 and tack.
3. Position the plate as shown in Figure 15C-2.
4. Adjust the welding current: Direct Current Reverse Polarity DCRP, electrode positive (+). Ampere range: 75 to 125.

PROCEDURE
1. *First pass:* Run a stringer bead vertically up the joint. Utilize the whipping motion, as in the previous lesson as shown in Figure 15C-3. Use an electrode work angle of 10 to 15° as shown in Figure 15C-4.
2. *Second pass:* Clean the first pass, then start the arc on the bottom left. Carry the puddle across to the right. **PAUSE.** Then swing the electrode, with a long arc length, over to the left side. The motion is shown in Figure 15C-5. Metal is deposited on each move from left to right.
3. Pause for a moment on the left, then move slowly to the right side. The pause at the left should be less than the pause at the right. There is more metal on the right side to absorb the heat.
4. *Final pass:* Start as with the second pass, but hold a slightly long arc. Do not move until the edge of the puddle touches the leading edge of the top plate.

172 VERTICAL WELDING OF PLATE

5. Move steadily to the right, pause, and continue as on the previous pass. Use the same movements shown in Figure 15C-5.
6. Compare with the sample weld shown in Figure 15C-6.

POINTS TO REMEMBER
1. Pause at the sides to fill in the undercut.
2. Pause longer on the right than on the left (especially on the final pass).
3. Be sure the legs of the fillet are equal, and the face is flat or slightly convex.
4. Hold a close arc when depositing metal and a long arc on the crossover.

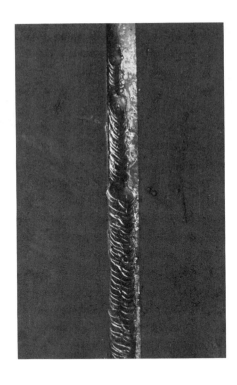

Figure 15C-6 Completed weld.

LESSON 15D
LAP JOINT FILLET, VERTICAL, 3-F POSITION, WEAVE BEADS (UPHILL) E-7018 ELECTRODES

OBJECTIVE
Upon completion of this lesson you should be able to correctly and safely weld a lap joint fillet in the 3-F position with low hydrogen E-7018 electrodes, using the uphill method in accordance with accepted industrial procedures.

NOTE
Although the lap joint in the vertical position is not in common use you may be called upon to weld one. Your practice to master this joint can only increase your overall welding skill.

MATERIAL AND EQUIPMENT
1. Personal welding equipment
2. E-7018 electrodes ⅛ in. (3.2 mm) diameter
3. 2 pcs. ⅜ in. × 3 in. × 12 in. long hot rolled steel plate

GENERAL INSTRUCTIONS
1. Thoroughly wire brush the intended joint area.
2. Overlap the plates ¾ in. as shown in Figure 15D-1.
3. Position the plate as shown in Figure 15D-2.
4. Adjust the welding current: Direct Current Reverse Polarity DCRP, electrode positive (+). Ampere range: 115 to 165.

PROCEDURE
1. *First pass:* Run a vertical stringer up the root utilizing the weave

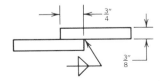

Figure 15D-1 Tacking procedure.

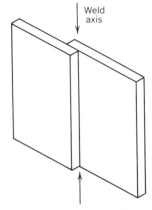

Figure 15D-2 Test position 3F.

motion shown in Figure 15D-3. Use an electrode work angle of approximately 0 to 5° upward as in Figure 15D-4. *Pause briefly at each side* and maintain a close arc.

2. *Second pass:* After cleaning the first pass, strike the arc in the center of the groove and move to the left. Pause at the lower left until a puddle forms.

3. Carry the puddle across to the right and pause to cover the first stringer.

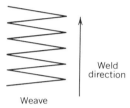

Figure 15D-3 Weave motion.

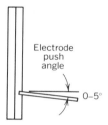

Figure 15D-4 Electrode push angle.

4. Weave back and forth as in Figure 15D-3. Hold a close arc and pause slightly longer on the right to fill in undercut.

5. *Finish pass—double cap:* Begin with the third bead on the right side of the joint. It should cover slightly more than half of the previous pass as shown in Figure 15D-5.

6. Run a fourth bead as wide as the third so it overlaps the third bead. (Refer to Figure 15D-5.) Be careful not to undercut or overlap the left-hand edge.

7. Compare your weld with the sample shown in Figure 15D-6.

POINTS TO REMEMBER

1. Pause at the sides with all passes to fill in the undercut.
2. Pause slightly longer at the right side of the joint.
3. Keep the legs of the fillet equal and the face flat or slightly convex.
4. Maintain a close arc at all times with E-7018. Do not lenghten it on the crossover, as you did in Lesson 15C with E-6011 electrodes.

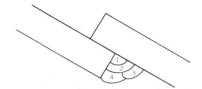

Figure 15D-5 Bead sequence.

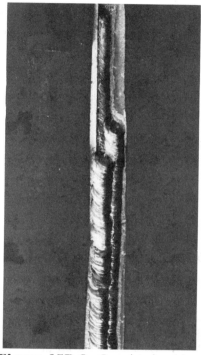

Figure 15D-6 Completed weld.

LESSON 15E
V-GROOVE BUTT JOINT, VERTICAL, 3-G POSITION, WITH BACK-UP BAR WEAVE BEADS (UPHILL) E-7018 ELECTRODE

OBJECTIVE
Upon completion of this lesson you should be able to correctly and safely weld a butt joint with back-up bar in the 3-G position, with E-7018 electrodes, using the uphill method to conform with QW-461.2 (C) ASME Section IX and AWS D1.1 Figure 5.8.1.1. C

NOTE
The V-groove butt joint, with a back-up bar, is in common use. It is needed where it is necessary to join heavy plate.

A welder who passes a test similar to the one in this lesson, qualifies to weld plate of limited thickness (as indicated in Figure 5.19 of the AWS Structural Welding Code). Passing the same test on 1 in. thick plate qualifies the welder for unlimited thicknesses.

MATERIAL AND EQUIPMENT
1. Personal welding equipment
2. E-7018 electrodes ⅛ in. (3.2 mm) diameter

3. 2 pcs. ⅜ in. × 3 in. × 12 in. long hot rolled steel plates
   1 pc. ⅜ in. × 1½ in. × 12 in. long hot rolled steel plates

GENERAL INSTRUCTIONS

1. *Plate preparation:* Bevel one long side on each plate to form a 45° included angle as shown in Figure 15E-1.
2. Clean the face of the bevel and the back and front of the plates in the intended weld area.
3. Place the plates, beveled edges together, upside down on a flat surface, spaced with a ¼ in. (6.4 mm) gap.
4. Center the ⅜ in. × 1½ in. × 12 in. back-up bar over the joint. Hold in position and place tacks on both sides, the ends, and the center as in Figure 15E-2.
5. Position the plates as shown in Figure 15E-3.
6. Adjust the welding current: Direct Current Reverse Polarity DCRP electrode positive (+). Ampere range: 115 to 165.

PROCEDURE

1. *First pass:* Strike the arc at the bottom of the groove. Hold the electrode straight into the groove as in Figure 15E-1, with about a 5° downward angle as shown in Figure 15E-4. Weld uphill, using a slight weaving motion. You can drag the electrode up the root opening as an alternative way.
2. Weave the electrode until it reaches the toes of the weld. With the correct amount of weave the coating should touch the sides of the joint. If the arc current (heat) is set properly, you will see and feel the electrode "bite" into the parent metal. Pause briefly at the sides to eliminate undercut.
3. Make sure both plates and the back-up bar are fused together. Do not overheat the sides of the bevel; this may cause back-undercut, as shown in Figure 15E-1.
4. Clean thoroughly and run the second pass. Weave the second pass using the same electrode angles, with the motion shown in Figure 15E-5. Pause at the edges of the weld.
5. Clean and run the third pass. The third pass is a double cap. Weave the first bead slightly wider than one-half the width of the joint. Clean the weld thoroughly, then run another uphill bead to cover the remainder of the joint. The second bead must overlap the first by about one-half bead. (See bead sequence Figure 15E-6.)
6. Keep the weld reinforcement less than ³⁄₃₂ in. above the surface of the plate. Excess reinforcement can cause stress risers or small notches. These can cause weld failures.
7. Clean and compare your weld with the sample in Figure 15E-7.

POINTS TO REMEMBER

1. Clean each bead thoroughly between passes.
2. Maintain a 0 to a 5° electrode angle downward on all passes.
3. Pause at the sides. Maintain a uniform weld face.

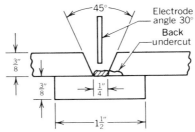

Figure 15E-1 Plate preparation.

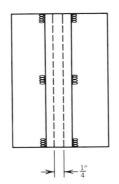

Figure 15E-2 Tacking procedure.

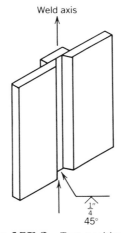

Figure 15E-3 Test position.

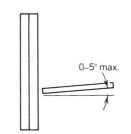

Figure 15E-4 Electrode angle.

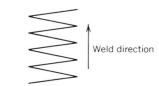

Figure 15E-5 Weave.

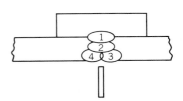

Figure 15E-6 Bead sequence.

4. Avoid overheating at the feather edge of the bevels. It can cause back-undercut.

CAUTION: **DO NOT QUICK QUENCH** (cool rapidly) specimens that are to be tested.

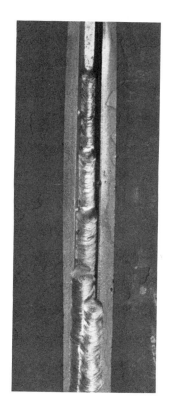

Figure 15E-7 Completed weld.

LESSON 15F
OPEN CORNER JOINT, VERTICAL 3-G POSITION, STRINGER AND WEAVE BEADS (UPHILL), E-6010 OR E-6011 ELECTRODES

OBJECTIVE
Upon completion of this lesson you should be able to correctly and safely weld an open corner joint in the vertical 3-G position with the E-6010 or E-6011 electrode using the "uphill method" in accordance with accepted industrial procedures.

NOTE
Open corner joints are used frequently in the fabrication of rectangular tanks, but rarely with a root opening. Most open corner joints are welded without full penetration. The lesson gives you a chance to practice first passes in open root groove welds. The open corner joint requires much less preparatory work, therefore you will have more time to spend welding.

The skill you develop on open roots will be helpful when you weld single bevel and V-groove joints.

MATERIAL AND EQUIPMENT
1. Personal welding equipment
2. E-6010 or E-6011 electrodes ⅛ in. (3.2 mm) diameter
3. 2 pcs. ⅜ in. × 3 in. × 12 in. long hot rolled steel

GENERAL INSTRUCTIONS
1. Remove all mill scale from the long edge and both sides of the intended weld area.
2. Prepare and tack the plates in the position shown in Figure 15F-1.
3. Position the weldment as shown in Figure 15F-2.
4. Adjust the welding current: DC reverse polarity DCRP, electrode positive (+). Ampere range: 75 to 125.

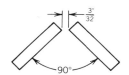

Figure 15F-1 Tacking procedure.

VERTICAL WELDING OF PLATE

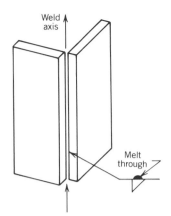

Figure 15F-2 Test position.

Figure 15F-3 Whip.

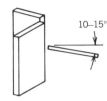

Figure 15F-4 Electrode push angle.

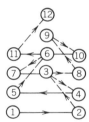

Figure 15F-5 Triangle.

Figure 15F-6 Weave.

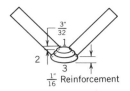

Figure 15F-7 Bead sequence.

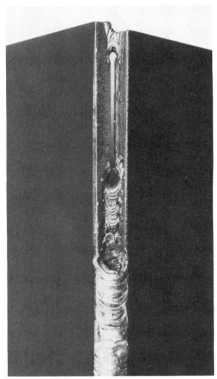

Figure 15F-8 Completed weld.

PROCEDURE

1. *First pass:* Weld uphill, using the whipping motion. Hold a close arc, then whip out of the puddle with an upward movement to the right, as illustrated in Figure 15F-3. Keep the electrode at the 10 to 15° upward angle, as shown in Figure 15F-4.

2. As you proceed upward, keep the "keyhole" open. If the keyhole becomes too large, whip out further on the upswing or lower your welding current.

3. Make sure you are obtaining penetration. Do not overdo it; 3/32 in. (2.4 mm) is ideal, and will provide proper **melt-thru.**

4. *Second pass and out:* Use either the triangle (Figure 15F-5) or the weave method (Figure 15F-6). Pause at the sides and keep the face flat and uniform.

5. The finish pass should have about 1/16 in. (1.6 mm) reinforcement. (See Figures 15F-7 and 15F-8.)

6. Cool, clean, and present to your instructor.

POINTS TO REMEMBER

1. Do not **overweld** the root pass. It will cause excessive root buildup.
2. Watch for incomplete penetration or excessive melt-thru.
3. Pause at the sides to eliminate undercut on second and outer passes.
4. Do not overweld the finish pass.

LESSON 15G
OPEN CORNER JOINT, VERTICAL, 3-G POSITION, WEAVE BEADS (DOWNHILL), E-6010 OR E-6011 ELECTRODES

OBJECTIVE
Upon completion of this lesson you should be able to correctly and safely weld an open corner joint in the vertical 3-G position with E-6010 or E-6011 electrodes utilizing the downhill method in accordance with accepted industrial procedures.

NOTE
This lesson is designed to develop your ability to weld plate and pipe with the downhill method.

It will also help you develop the ability to "drag" the first pass just like a pipeline welder.

Dragging requires practice, but after mastering the method you will be able to make out-of-position open root butt welds much easier and faster.

MATERIALS AND EQUIPMENT
1. Personal welding equipment
2. E-6010 or E-6011 electrodes ⅛ in. (3.2 mm) diameter
3. 2 pcs. ⅜ in. × 3 in. × 12 in. long hot rolled steel
4. 2 spacers ³⁄₃₂ in. × ⅝ in. × 6 in. long

GENERAL INSTRUCTIONS
1. Remove all mill scale from the long edge and both sides of the intended weld area.
2. Prepare and tack the plates as in Figure 15G-1. Be sure the root opening is uniform and the correct width.
3. Position the weldment as shown in Figure 15G-2.
4. Adjust the welding current: DC reverse polarity, DCRP electrode positive (+). Ampere range: 75 to 125.

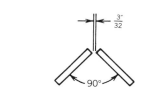

Figure 15G-1 Tacking procedure.

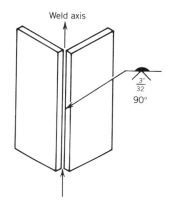

Figure 15G-2 Test position 3G.

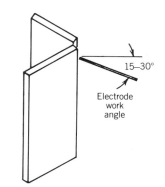

Figure 15-G-3 Electrode work angle.

PROCEDURE
1. The root opening is ³⁄₃₂ in. (2.4 mm), as shown. The first pass will be dragged. Start at the top and weld toward the bottom.
2. *First Pass:* Point the electrode directly into the center of the joint. Use an upward angle of about 15 to 30° as shown in Figure 15G-3.
3. Strike the arc and weave down until you pass over the tack and until the end of the tack is reached. Hold a fairly long arc.
4. When you pass the tack, gently push the electrode into the root. The arc should blow out the opposite side of the joint, and the electrode should touch the workpiece.
5. Slowly drag the electrode down the joint making sure that the electrode angle remains the same. If you are not getting enough penetration, you must increase the amperes, increase the electrode angle, or decrease the rate of travel.
6. If the arc blows through excessively, decrease the electrode angle, lower the amps, or increase the rate of travel. Listen for a harsh, blowing sound and you will know that the arc is penetrating.
7. *Second and outer passes:* Clean the bead thoroughly. Start downhill at the top of the joint. Weave slightly from side to side, with an upward-over motion as shown in Figure 15G-4. Maintain a close arc and the electrode drag angle shown in Figure 15G-3.
8. If the face of the weld is slightly

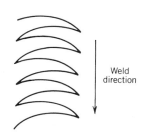

Figure 15G-4 Up and over.

concave on the next-to-last pass, run a stringer, about 5/16 in. (8.0 mm) wide down the center before applying the finish pass, which is another weave. (See Figure 15G-5 for the bead sequence.)

9. Cool, clean, and compare your weld with Figure 15G-6.

POINTS TO REMEMBER

1. Maintain the same electrode angle throughout the procedure.
2. Clean thoroughly between passes.
3. Use the up and over motion. Be sure you get fusion on the sides.
4. Listen for a harsh sound on the first pass; it is your signal that you are getting penetration.
5. Guard against excessive overlap on the finish pass.

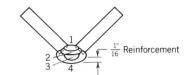

Figure 15G-5   Bead sequence.

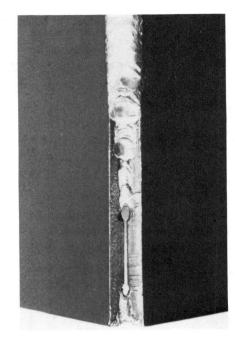

Figure 15G-6   Completed weld.

LESSON 15H
V-GROOVE BUTT JOINT, VERTICAL, 3-G POSITION, STRINGER AND WEAVE BEADS (UPHILL) E-6010 OR E-6011 ELECTRODES

OBJECTIVE
Upon completion of this lesson you should be able to correctly and safely weld a V-groove butt joint in the vertical 3-G position with E-6010 or E-6011 electrodes, utilizing the uphill method, while conforming to the requirements of QW-461.2 (C) ASME Section IX Boiler and Pressure vessel code and AWS Structural Code D1.1 Figure 5.8.1.1.C

NOTE
One of the more difficult joints to weld is the open root V-joint. This joint is in wide use and is often used to test a welder's ability.

Quite a bit of skill is necessary to weld root passes with the correct penetration. Excessive burn through is not acceptable, just as lack of penetration is not acceptable.

Be sure to keep the "keyhole" open as the weld progresses during the first pass.

MATERIAL AND EQUIPMENT
1. Personal welding equipment
2. E-6010 or E-6011 electrodes 1/8 in. (3.2 mm) 5/32 in. (4.0 mm) diameter
3. 2 pcs. 3/8 in. × 5 in. × 12 in. long hot rolled steel plate
4. 2 spacers 3/32 in. × 5/8 in. × 6 in. long

GENERAL INSTRUCTIONS
1. Bevel the long side of the plates with a 30° angle. This provides a final included angle of 60°.
2. Remove the scale on the intended weld areas by grinding or sanding.
3. File a 3/32 in. (2.4 mm) root face land or shoulder on each bevel.
4. Fit up with a 3/32 in. (2.4 mm) root gap and tack at each end as shown in Figure 15H-1.
5. Preposition the plates slightly as in Figure 15H-2 so you can fit a 1/8 in. (3.2 mm) electrode beneath the joint.

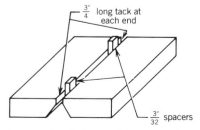

Figure 15H-1   Tacking procedures.

6. Position the joint at a comfortable height, in the position shown in Figure 15H-3.
7. Adjust the welding current: DC reverse polarity, DCRP, electrode positive (+). Ampere range: 1/8 in. (3.2 mm)—75 to 125; 5/32 in. (4.0 mm)—110 to 170.

PROCEDURE

1. Check your welding condition on scrap so that you can easily whip the electrode uphill, as in Figure 15H-4.
2. Start the arc on the bottom tack. Hold the arc at the beginning of the root opening until a keyhole appears. Use an electrode angle 10 to 15° as shown in Figure 15H-5.
3. Move the electrode upward in a slight whipping motion, then return to the keyhole. Pause there until the keyhole fills and another opens above it. Continue to whip uphill.
4. If the keyhole does not appear, you are:
   (a) Moving upward too rapidly
   (b) Do not have sufficient arc current to penetrate
   (c) Do not have the proper root opening
5. If the keyhole is too large, you are:
   (a) Moving upward too slowly
   (b) Have the current set too high
   (c) Holding too long an arc
6. *Clean and run the second pass.* Start at the bottom of the joint. Hold a normal arc and weld uphill using either the weave shown in Figure 15H-6, or the triangle motion shown in Figure 15H-7. Pause at the sides to eliminate undercut.
7. *Third and out:* Switch to 5/32 in. diameter electrodes and continue to weld the same way. The next to last pass should be 1/32 to 1/16 in. below the plate surface. This gives you a guide for the last pass.
8. The final pass must be about 1/16 in. above the surface of the plate. Excess reinforcement can create a stress riser or notch which can lead to weld failure.
9. Cool, clean, and compare your weld with Figure 15H-8.

POINTS TO REMEMBER

1. Check for smooth bead appearance and undercut. If you have undercut, you neglected to *pause* at sides.
2. The prepositioning is important; without it you will have a distorted weldment.
3. The first pass must have a keyhole at all times, or you will not have penetration.

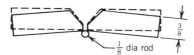

Figure 15H-2 Prepositioning.

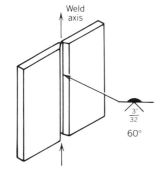

Figure 15H-3 Test position 3G.

Figure 15H-4 Whipping motion.

Figure 15H-5 Push angle.

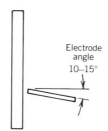

Figure 15H-6 Weave.

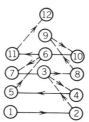

Figure 15H-7 Triangle.

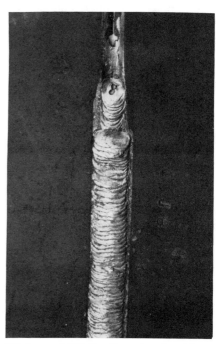

Figure 15H-8 Completed weld.

# LESSON 15I
SQUARE GROOVE BUTT JOINT, VERTICAL, 3-G POSITION, WITH THE BACK-UP BAR (UPHILL), 3-G POSITION, E-7018 ELECTRODES, AND (DOWNHAND) 1-G POSITION, E-7024 ELECTRODES

OBJECTIVE
Upon completion of this lesson you should be able to correctly and safely weld a square groove butt with back-up bar. The fillets shall be welded in the vertical position with E-7018. The remainder of the groove may be welded in the 1-G position with E-7024, while conforming to the requirements of QW-461.4 (A) ASME Boiler and Pressure vessel code and AWS Structural Code D1.1 Figure 5.8.1.3 C.

NOTE
Some construction companies that erect buildings and bridges often use a test based on a joint similar to the one in this lesson. The vertical passes determine your ability to handle low hydrogen electrodes in out-of-position welding.

The critical part of the uphill test is the amount of first-pass penetration.

After the first pass on both fillets is completed, the plate is moved to the flat position. The weld is completed with iron powder electrodes or in some instances low hydrogen electrodes.

The purpose of the downhand portion of the weld is to test your ability to deposit filler metal without defects.

Many times this is the first test you receive when applying for a job. Successful completion usually means an offer of employment.

MATERIAL AND EQUIPMENT
1. Personal welding equipment
2. E-7018 electrodes 5/32 in. (4.0 mm) diameter
3. E-7024 electrodes 5/32 in. (4.0 mm) diameter
   2 pcs. 1/2 in. × 3 in. × 12 in. long hot rolled steel plate
   1 pc. 1/2 in. × 1 1/2 in. × 12 in. long hot rolled steel plate
4. 2 spacers 15/16 in. × 15/16 in. × 6 in. long

GENERAL INSTRUCTIONS
1. Clean the mill scale and slag from the intended root and face areas of the 12 in. sides of the plate.
2. Place the plates, face down, with a 15/16 in. space between them as in Figure 15I-1.
3. Cover the space with the back-up bar and tack weld at both ends as shown in Figure 15I-2.
4. Position the joint as shown in Figure 15I-3.

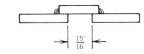

Figure 15I-1

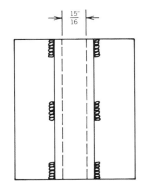

Figure 15I-2 Tacking procedure.

5. Adjust the welding current as follows: E-7018 electrodes, DC reverse polarity, electrode positive (+). Ampere range: 150 to 220. E-7024 electrodes AC or DC straight polarity, electrode negative (−). Ampere range: 180 to 250.

PROCEDURE
1. *Uphill fillet welds:* Use the weave method to make two filets in the 3-G positions with 5/32 in. E-7018. Use an electrode angle of approximately 5° downward as shown in Figure 15I-4. These are welds 1 and 2 in Figure 15I-5.

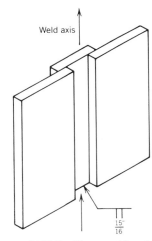

Figure 15I-3 Test position 3G.

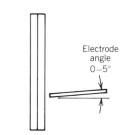

Figure 15I-4

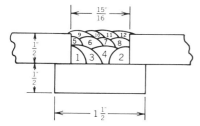

Figure 15I-5 Bead sequence.

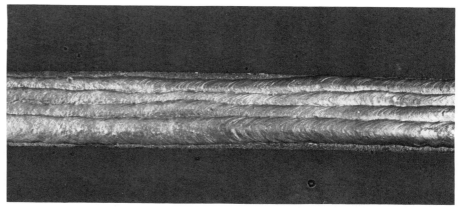

Figure 15I-6 Completed weld.

2. Clean thoroughly and place the plate in the 1-G or downhand position as in Figure 15I-5.
3. Fill in the unwelded area between the fillet welds with two stringer beads (3 and 4) using 5/32 in. E-7024 drag or hold a close arc.
4. Fill in the rest of the groove as in square groove welding in the flat or downhand position.
5. Do not overlap more than 1/8 in. on the finish pass. Be sure to overlap half of each stringer as in Figure 15I-5.
6. Cool, clean, and compare with completed sample in Figure 15I-6 before presenting to your instructor.

POINTS TO REMEMBER
1. The fillet welds are critical. Be sure you have good fusion and penetration.
2. Clean thoroughly. Pay special attention to the edges of the square groove.
3. Avoid undercut on the finish pass.

LESSON 15J
V-GROOVE BUTT JOINT, VERTICAL, 3-G POSITION WEAVE BEADS, E-6010 or E-6011 ELECTRODE. FIRST PASS DRAGGED (DOWNHILL) SECOND AND OUTER PASSES WEAVED (UPHILL)

OBJECTIVE
Upon completion of this lesson you should be able to correctly and safely weld a V-groove butt joint in the 3-G position with E-6010 or E-6011 electrode, using the downhill method for the first pass and the uphill method for the remainder, while conforming to the requirements of QW-461.2 (C) ASME Boiler and Pressure vessel code and AWS Structural Code D1.1 Figure 5.8.1 A

NOTE
Open root V-groove joints are one of the most difficult to weld. It naturally follows that this joint is often used as a test of a welder's skill and ability.

This lesson uses your skills in both the downhill and uphill methods of welding. You will use a grinder to remove excess buildup from the first pass and keep slag from being trapped at the **toes** of the bead. If the first pass is not cleaned and ground properly, the second pass wagon tracks will occur. Wagon tracks show up on an x-ray as two lines at the toes of the first pass. The tracks indicate that there is a *lack of fusion* due to slag or lack of second pass penetration. If not discovered, this condition will eventually cause weld failures.

MATERIAL AND EQUIPMENT
1. Personal welding equipment
2. E-6010 or E-6011 electrodes 1/8 in. (3.2 mm) diameter
3. 2 pcs. 3/8 in. × 4 in. × 12 in. long hot rolled steel plate
4. 2 spacers 3/32 in. × 5/8 in. × 6 in. long

GENERAL INSTRUCTIONS
1. Bevel the 12 in. wide side of the plates to a maximum of 37½°.

2. Remove all mill scale and clean the intended weld areas.
3. Sand, or file, a land 3/32 in. (2.4 mm) wide on the bevel.
4. Place the plates face down with a 3/32 in. spacer between the lands and tack each end as shown in Figure 15J-1.
5. Preposition the plates as shown in Figure 15J-2. You can use a 1/8 in. diameter electrode placed under the root.
6. Position the joint as shown in Figure 15J-3.
7. Adjust the welding current. DC reverse polarity, DCRP electrode positive (+). Ampere range: 75 to 125

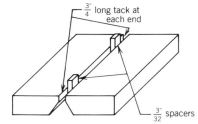

Figure 15J-1 Tacking procedure.

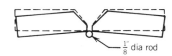

Figure 15J-2 Prepositioning.

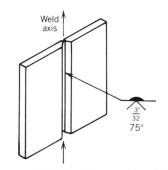

Figure 15J-3 Test position 3G.

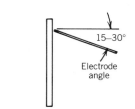

Figure 15J-4

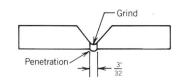

Figure 15J-5

Figure 15J-6 Weave.

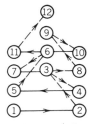

Figure 15J-7 Triangle.

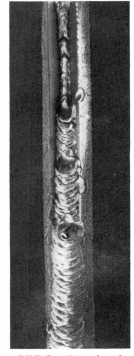

Figure 15J-8 Completed weld.

PROCEDURE

1. Check your bead by dragging the electrode on scrap. You should be able to lay the electrode against the metal at about a 15 to 30° drag angle and not stub out as in Figure 15J-4.
2. *First pass:* Strike the arc on the top tack. Hold a long arc and weld over the tack until you reach the open root.
3. Gently press the electrode into the root with an electrode angle of 15 to 30°, and drag it steadily along. Listen carefully. If you hear a harsh, blowing sound, you are getting good penetration.
4. If you are not getting penetration, you are not getting enough heat into the joint. To correct the problem, you should:
   (a) Increase the electrode angle.
   (b) Decrease the travel speed.
   (c) Increase the amperage.
5. *Second pass and outer passes.* Clean thoroughly and grind out any excess weld metal in the root pass as shown in Figure 15J-5.
6. Weave or use the triangle motion as in Figures 15J-6 and 15J-7. Pause at the sides and keep the width of the pass uniform.
7. Examine the weld for evenness, contour, and undercut.
8. Cool, clean, and compare with sample Figure 15J-8 before presenting to your instructor.

POINTS TO REMEMBER

1. One-hundred percent penetration on the first pass is essential. Attempt to obtain a minimum of 3/32 in. (2.4 mm) reinforcement.
2. Grind off the excess buildup of the first pass before continuing.
3. Check constantly for undercut.
4. The finish pass should be 3/32 in. above the plate surface and have a flat or slightly convex contour.

LESSON 15K
V-GROOVE BUTT JOINT, VERTICAL, 3-G POSITION, AWS STRUCTURAL WELDING CODE D1.1 WELDER QUALIFICATION TEST FOR LIMITED PLATE THICKNESS (E-6010 AND E-7018 ELECTRODES)

OBJECTIVE
Upon completion of this lesson you should be able to safely and correctly weld an open root, V-groove, butt joint in the 3-G position, utilizing E-6010 and E-7018 electrodes while conforming to Figure 5.20A of the AWS Structural Welding Code.

NOTE
There are a number of plate tests that are used to qualify welders. This test is the most demanding because of the degree of skill required to pass the test.

Successful completion of this test qualifies the welder Part C Welder Qualification, D1.1 AWS Structural Welding Code which states:

The welder who makes a complete joint penetration plate groove weld procedure qualification test that meets the requirements, is thereby qualified for that process and test position for plates and square or rectangular tubing equal to or less than the thickness of the test plate welded. If the test plate is 1 inch (25.4 mm) or greater in thickness, the welder will be qualified for all thicknesses.

In addition the 5.23.1.3 states that:

Qualification in the 3 G (vertical) position qualifies for flat, horizontal, and vertical position groove and flat, horizontal, and vertical position fillet welding of plate, and flat and horizontal position fillet welding of pipe and tubing.

The welder who successfully completes this test demonstrates a high degree of skill in shielded metal arc welding.

MATERIAL AND EQUIPMENT
1. Personal welding equipment
2. E-7018 electrodes ⅛ in. (3.2 mm) E-6010 and ⁵⁄₃₂ in. (4.0 mm) 2 pcs. ⅜ in. × 5 in. × 12 in. long hot rolled steel plates
3. 2 spacers ³⁄₃₂ in. × ⅝ in. × 6 in. long

GENERAL INSTRUCTION
1. Bevel the long side of the plates (max. 30° angle).
2. Remove the mill scale and clean the intended weld by grinding or sanding.
3. File a ³⁄₃₂ in. (2.4 mm) root face on each bevel. Maximum allowed is ⅛ in. (3.2 mm).
4. Fit up with a ³⁄₃₂ in. (2.4 mm) root gap and tack at each end as shown in Figure 15K-1. Maximum root gap allowed is ⅛ in. (3.2 mm).
5. Preposition the plates slightly as shown in Figure 15K-2. If you can fit a ⅛ in. (3.2 mm) electrode beneath the root, the amount of prepositioning is sufficient.
6. Position the test piece at a comfortable height in the position as shown in Figure 15K-3.
7. Adjust the welding current: DC reverse polarity, electrode positive (+). Ampere range: ⅛ in. (3.2 mm) E-6010 DCRP—75 to 125; ⁵⁄₃₂ in. (4.0 mm) E-7018 DCRP—150 to 220.

PROCEDURE
1. Check your amperage using the ⅛ in. E-6010 electrode so that you can easily weld uphill, using the whipping method. Hold the electrode pointed slightly upward as in Figure 15K-4.
2. Start the arc on the lower tack weld. Hold a long arc as you pass over the tack.
3. When you reach the open root,

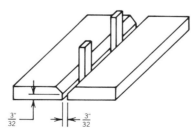

Figure 15K-1 Tacking procedure.

Figure 15K-2 Prepositioning.

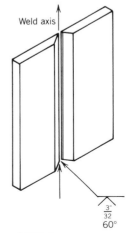

Figure 15K-3 Test position 3G.

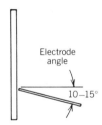

Figure 15K-4

pause briefly until a keyhole forms as in Figure 15K-5.

4. Move the electrode upward with a slight whipping motion as in Figure 15K-6. Pause on the downstroke until the keyhole appears.
5. If the keyhole does not appear:
   (a) Hold a fairly long arc until the root is very hot, then,
   (b) Gently force the electrode into the puddle and through to the other side of the joint.
6. Pause slightly until the keyhole opens. Then proceed upward, making sure that you do not lose the keyhole. Remember if the keyhole is too large you are:
   (a) Moving upward too slowly
   (b) Working with too high an amperage setting
   (c) Holding too long an arc
7. *Second pass:* Using the E-7018 electrode start at the bottom of the joint and weld uphill. Use the weave motion as shown in Figure 15K-7. Pause at the sides.
8. *Third pass:* Weave uphill with E-7018 electrodes. Pay close attention to the edges of the bevel. The weld should extend about 3/32 in. (2.4 mm) beyond the edge of the bevel. Be sure to move steadily across the middle. This insures a solid flat or slightly convex surface.
9. Remove two coupons 1½ in. wide—one from each side of the center line.
10. Prepare them for testing by filing a radius as in Figure 15K-8.
11. Subject one of them to a root bend test and the other to a face bend.
12. Present the bend specimens to the instructor for evaluation.

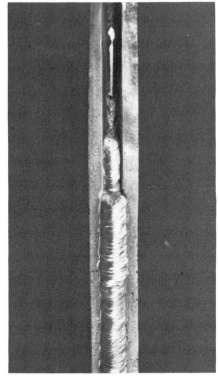

Figure 15K-5 Completed weld.

Figure 15K-6 Whip.

Figure 15K-7 Weave.

Figure 15K-8 Proper edge preparation.

POINTS TO REMEMBER

1. Undercut is caused by failure to pause at the sides.
2. Without a keyhole you have no penetration. Cut two test coupons 1½ in. wide from each side of the center line of the plate.

CHAPTER 16
OVERHEAD WELDING OF PLATE

Many people feel the overhead position is the most difficult of all welding positions to master. This is not necessarily true, but it takes a lot of practice and is physically tiring.

The most difficult part of learning overhead welding is that your arms get tired. It takes time to develop the strength to hold the electrode steady in that position for the length of time required while you run a pass. This is not as difficult as it seems because it takes less than 2 minutes to burn an electrode.

Welding procedures for the overhead position are similar to those used for the horizontal position. With fillet weld test plates the fillet weld is deposited on the underside of the horizontal surface and against the vertical surface as shown in Figure 16-1.

For the overhead position, groove welded plates should be approximately horizontal and the weld metal deposited on the underside as shown in Figures 16-2 and 16-3.

As in other welding positions electrode angles, arc length, current settings, and travel speed are extremely important. Arc length is very important in overhead welding. An arc length that is too long will cause the molten pool to become too large and fluid, and it will become difficult to control the puddle. The puddle will sag badly or even fall away. Maintain a close arc at all times when welding overhead.

Proper protective clothing and equipment are required when welding in any position. They are absolutely essential for your safety when you weld in the overhead position. Always use and wear *all* the required protective clothes and equipment.

Industrial use of overhead welding is kept to a minimum for a number of reasons. Welding overhead is slow. Only the smaller diameter electrodes, such as 3/32 in. (2.4 mm), 1/8 in. (3.2 mm), 5/32 in. (4.0 mm), and 3/16 in. (4.8 mm) electrodes are normally used. This increases the time required to complete a job. Muscle fatigue tends to decrease the welder's output.

Skilled overhead position welders are difficult to find. The welder who can produce quality welds in the overhead position is not an ordinary welder. Such an individual has a fairly high degree of ability and is sought after by those who require welders for out-of-position welding.

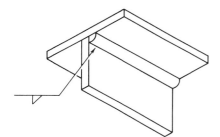

Figure 16-1 T-joint fillet weld—overheal position 4F.

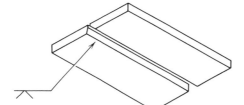

Figure 16-2 Groove welds—overhead position 4G.

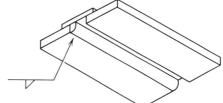

Figure 16-3 Square groove butt—overhead position 4G.

LESSON 16A
T-JOINT FILLET, OVERHEAD, 4-F POSITION, STRINGER BEADS (E-6010 OR E-6011 ELECTRODES)

OBJECTIVE
Upon completion of this lesson you should be able to safely and correctly weld a T-joint fillet in the overhead position using E-6010 or E-6011 electrodes.

NOTE
The T-joint fillet is one of the most widely used overhead weld joints found in industry. Normally, it requires very little joint preparation and it lends itself to many applications. It is used in such applications as constructing bridges, buildings, ships, trucks, and machinery.

MATERIAL AND EQUIPMENT
1. Personal welding equipment

2. E-6010 or E-6011 electrodes, ⅛ in. (3.2 mm) diameter
3. 1 pc. carbon steel plate, ⅜ in. × 4 in. × 12 in. long
   2 pcs. carbon steel plate, ⅜ in. × 1¾ in. × 12 in. long

GENERAL INSTRUCTIONS

1. Wire brush the intended weld area of the plate.
2. Tack the plates together as shown in Figures 16A-1 and 16A-2.
3. Position the practice piece in the fixture as in Figure 16A-3.
4. Be sure the workpiece and workpiece clamp are secure.
5. Adjust the welding current: DC reverse polarity, electrode positive (+) DCRP. Ampere range: 75 to 125.

PROCEDURE

1. Place the electrode in the holder as shown in Figure 16A-4.
2. Right-handed welders should start at the left side of the joint. Use a 30° work angle off the vertical plate as shown in Figure 16A-5. Work with a drag angle of approximately 10° as shown in Figure 16A-6.
3. Strike the arc and hold a fairly long arc. Begin a stringer and weld over the tack using the "C" motion shown in Figure 16A-7. When the end of the tack is reached, shorten the arc. Then complete the stringer pass using the "C" motion. Pause at the upper left of the "C" to fill in the undercut. Direct the arc at the overhead plate as shown in Figure 16A-5.
4. Clean the first pass thoroughly. Use a 5 to 10° work angle as in Figure 16A-5. Run the next stringer so it covers about ⅛ in. of the vertical plate and about two-thirds of the first stringer.
5. After cleaning, run the third stringer. Use a 5 to 10° work angle, as shown in Figure 16A-5, favoring the top plate. About one-half to two-thirds of bead 2 should be seen. The legs of the weld should be equal, with the face of the weld at a 45° angle to the plate surface.
6. Cool, clean, and show the joint to the instructor. If it is satisfactory, weld a third pass. The third pass should have three beads as shown in Figure 16A-5. Clean thoroughly after each stringer. Compare your weld with the sample in Figure 16A-8.

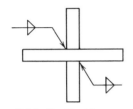

Figure 16A-1 Tacking procedure.

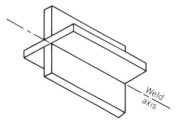

Figure 16A-2

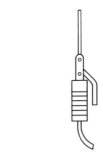

Figure 16A-4

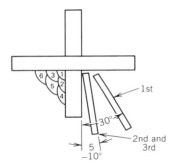

Figure 16A-5 Bead sequence for 3 passes.

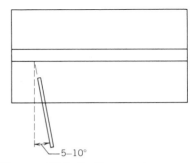

Figure 16A-6 Drag angle.

Figure 16A-7 "C" motion.

Figure 16A-3 Test position.

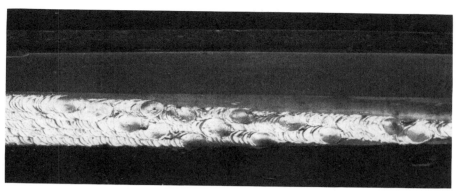

Figure 16A-8 Completed weld.

7. Use the other three joints for practice.

POINTS TO REMEMBER
1. Secure the workpiece and workpiece clamp.
2. Maintain correct electrode angles.
3. Hold a close arc when welding in the overhead position.
4. Pause at the upper left of the "C" or you will leave undercut.

LESSON 16B
T-JOINT FILLET, OVERHEAD, 4-F POSITION, STRINGER BEADS (E-7018 ELECTRODES)

OBJECTIVE
Upon completion of this lesson you will be able to safely and correctly weld a T-joint fillet, in the overhead position, using E-7018 low hydrogen electrodes.

NOTE
Welding overhead T-joint fillets with low hydrogen electrodes is similar to welding with E-6010 electrodes. When using low hydrogen electrodes, however, you must *never* use a whipping motion. The electrode should never move outside the edge of the puddle.

In the more critical applications the weld bead width should be less than *three* times the diameter of the core wire. Some applications allow a width of up to five core wire diameters. But electrode oscillation and bead width should be minimized if quality welds are expected.

MATERIAL AND EQUIPMENT
1. Personal welding equipment
2. E-7018 low hydrogen electrodes 1/8 in. (3.2 mm) diameter
3. 1 pc. carbon steel plate, 3/8 in. × 4 in. × 12 in. long
   2 pcs. carbon steel plate, 3/8 in. × 1 3/4 in. × 12 in. long

GENERAL INSTRUCTIONS
1. Wire brush the intended weld area of the plates
2. Tack the plates as shown in Figures 16B-1 and 16B-2.
3. Clamp the practice plates in position as shown in Figure 16B-1.

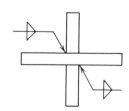

Figure 16B-1 Tacking procedure.

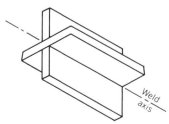

Figure 16B-2 Test position 4-F.

4. Be sure the workpiece and workpiece clamp are secure.
5. Adjust the welding current: AC or DCRP/electrode positive. Ampere range: 115–165.

PROCEDURE
1. Place the electrode in the holder as shown in Figure 16B-3.
2. Strike an arc using the tapping method. Start the first bead at the left side of the joint with the electrode at a 30° work angle off the vertical plate as in Figure 16B-4. Use a drag angle of approximately 10° as in Figure 16B-5.
3. Hold a slightly long arc and weld over the tack using the "C" mo-

Figure 16B-3

188 OVERHEAD WELDING OF PLATE

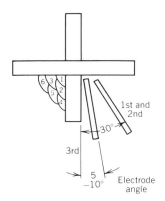

Figure 16B-4 Bead sequence.

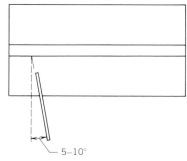

Figure 16B-5 Drag angle.

tion. Make sure that you do not oscillate excessively or leave the puddle.

4. When the end of the tack is reached, hold as short an arc as possible, without stubbing out, and complete the stringer. Move the electrode along, steadily pulling, using the "C" motion. But never allow the tip of the electrode to leave the puddle. "Play" the arc on the overhead plate by pausing slightly at the top left of the "C" motion.

5. Clean thoroughly. Use the slag hammer to remove the slag, then brush the joint thoroughly. After brushing inspect carefully for any slag you might have missed. The slag left by low hydrogen electrodes is sometimes difficult to remove, especially if there is an undercut area where it can adhere.

6. Run the second stringer, using the same electrode angles as with the first pass. Cover about two-thirds of the first pass and about 1/8 in. of the vertical plate as shown in Figure 16B-4.

7. Clean thoroughly and run the third stringer. Use a work angle of 5 to 10° off the vertical plate as shown in Figure 16B-4. About one-half to two-thirds of the second stringer should be uncovered. The weld legs should be equal. The face of the weld should make a 45° angle with the plate surfaces.

8. Cool, clean thoroughly, and compare your weld with the sample shown in Figure 16B-6. Show the joint to your instructor. Your instructor may tell you to run a third pass, consisting of three stringers as shown in Figure 16B-4. Practice welds often use more weld metal than is allowable in industry. It is good practice to keep your weld deposits less than the thickness of the base metal. However, there are practical applications where you may be given directions to make thicker welds.

9. Use the other three joints for more practice.

POINTS TO REMEMBER

1. Secure the workpiece and workpiece clamp.
2. Hold a close arc when welding overhead.
3. Do not oscillate excessively, stay inside the puddle.
4. Pay strict attention to cleaning between passes.

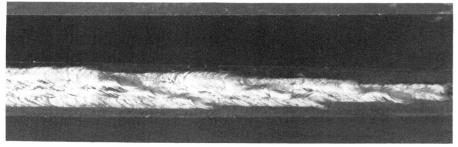

Figure 16B-6 Completed T with E-7018.

LESSON 16C
LAP JOINT FILLET, OVERHEAD, 4-F POSITION, STRINGER BEADS
(E-6010 and E-7018 ELECTRODES)

OBJECTIVE
Upon completion of this lesson you should be able to safely and correctly weld a lap joint fillet in the overhead position, using both E-6010 and E-7018 electrodes.

NOTE
The lap joint is welded in much the same manner as the T-joint. How-

ever, there is a tendency to overweld the lower stringer on the finish pass. The overwelding causes the leading edge of the bottom plate to burn away. This, in turn, gives the weld a sagging appearance, and eliminates the possibility of having a finished weld with equal legs.

MATERIAL AND EQUIPMENT
1. Personal welding equipment
2. E-6010 and E-7018 electrodes, ⅛ in. (3.2 mm) diameter
3. 2 pcs. carbon steel plate, ½ in. × 4 in. × 12 in. long

GENERAL INSTRUCTIONS
1. Wire brush the intended weld area of the plates.
2. Tack the plates as shown in Figure 16C-1.
3. Position the plates as shown in Figure 16C-2.
4. Be sure the workpiece and workpiece clamp are secure.
5. Adjust the welding current. DC reverse polarity, electrode positive (+). DCRP. Ampere range: 78 to 125 for ⅛ in. (3.2 mm) E-6010 electrodes; 115 to 165 for ⅛ in. (3.2 mm) E-7018 electrodes.

PROCEDURE
1. Place the electrode in the holder as in the previous lesson.
2. Start with the E-6010 electrode at the left side of the joint. Position the electrode with a 30° work angle off the vertical surface of the lower plate as shown in Figure 16C-3. Use a drag angle of approximately 10°.
3. Hold a long arc until you reach the end of the tack. Then, hold a close arc and run the stringer, as shown in Figure 16C-4, using the "C" motion. Do not leave the puddle and favor the top plate.
4. Clean the first pass, then run the second stringer. Use an E-7018 electrode with the same angles as

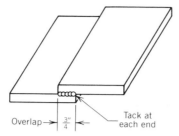

Figure 16C-1 Tacking procedure.

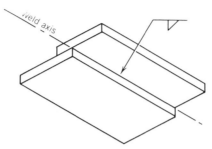

Figure 16C-2 Test position 4-F.

for the first pass. Refer to Figures 16C-3 and 16C-4.
5. After cleaning the second pass run the third stringer. Use an E-7018 electrode held at about a 5 to 10° work angle off the vertical. Refer to Figures 16C-3 and 16C-4.
6. The next stringer is critical. After cleaning the third stringer run the fourth one. Use the E-7018 at a 30° work angle. Point it at the lower edge of the second stringer as shown in Figure 16C-4. Pull the electrode along steadily, or use the "C" motion. Guard against sag or excessive overlapping of the toe of the vertical leg. Watch the bottom edge of the puddle. You will be able to tell when the leading edge is correctly filled.
7. Run stringers 5 and 6 as in the

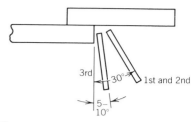

Figure 16C-3 Electrode angles.

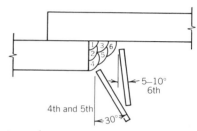

Figure 16C-4 Bead sequence.

welding of T-joints as shown in Figure 16C-4. Pay close attention to the last stringer. A slight pause at the top left of the "C" will prevent undercut. Be sure to clean each stringer before welding the next one. Cool, clean, then compare with the sample weld shown in Figure 16C-5. Present the weld to your instructor.

POINTS TO REMEMBER
1. Secure the workpiece and work lead.
2. Pay close attention to your electrode angles.
3. Hold a close arc.
4. Do not oscillate excessively or leave the puddle.
5. Pay strict attention to cleaning between passes.

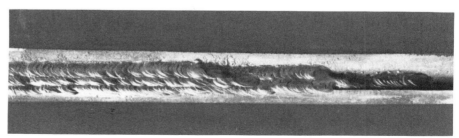

Figure 16C-5 Completed weld.

LESSON 16D
OPEN CORNER JOINT, OVERHEAD, 4-G POSITION, STRINGERS AND WEAVE (E-6010 OR E-6011 ELECTRODES)

OBJECTIVE
Upon completion of this lesson you should be able to safely and correctly weld an open corner joint, in the overhead position with E-6010 and E-6011 electrodes utilizing both the stringer and weave methods.

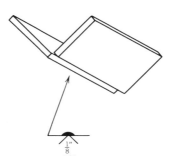

Figure 16D-1 Tacking procedure.

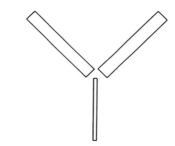

Figure 16D-2 Test position 4-G.

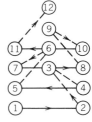

Figure 16D-4 Whipping motion.

Figure 16D-5 Triangle.

NOTE
This joint will provide you with experience in welding first passes of open root groove welds. It will do this without all the preparatory work associated with beveled V-groove joints. It will also offer practice in filling and finishing a groove weld.

The first pass is slightly different from that of a V-groove which has been properly prepared. It is not so different, however, that you will not gain worthwhile experience from this lesson.

In addition to being an excellent weld for your training, this joint is frequently used in certain areas of industry.

MATERIAL AND EQUIPMENT
1. Personal welding equipment
2. E-6010 or E-6011 electrodes, 1/8 in. (3.2 mm) diameter
3. 2 pcs. carbon steel plate, 3/8 in. or 1/2 in. thick × 4 in. × 12 in. long

GENERAL INSTRUCTIONS
1. Remove all scale and slag from the edges of the plates. Wire brush the intended weld area of the plates.
2. Tack the plates as in Figure 16D-1.
3. Position the plates as in Figure 16D-2.

Figure 16D-3 Electrode work angle.

4. Be sure the workpiece and workpiece clamp are secure.
5. Adjust the welding current: DC reverse polarity, electrode positive (+) DCRP. Ampere range: 75 to 125.

PROCEDURE
1. Strike an arc on the tack at the left side of the joint. Hold the electrode perpendicular to the underside of the groove as in Figure 16D-3. Use a drag angle of 5 to 15°.
2. Carry a slightly long arc until you reach the end of the tack. Then pause to open a keyhole. When the keyhole appears, use the whipping motion shown in Figure 16D-4. The whipping motion is used to keep the keyhole the correct size. If it gets too large, whip out further. If it tends to close on you, stay in the puddle longer. Don't whip out so far. Raising or lowering the amperage will also help.
3. The second pass can be a wide stringer or a small weave. Use the triangle motion shown in Figure 16D-5. Make the second bead about 1/8 in. wider on each side of the first weld. Remember to pause at the sides to eliminate undercut.
4. If you are using 3/8 in. plate, the next pass will be the finished pass. Use the triangle method. Make sure you whip the electrode toward the center of the groove. This way you will not scar or undercut

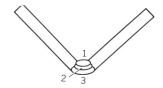

Figure 16D-6 Bead sequence.

the plate at the toes of the weld. See the bead sequence in Figure 16D-6.

5. If you are using ½ in. plate, you will require two more passes. Use the triangle method for both. The finish pass should have about 3/32 in. (2.4 mm) reinforcement, both above the parent metal, at the edges of the groove, and on the other side or root side of the piece.
6. You can use the weave you learned in vertical welding.

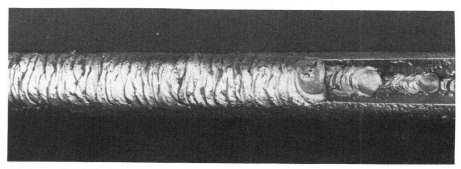

Figure 16D-7 Completed weld.

7. Cool, clean, and compare with the weld sample in Figure 16D-7. Present it to your instructor.

POINTS TO REMEMBER
1. Secure the workpiece and workpiece clamp.
2. Hold a close arc.
3. Pause at the sides to eliminate undercut.
4. Do not deposit excessive weld metal.

LESSON 16E
V-GROOVE BUTT JOINT, OVERHEAD, 4-G POSITION, OPEN ROOT, FIRST PASS DRAGGED, SECOND AND OUTER PASSES STRINGER (E-6010 OR E-6011 ELECTRODES)

OBJECTIVE
Upon completion of this lesson you should be able to safely and correctly weld an open root V-groove butt joint, in conformance with Figure 5.20a, D1.1 AWS Structural Code, in the overhead position using E-6010 or E-6011 Electrodes.

NOTE
The overhead open root V-groove butt joint is probably the most difficult plate joint to weld. The first pass is the most critical. Much can be learned about a welder's ability by the weld preparation and by the quality of the resulting weld.

Proper penetration and fusion of the parent metal is a must. The welder who qualifies on this overhead joint also qualifies on the flat position groove joints, or flat horizontal and overhead fillet joint with plates, square, or rectangular tubing up to ¾ in. in thickness.

MATERIAL AND EQUIPMENT
1. Personal welding equipment
2. E-6010 or E-6011 electrodes, ⅛ in. (3.2 mm) diameter
3. 2 pcs. carbon steel plate, ½ in. × 4 in. × 12 in. long

GENERAL INSTRUCTIONS
1. Clean all intended weld areas, then tack the plates as shown in Figure 16E-1.

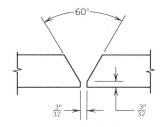

Figure 16E-1 Configuration of workpiece.

2. Position the plates as shown in Figure 16E-2.
3. Be sure the workpiece and workpiece clamp are secure.
4. Adjust the welding current. DC reverse polarity, electrode positive (+) DCRP. Ampere range: 75 to 125.

OVERHEAD WELDING OF PLATE

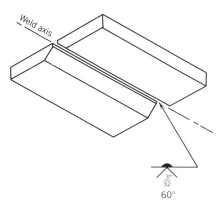

Figure 16E-2 Test position 4-G.

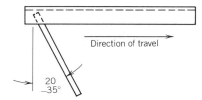

Figure 16E-3

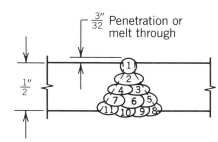

Figure 16E-4 Bead sequence.

PROCEDURE

1. Place the electrode in the holder as in the previous lesson in the overhead position.
2. Start at the left side of the joint. Hold the electrode perpendicular to the plate. Use a drag angle of approximately 20 to 35° as in Figure 16E-3. Then weld over the tack, holding a slightly long arc.
3. When you reach the end of the tack, close the arc gap. Do this until the electrode coating touches the edges of the root face. Pull the electrode steadily along the joint. You get good penetration when the arc is almost invisible and protruding out the opposite side of the plate. The desired amount of penetration is shown in Figure 16E-4. During this pass, the electrode must never lose contact with the workpiece.
4. Another way to know the arc is penetrating sufficiently is to listen for a "harsh" sound. The harsh sound indicates the arc is forcing molten metal through the gap onto the opposite side of the plate.
5. You can increase penetration by:
   (a) increasing the arc current
   (b) slowing the rate of travel
   (c) increasing the electrode angle
6. You can decrease penetrations by:
   (a) decreasing the arc current
   (b) increasing the rate of travel
   (c) decreasing the electrode angle
7. Clean the bead thoroughly. Pay close attention to the toes of the weld. If you are not careful, slag will remain in the ridges at the toes of the root pass. It is extremely difficult to "burn out" the slag.
8. Use the "C" motion or a whipping motion; run a second stringer as shown in Figure 16E-4. Be sure to cover the root pass completely. The crater must be deep enough to burn out the ridges (known as "wagon tracks") at the toes of the weld.
9. Complete the remainder of the joint with stringer beads. Make sure the finish pass has about $3/32$ in. (2.4 mm) reinforcement as in Figure 16E-4. Be careful that you do not get undercut on the finish pass.
10. Cool, clean, and compare your weld with the sample shown in Figure 16E-5.
11. Present it to your instructor

POINTS TO REMEMBER

1. Be sure the workpiece is properly prepared before welding.
2. Make sure that you preposition the plates as in previous lessons using open root V-grooves, or they will distort and not be on the same plane.
3. Maintain a close arc.
4. Drag the first pass. Maintain contact between the electrode and the full length of the root.
5. Limit the weld reinforcement to $3/32$ in. (2.4 mm).

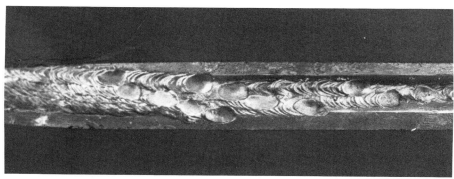

Figure 16E-5 Completed weld.

LESSON 16F
V-GROOVE BUTT JOINT, OVERHEAD, 4-G POSITION, OPEN ROOT, IN ACCORDANCE WITH SECTION IX ASME BOILER AND PRESSURE VESSEL CODE (E-6010 AND E-7018 ELECTRODES)

OBJECTIVE
Upon completion of this lesson you should be able to safely and successfully pass an open root V-groove butt joint in the overhead position.

The student will remove, prepare, and subject coupons to a guided bend test as described in Figure QW-466.2 Section IX ASME Boiler and Pressure Vessel Code.

NOTE
The QW numbers listed above are for reference purposes only. The coupons will be removed as indicated by Figure 16F-1. Use the bending jig available in your shop.

In the previous lessons you developed your overhead welding skills. Overhead welding is not really difficult. The joint in this lesson is probably the most challenging of all. You will use the whipping motion for the first pass. It is important to obtain excellent penetration for the entire joint length.

Successful completion of this test joint will qualify you to weld V-groove and fillet welds, on plate in the 4-G and 1-G positions. It also qualifies you for pipe over 24 in. in diameter.

MATERIAL AND EQUIPMENT
1. Personal welding equipment
2. E-6010 and E-7018 electrodes, 1/8 in. (3.2 mm) diameter
3. 2 pcs. carbon steel plate, 1/2 in. × 5 in. × 10 in. long

GENERAL INSTRUCTIONS
1. Be sure that the bevel angle and root face are prepared according to Figure 16F-2. Clean all intended weld areas.
2. Tack the plates as shown in Figure 16F-2.
3. Position the test piece as shown in Figure 16F-3.
4. Be sure the test piece and workpiece clamp are secure.
5. Adjust the welding current: DC reverse polarity, electrode positive (+) DCRP. Ampere range: E-6010 75 to 125; E-7018 115 to 165.

PROCEDURE
1. Use the E-6010 electrode and a drag angle of approximately 10° to whip in the first pass. Make sure that you open a keyhole. Keep it open the entire length of the joint.
2. Hold a very close arc. Push the electrode slightly through the keyhole. This will force the molten metal to pile up on the top side of the joint (approximately 1/16 to 3/32 in. in height) as shown in Figure 16F-4.
3. Clean the first pass thoroughly. Run a stringer by drawing the electrode along the joint for the second pass. Use E-7018 electrodes for this and the rest of the passes. Make sure the arc is running fairly "hot." The "heat" is needed to "burn out" any wagon tracks or deep grooves at the weld toes. Stay in the puddle.
4. The third pass will be made with two stringers. Remember to favor the toe area of the weld. Point the electrode slightly in the direction

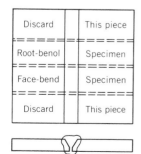

Figure 16F-1 QW-463.2(a) Plates 1/16 to 3/4 in. Performance qualification.

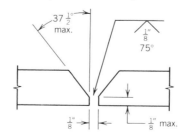

Figure 16F-2 Joint configuration.

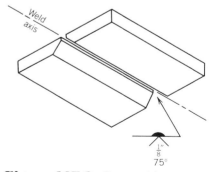

Figure 16F-3 Test position.

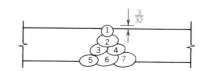

Figure 16F-4 Bead sequence.

you want the metal to flow toward, or the area you want to "burn out." (See Figure 16F-5.) Place beads 3 and 4 as shown in Figure 16F-4.

5. Complete the remainder of the joint using the stringer method. Three more beads are needed for this pass. Be careful of producing undercut or excessive reinforcement on the finish pass. See Figure 16F-4 for the bead sequences and amount of weld reinforcement.

6. Allow the workpiece to cool at room temperature.

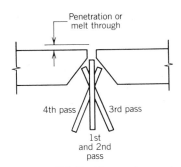

Figure 16F-5 Electrode work angles.

7. Remove the test coupons as shown in Figure 16F-1, and prepare them for testing. Flame cut coupons and cool them slowly. Do not rapid quench. Make one root bend and one face bend.

8. Present your results to your instructor.

POINTS TO REMEMBER

1. Secure the test piece and work lead.
2. Preposition the test plates as in previous open root butt welding before welding.

CHAPTER 17
WELDING SHEET METAL IN THE FLAT POSITION

Welding sheet metal (especially below 16 gauge) is extremely difficult with the shielded metal arc welding process. Thin sheet metal is better suited for welding by Gas Metal Arc (GMAW), Gas Tungsten Arc (GTAW), or the Oxy-acetylene (OAW) welding processes. Gas metal arc welding is a fast and economical process. Gas tungsten arc welding, although slower, has the ability to produce exceptionally sound welds in the "hard to weld" metals such as stainless steel, aluminum, magnesium, titanium, copper, and so on.

The oxy-acetylene process is an old standby, but it lacks the speed needed for most types of production welding. The oxy-acetylene method also increases the heat input and causes the metal to warp.

A skilled shielded metal arc welder can successfully weld some sheet metal gauges. However, 16 gauge is the thinnest that can be welded successfully without too much distortion or burn through.

Joint preparation and fit up are of great importance in sheet metal welding. Slight openings in fillet joints or butt joint faces will cause the arc to burn away the sheet metal edges during the welding operation.

For good joint preparation prior to fit up it is important to shear or saw all metal to be joined. Flame cutting is a poor choice. Even if the welder is skillful enough to eliminate the slag from the cutting operation, the metal will warp from the heat. Good fit up is difficult, if not impossible, to obtain with warped sheet metal.

A good fit also depends on the tacking procedure. Tacks on sheet metal should be close together. They should be placed no more than 1 in. to 1½ in. apart, depending on the gauge of the metal and type of joint. Tacks that are further apart will allow the joint to open from the heat. A poor joint will be the result.

Even if you are able to weld the joint, the weld will have to be excessively wide whenever there is an open space between the two sheets of the joint.

You can increase your chance of making acceptable sheet metal welds by knowing your electrodes. You should be able to choose the correct electrode for the job.

The lessons in this chapter will give you practice with some medium to shallow penetration electrodes especially suited for welding sheet metal.

LESSON 17A
WELDING SHEET METAL, FLAT (DOWNHAND) POSITION, OPEN CORNER JOINT (E-6013 ELECTRODES)

OBJECTIVE
Upon completion of this lesson you should be able to safely and correctly weld an open corner, 16 gauge hot rolled sheet metal joint in the downhand position. The joint will be using E-6013 electrodes inclined about 30° in the direction of travel.

NOTE
The E-6013 electrode is probably the best all-around choice for use on light-gauge metal. It has a shallow penetration characteristic, which helps to minimize the chance of burn through. It can be used on AC or any DC polarity. It is an all-position electrode. Its slag is easily removed; in fact, the slag sometimes peels off in large sections as the weld metal cools.

This electrode can be dragged, with the electrode coating touching the surface of the joint, or used with a normal arc length.

The E-6013 electrode performs well with any type of welding current. However, for the best results on light-gauge metal, DC straight polarity-electrode negative (−) is the best choice. Penetration is minimum and deposition rate is highest with DCSP.

The open corner joint is widely used in tank fabrication and many other sheet metal structures. The joint in this lesson requires little preparation. It can be welded at a moderately fast speed, which makes it a very economical operation.

MATERIAL AND EQUIPMENT
1. Personal welding equipment
2. Pliers
3. 2 pcs. 16-gauge sheet metal 2 in. × 10 in. long
4. E-6013 electrodes 3/32 in. (2.4 mm) diameter

WELDING SHEET METAL IN THE FLAT POSITION

GENERAL INSTRUCTIONS

1. Adjust the welding current on 16-gauge scrap sheet. DC straight polarity, electrode negative (−). Ampere range: 45 to 90.
2. Tack the two pieces of sheet metal to form the open corner joint shown in Figure 17A-1. Tacks on sheet metal differ from those used on plate. Sheet metal tacks should be small; they are no larger than arc strikes. Apply them by striking and maintaining an arc for a two or three counts and then breaking the arc. A sample tack is shown in Figure 17A-2.
3. After tacking, secure the weldment as shown in Figure 17A-3.

PROCEDURE

1. Remove all slag from the tacks and wire brush the intended weld areas.
2. Place the electrode in the holder at a 90° angle.
3. Starting at the highest point of the joint, strike the arc and hold a normal arc gap. Move steadily along the joint with a perpendicular work angle as in Figure 17A-4. Use a 5 to 10° drag angle as shown in Figure 17A-5. The electrode can also be dragged along in contact with the metal. However, the electrode drag angle must be increased to approximately 25 to 30°.
4. Stop midway along the joint. Remove the slag and inspect the bead. It should be slightly convex with no overlap at the toes of the weld. It should be free of surface holes and slag inclusion. Figure 17A-6 shows the desired bead profile.
5. If the face of the bead is convex and there is overlap, there is too much filler metal. This happens when your rate of travel is too slow, your heat is set too high, or you have too much drag angle. Change your welding conditions accordingly.

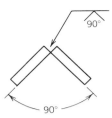

Figure 17A-1 Tacking procedure.

6. If the bead is flat or concave, there is too little metal. This happens when your rate of travel is too fast or your amperage is too high. Change your welding conditions accordingly.
7. If the bead is high, just setting on top of the joint, your amperage is too low.
8. After you make the necessary adjustments, pick up the weld at the crater. Hold a long arc to start, then complete the joint. Inspect as before and continue to weld practice pieces until you can produce acceptable welds.
9. Cool, clean, and present to your instructor.

POINTS TO REMEMBER

1. Obtain a good fit.
2. Keep tacks approximately 1 in. apart.
3. If the flux gets ahead of the puddle, stop, clean, and restart.
4. Do not overweld.
5. Try both the normal arc gap and the drag technique.

Figure 17A-2 Sample of tacks.

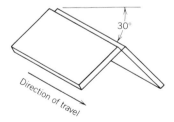

Figure 17A-3 Test position.

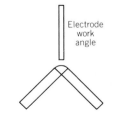

Figure 17A-4 Electrode angle.

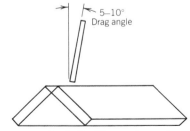

Figure 17A-5 Electrode angle.

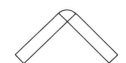

Figure 17A-6 Bead contour.

CHAPTER 18
WELDING SHEET METAL IN THE HORIZONTAL POSITION

Welding sheet metal in the horizontal position is slightly more challenging than welding in the flat or downhand position, but it is not really difficult.

As with other joints care must be taken in fitting and tacking the joint. You must always be on guard against burning through, and avoid undercut and excess weld puddle sag. The proper welding current setting is of great importance. You should not have too much heat input.

The lessons in this chapter will give you practice in welding various horizontal joint configurations. You will use E-6011, E-6012, and E-6013 electrodes for welding the sheet metal.

LESSON 18A
WELDING SHEET METAL, T-JOINT FILLET, 2-F HORIZONTAL POSITION (E-6012 ELECTRODES)

OBJECTIVE
Upon completion of this lesson you should be able to safely and correctly weld a 10- or 12-gauge sheet metal T-joint fillet, in the 2-F horizontal position, using E-6012 electrodes.

NOTE
The E-6012 electrode, due to its moderate arc, produces less penetration than the E-6010 or E-6011. It also produces a much quieter and softer arc, with less spatter and a smoother surface than the E-6010 or E-6011. The deposition rate is slightly higher, due in part to the use of DC straight polarity. Compared with DCRP, DCSP puts more heat into the electrode causing it to melt rapidly.

The E-6012, because of its moderate penetration and ability to weld poorly fit-up joints, is a good choice for use on sheet metal application.

MATERIAL AND EQUIPMENT
1. Personal welding equipment
2. Pliers
3. 2 pcs. 10 gauge or 12 gauge hot rolled sheet metal, 2 in. × 10 in. long
4. 1/8 in. (3.2 mm) diameter E-6012 electrodes

GENERAL INSTRUCTIONS
1. Adjust the welding current: AC or DC straight polarity DCSP electrode negative (−). Ampere range: 80 to 140.
2. Adjust the current on 10 gauge or 12 gauge sheet metal.
3. Clean the intended weld areas.
4. Tack weld the two pieces to form a T-joint as in Figure 18A-1. Make sure the vertical piece is perpendicular.
5. Place one tack at each end as in Figure 18A-2. Place smaller tacks between the end tacks, at about 1 in. intervals. These tacks are sometimes referred to as bubble tacks. They should be round and slightly larger in diameter than the electrode.
6. Secure the tacked pieces in position as shown in Figure 18A-3. Tilt it, up to 15° in the direction of travel. This will let you weld

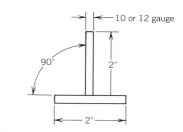

Figure 18A-1 Fit up.

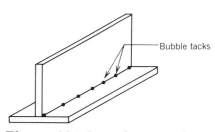

Figure 18A-2 Tacking procedure.

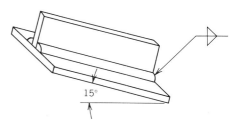

Figure 18A-3 Weldment position 2-F.

faster and reduce the chance of undercut.

PROCEDURE

1. Clean the slag left by the tacking procedure and start at the left side.
2. Place the electrode in the holder, at a 45° angle as you did when welding horizontal fillets on plate. The electrode work angle should be between 30 and 45 degrees as shown in Figure 18A-4. Use a 10 to 15° lead or drag angle.
3. Maintain a puddle slightly larger than the electrode diameter, (maximum 3/16 in.), by holding a close arc with the electrode pointed into the heel of the joint. Favor the bottom piece slightly.
4. You can either hold a close arc or drag the electrode. If you drag the electrode, touch the leading edge of the coating to the surface of the joint. Watch the upper trailing edge of the puddle. If it does not fill in completely, you should travel slower. Otherwise, the vertical leg will be undercut.
5. When you reach the halfway point, stop. Clean the weld and remove the slag. Examine the bead to be sure you are fusing both pieces. The weld face should be at a 45° angle. The legs should be equal and the surface flat or slightly convex. This is illustrated in Figure 18A-5.
6. Restart the arc using the same electrode, and complete the weld.
7. When you have completed the weld, cool and clean it thoroughly. Then weld the other side as indicated in Figure 18A-1.
8. Present the weld to your instructor.

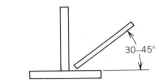

Figure 18A-4 Work angle.

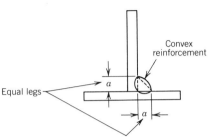

Figure 18A-5 Desired weld profile.

POINTS TO REMEMBER

1. Be sure the fit up is good. Use plenty of tacks.
2. Use a work angle of between 30 and 45°.
3. Clean thoroughly before starting to weld and at the midpoint.
4. Maintain a close arc or use the drag technique.
5. Watch out for undercut or overwelding.

LESSON 18B
WELDING SHEET METAL, LAP JOINT FILLET, 2-F HORIZONTAL POSITION (E-6013 ELECTRODES)

OBJECTIVE
Upon completion of this lesson you should be able to safely and correctly weld a 16-gauge hot rolled sheet metal lap joint fillet in the 2-F horizontal position, using E-6013 electrodes.

NOTE
From your experience in Chapter 17 you should have gained an understanding of the characteristics of the E-6013 electrode. Also, you should have developed a fair degree of skill in the use of the electrode. In this lesson you will improve on those skills by using the E-6013 electrode in the horizontal position.

Keep in mind that this electrode has a fairly heavy slag deposit. The slag is needed to help the weld bead contour and surface. Because of this it is not necessary for you to whip or oscillate the electrode. Electrode movement may be kept to a minimum. This does not mean you cannot manipulate the electrode if it is necessary or more comfortable.

MATERIAL AND EQUIPMENT
1. Personal welding equipment
2. Pliers
3. 2 pcs. 16 gauge hot rolled sheet metal, 2 in. × 10 in. long
4. 3/32 in. (2.4 mm) diameter E-6013 electrodes

GENERAL INSTRUCTIONS
1. Adjust the welding current: AC or DC straight polarity DCSP electrode negative (−). Ampere range: 45 to 90.
2. Clean all intended weld areas. Then tack the two pieces on the ends as shown in Figure 18B-1. Overlap the pieces 3/4 in.

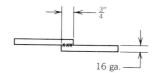

Figure 18B-1 Fit up.

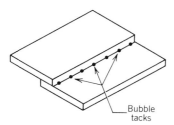

Figure 18B-2 Tacking procedure.

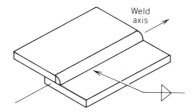

Figure 18B-3 Weldment position 2-F.

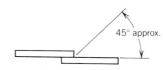

Figure 18B-4 Work angle.

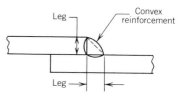

Figure 18B-5 Desired weld profile.

3. Make sure there are no openings along the joint and place bubble tacks as in Figure 18B-2. The closely spaced tacks will keep the joint from opening up as the weld progresses.

4. Secure the tacked pieces in position as in Figure 18B-3. You may tilt the pieces slightly (a maximum of 15°) toward the direction of travel. If you do, welding will proceed at a slightly faster pace and you will minimize the chance of undercut.

PROCEDURE

1. Remove the slag left by the tacking procedure. Place the electrode in the holder at a 45° angle as in Lesson 18A.

2. Start at the left side of the joint. The electrode work angle should be approximately 45° as shown in Figure 18B-4. Use about a 10 to 15° lead or drag angle. Maintain the puddle slightly larger than the legs of the joint. Take care: Do not burn away an excessive amount of the toe at the top of the vertical leg.

3. Pull the puddle, with a uniform width, along the entire length of the joint. With a 3/32 in. diameter electrode it is very difficult to keep the weld small enough for metal of this thickness. If you have too much trouble, try using 1/16 in. (1.6 mm) diameter electrodes.

4. Check the completed weld for surface holes, slag inclusion, and discontinuities, including spaces left unwelded. The weld should be completed in a single pass. It should have equal legs and a flat or slightly convex face as shown in Figure 18B-5.

5. Cool and clean the sample. Then weld the opposite side as indicated in Figure 18B-3.

6. Present the completed weld to your instructor.

POINTS TO REMEMBER

1. Be sure the fit is good. Use plenty of tacks.
2. Use a work angle of approximately 45°.
3. Clean thoroughly before starting to weld, and when you are finished.
4. Keep the electrode manipulation to a minimum.
5. Do not overweld.

CHAPTER 19
WELDING SHEET METAL IN THE VERTICAL POSITION

The previous chapters describe the welding of many joint configurations, in a variety of positions, using the E-6010, E-6013, E-7014, E-7024, and E-7018 electrodes. Until you mastered them each position, joint configuration, and electrode seemed difficult at first. Welding sheet metal in the vertical position is not any different. Follow the directions in this lesson and with your instructor's help you will soon master this lesson.

You will learn to weld the "downhill" technique. This means welding from the top of the joint to the bottom. You will use E-6011 electrodes and DCSP. There are some welders who will tell you this is not possible, but don't believe them. Direct current straight polarity produces a forceful arc stream. If you maintain a slightly long arc and move rapidly, in a short time you will be able to make a good weld.

The downhill technique, in addition to being a fast method of welding, produces very little spatter. Most of the spatter can be rubbed off with your welder's glove. However, it is better practice to use a wire brush; your gloves will last longer.

LESSON 19A
WELDING SHEET METAL, OPEN CORNER JOINT, 3-G VERTICAL POSITION (DOWNHILL) (E-6011 ELECTRODES)

OBJECTIVE
Upon completion of this lesson you should be able to safely and correctly weld a 10- or 12-gauge sheet metal open corner joint in the 3-G vertical position, using the downhill technique and E-6011 electrodes.

NOTE
Vertical downhill welding of sheet metal is no more difficult than welding plate downhill. However, you must pay attention or you might burn through the metal. You should not even consider welding sheet metal in the uphill mode. All vertical sheet metal welding works better with the downhill technique.

MATERIAL AND EQUIPMENT
1. Personal welding equipment
2. Pliers
3. 2 pcs. 10 gauge or 12 gauge × 2 in. × 10 in. long hot rolled sheet metal
4. ³⁄₃₂ in. (2.4 mm) diameter E-6011 electrode

GENERAL INSTRUCTIONS
1. Adjust the welding current: Direct current straight polarity DCSP electrode negative (−). Ampere range: 40 to 80.
2. Tack weld the joint as in Figures 19A-1 and 19A-2. Place the tacks 1 to 1½ in. apart.
3. Be sure there are enough tacks; otherwise the joint will open during welding.
4. Secure the weldment in the 3-G vertical position as in Figure 19A-3.
5. Remove all slag from the tacking procedure and clean the intended weld area. The welding will be from the top of the joint toward the bottom.

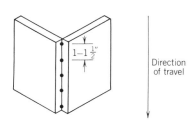

Figure 19A-2 Tacking procedure.

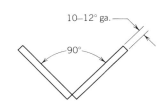

Figure 19A-1 Fit up.

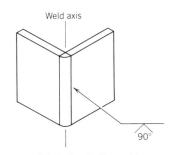

Figure 19A-3 3-G position.

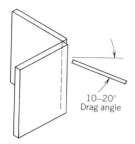

Figure 19A-4 Drag angle.

Figure 19A-5 Desired weld profile.

PROCEDURE

1. Place the electrode in the holder at a 45° angle.
2. Start at the top of the joint. Strike an arc and move steadily toward the bottom using about a 10 to 20° drag angle as shown in Figure 19A-4. Point the electrode straight into the joint.
3. As you follow the groove it may be necessary to weave the electrode slightly from side to side. This will fuse the toes of the weld. If the puddle is wide enough, the weaving is not necessary.
4. Hold a fairly long arc. Do not be concerned by any flux that moves ahead of the puddle. Whatever flux is present will be light, and it will have no effect on the advancing weld.
5. Concentrate on completing the weld without stopping. If a pause is necessary, be sure to clean the crater area thoroughly before continuing.
6. The desired weld can range from a flat face to a slightly convex face as in Figure 19A-5.
7. You can continue to practice this joint by adding pieces to the weldment. Be sure to clean all plate edges and intended weld areas before you begin. Refer to Figure 19A-1.
8. Cool and clean your completed welds and present them to your instructor.

POINTS TO REMEMBER

1. Take your time when fitting and tacking.
2. Maintain the suggested electrode angles.
3. Clean after tacking and after all pauses.
4. Keep the rate of travel constant.

LESSON 19B
WELDING SHEET METAL, OPEN CORNER JOINT, 3-G VERTICAL POSITION (DOWNHILL) (E-6013 ELECTRODES)

OBJECTIVE
Upon completion of this lesson you should be able to safely and correctly weld a 10- or 12-gauge sheet metal open corner joint in the 3-G vertical position, using the downhill technique and E-6013 electrodes.

NOTE
Welding an open corner joint, in the 3-G vertical position, with E-6013 electrodes is similar to doing it with E-6011 electrodes.

The main difference is you must concentrate on not allowing the slag to get ahead of the puddle. The slag from the E-6013 electrode is much heavier than the slag from the E-6011, and will cause voids in the weld.

You can reduce the chances of voids occurring if you keep your arc length and rate of travel constant.

MATERIAL AND EQUIPMENT
1. Personal welding equipment
2. 2 pcs. 10 gauge or 12 gauge × 2 in. × 10 in. long hot rolled sheet metal
3. 1/8 in. (3.2 mm) diameter E-6013 electrodes

GENERAL INSTRUCTIONS
1. Adjust the welding current: DC straight polarity DCSP electrode negative (−). Ampere range: 80 to 130.
2. Clean all intended weld areas.
3. Tack two pieces to form an open corner joint as in Figures 19B-1 and 19B-2. Make sure the pieces are at a right angle as indicated.
4. Place a tack at each end and enough tacks in between. Make sure the joint will not open up during the welding procedure.
5. Clean the slag left by the tacking procedure.

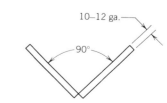

Figure 19B-1 Fit up.

WELDING SHEET METAL IN THE VERTICAL POSITION

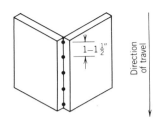

Figure 19B-2 Tacking procedure.

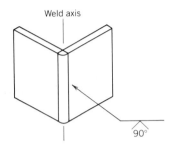

Figure 19B-3 3-G position.

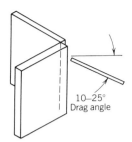

Figure 19B-4 Drag angle.

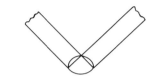

Figure 19B-5 Convex weld profile.

6. Secure the weldment in position, as shown in Figure 19B-3.

PROCEDURE

1. Place the electrode in the holder at a 45° angle.
2. Start to weld from the top. Aim the electrode straight into the joint. The drag angle should be between 10 to 25° as shown in Figure 19B-4.
3. Using the downhill technique drag the electrode downward using a slight weaving motion if necessary. Spread the weld pool to cover the entire joint.
4. Do not oscillate the electrode excessively. Excessive oscillation will cause the weld to be too wide. The weld width should be slightly more than the joint width.
5. If the slag runs ahead of the molten puddle, you should stop. Chip the weld at the crater, clean, and start over. Remember to hold a long arc when restarting. Completely fill in the crater before continuing the bead.
6. Experiment a little with the length of the arc. Try a short length arc and a medium arc. Note the difference, then use the arc length you feel is the best for you. This holds true for electrode angles. Some angles improve the appearance of the weld and the ease with which it is welded.
7. Clean, cool, and evaluate the weld. Present it to your instructor for comments. It should have an even, smooth appearance, without surface holes or slag inclusions. It should have a convex weld profile as shown in Figure 19B-5.

POINTS TO REMEMBER

1. Take the time to fit and tack properly.
2. Maintain the suggested electrode angles.
3. Clean after tacking and after all pauses.
4. Maintain a constant arc gap and rate of travel.
5. Do not overweld the joint.

LESSON 19C
WELDING SHEET METAL, LAP JOINT FILLET, 3-F VERTICAL POSITION (DOWNHILL) (E-6013 ELECTRODES)

OBJECTIVE

Upon completion of this lesson you should be able to safely and correctly weld a 16-gauge sheet metal lap joint fillet in the 3-F vertical position, using the downhill technique and E-6013 electrodes.

NOTE

Sixteen-gauge sheet, because it is fairly thin, is slightly more difficult to weld. The downhill technique keeps the heat input to a minimum. Use of a smaller diameter electrode, such as the 3/32 in. (2.4 mm) diameter electrode, also reduces the heat input.

The shape of this joint requires the use of a different work angle than those used with the previous joints. You must pay closer attention to the top edge of the overlapping sheet, otherwise it will melt away.

MATERIAL AND EQUIPMENT

1. Personal welding equipment
2. 2 pcs. 16 gauge × 2 in. × 10 in. hot rolled sheet metal.
3. 3/32 in. (2.4 mm) diameter E-6013 electrodes

GENERAL INSTRUCTIONS

1. Adjust the welding current: AC or Direct current, straight polarity DCSP electrode negative (−). Ampere range: 45 to 90.

2. Tack the joint on the ends as in Figure 19C-1. Place enough tacks between so the joint will not open from the heat of the welding.
3. Remove all slag from the tacking procedure and clean all intended weld areas.
4. Secure the weldment as shown in Figure 19C-2. The direction of welding will be downward, from the top of the joint.

PROCEDURE
1. Place the electrode in the holder at a 45° angle.
2. Start at the top of the joint using a 10 to 20° drag angle, a work angle of 20 to 25°, off the vertical edge of the overlapping sheet, as shown in Figure 19C-3.
3. Use the drag technique. Hold a fairly short arc. The puddle should be just large enough in diameter to fuse the entire joint.
4. Favor the surface of sheet #2, Figure 19C-3. Watch the left side of the puddle closely. Make sure it fuses the leading edge of sheet #1 shown in Figure 19C-4. If you move too slowly or weave excessively, you will burn away the leading edge of sheet #1. This is not acceptable.
5. It is very difficult to obtain a finished weld with equal legs on thin sheet metal while using a 3/32 in. (2.4 mm) diameter electrode. A better electrode choice would be 1/16 in. (1.6 mm) diameter. If 1/16 in. electrodes are available, use them to practice this joint.
6. Usually the leg of the fillet on sheet #2 is slightly longer. This is acceptable as long as the other leg is fused properly.
7. Clean, cool, and check for places on the leading edge of sheet #1 that were left unwelded. Compare with the sample weld in Figure 19C-5. Present your weld to your instructor.

POINTS TO REMEMBER
1. Fit and tack properly.
2. Maintain the suggested electrode angles.
3. Clean after tacking.
4. Maintain a constant rate of travel.
5. Oscillate just enough to fuse the toes of the joint.
6. Do not overweld.

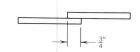

Figure 19C-1 Fit up.

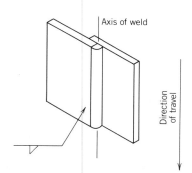

Figure 19C-2 3-F position.

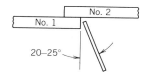

Figure 19C-3 Top view of work angle.

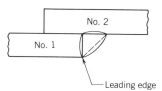

Figure 19C-4 Desired weld profile.

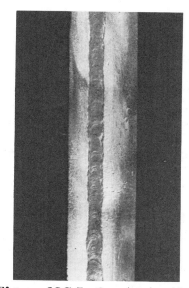

Figure 19C-5 Completed weld.

LESSON 19D
WELDING SHEET METAL, LAP JOINT FILLET, DISSIMILAR METAL THICKNESS, 3-F VERTICAL POSITION (DOWNHILL) (E-6013 ELECTRODES)

OBJECTIVE
Upon completion of this lesson you should be able to safely and correctly weld a lap joint fillet, consisting of a 16-gauge piece of sheet metal welded to a 3/16 in. thick piece of strip, in the 3-F Vertical position, utilizing the downhill technique and E-6013 electrodes.

WELDING SHEET METAL IN THE VERTICAL POSITION

NOTE

When welding thin sheet metal to heavier thickness metal, you must concentrate the arc heat on the heavier piece. The electrode must be pointed toward the heavy metal at an angle that will not let the arc melt away the thin section.

There are many applications in industry where welding of this type is necessary. Developing an expertise in welding light sheet to heavier sheet or plate will improve your welding skills.

MATERIAL AND EQUIPMENT

1. Personal welding equipment
2. 2 pcs. 16 gauge × 2 in. × 10 in. long hot rolled sheet metal
3. 2 pcs. 3/16 in. × 2 in. × 10 in. long hot rolled strip
4. 3/32 in. (2.44 mm), 1/8 in. (3.2 mm) diameter E-6013 electrode

GENERAL INSTRUCTIONS

1. Adjust the welding current: DC straight polarity, DCSP electrode negative (−). Ampere range: 3/32 in. (2.4 mm) diameter 45 to 90; 1/8 in. (3.2 mm) diameter 80 to 130.
2. Tack the joint as in Figures 19D-1 and 19D-2.
3. Remove all slag from the tacking procedure. Clean all intended areas.
4. Secure the weldment in the 3-F vertical position, as shown in Figure 19D-2. The direction of welding will be downhill from the top of the joint.

PROCEDURE

1. Place the electrode in the holder at a 45° angle.
2. Start at the top of the joint using a 10 to 20° work angle off the vertical edge of the overlapping sheet as shown in Figure 19D-3. Use a drag angle of 10 to 20° as shown in Figure 19D-4.
3. Use the drag technique. Hold a normal arc and favor the heavier metal. Cause the molten pool to fuse the edge of the light-gauge metal by oscillating the electrode slightly. This will wash the metal from the heavy section toward the lighter section.
4. Do not pause too long on the light-gauge leg of the joint. It will burn or melt away.
5. Maintain a steady rate of travel. Make sure you are fusing the root of the joint as in Figure 19D-5.
6. Clean, cool, and check for weld defects of any kind. Weld the other three fillets as indicated in Figure 19D-5.
7. Compare the welds with the sample shown in Figure 19D-6.
8. Present to your instructor.

POINTS TO REMEMBER

1. Fit and tack properly.
2. Maintain the suggested electrode angles.
3. Clean after tacking.
4. Maintain a constant rate of travel.
5. Be sure to fuse the leg of the lighter-gauge metal.
6. Be sure to fuse the root of the joint.

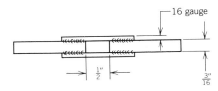

Figure 19D-1 Fit up.

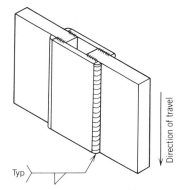

Figure 19D-2 3-F position.

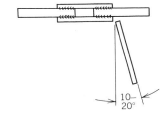

Figure 19D-3 Work angle.

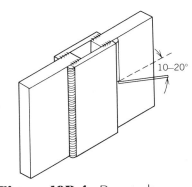

Figure 19D-4 Drag angle.

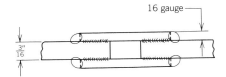

Figure 19D-5 Desired weld.

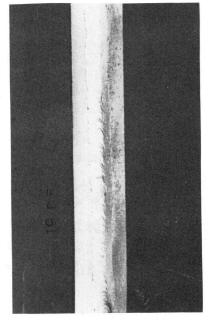

Figure 19D-6 No legend.

CHAPTER 20
WELDING SHEET METAL IN THE OVERHEAD POSITION

You may be called upon to weld light-gauge sheet metal in the overhead position. However, it is impractical to do and should be avoided if at all possible.

There always are exceptions to any rule, so therefore you should learn to weld sheet metal in the overhead position. This way you will be prepared to fulfill your obligations when the exceptions arise.

The arc current should be increased slightly when welding in the overhead position. Overhead welding requires a more forceful arc stream to obtain good fusion. Also you will have to pay close attention to the electrode work angles because they control the shape of the bead.

LESSON 20A
WELDING SHEET METAL, LAP JOINT FILLET, 4-F OVERHEAD POSITION (E-6013 ELECTRODES)

OBJECTIVE
Upon completion of this lesson you should be able to safely and correctly weld a 16-gauge hot rolled sheet metal lap joint fillet, in the 4-F overhead position, with E-6013 electrodes.

NOTE
The American Welding Society definition of the 4-F position is: "The test plates shall be so placed that each fillet is deposited on the underside of the horizontal surface and against the vertical surface." This is illustrated in Figures 20A-1 and 20A-2.

Welding sheet metal is almost the same as plate welding, except you must be careful not to burn through the overlapping section. You should be sure the end tacks, particularly the one where the weld will finish, are substantial. The end tacks should be at least ½ in. long. Otherwise, they burn away and leave an unwelded section at the end of the joint.

Depending on the amount of excessive heat input, you may have to stop welding near the end. Then you would finish the weld by adding filler metal a "dab" at a time.

MATERIAL AND EQUIPMENT
1. Personal welding equipment
2. 4 pcs. 16 gauge × 2 in. × 10 in. long hot rolled sheet metal
3. E-6013 electrodes 3/32 in. (2.4 mm) diameter
4. Fixture for positioning the specimen in the overhead position

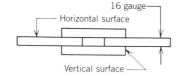

Figure 20A-1

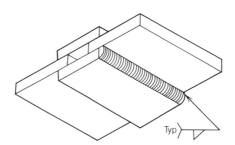

Figure 20A-2 4-F position.

GENERAL INSTRUCTIONS
1. Adjust the welding current: Alternating current or direct current straight polarity, electrode negative (−). DCSP. Ampere range: 45 to 90.
2. Tack the weldment in the position shown in Figure 20A-1.
3. Clean the tacks and all intended weld areas.
4. Secure the plates in the 4-F overhead position as shown in Figure 20A-2.
5. Place the electrode in the holder as shown in Figure 20A-3.

PROCEDURE
1. Use DCSP for this lesson. Start at the left side of the joint with the

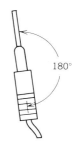

Figure 20A-3

electrode held at a 10 to 20° work angle off the vertical face of the joint as shown in Figure 20A-4.

2. Favor the horizontal surface, making sure the leading edge and the vertical leg are fused as shown in Figure 20A-5. Do not burn the leading edge of the bottom sheet back more than approximately $1/16$ in.

3. Use a slight "C" motion, or the whipping method, to control the puddle. Do not overweld by making too wide a bead. Do not pile excess weld metal on the surface. Guard against burn through in the last few inches.

4. Clean, cool, and examine the joint for appearance and soundness. The weld contour should be flat or slightly convex as in Figure 20A-5.

5. Weld the remaining three fillets as indicated by the symbol in Figure 20A-2. "Typ" means all joints of the type indicated should be welded the same.

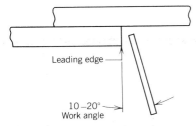

Figure 20A-4 Work angle.

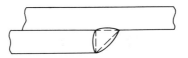

Figure 20A-5 Desired weld profile.

POINTS TO REMEMBER

1. Set your amperage correctly. If the arc current is too low, it will be difficult to maintain an arc and fuse the metal properly. You will have to use a very slow travel speed. If the amperage is too high, it can cause burn through, undercut, and require you to weld too fast to properly make the joint.

2. Try to keep the arc on the bottom of the horizontal surface as shown in Figure 20A-4.

3. Do not allow the arc stream to contact and overheat the lower edge of the vertical leg. The edge of the puddle has enough heat in it to fuse the vertical leg just by touching it.

4. Pay close attention to the last inch or two. Guard against burn through.

LESSON 20B
WELDING SHEET METAL, OPEN CORNER JOINT, 4-G OVERHEAD POSITION, E-6012 ELECTRODE

OBJECTIVE
Upon completion of this lessson you should be able to safely and correctly weld a 10- or 12-gauge hot rolled sheet metal open corner joint in the 4-G overhead position, with E-6012 electrodes.

NOTE
The E-6012 electrode is a moderate penetration electrode. When used with DCSP, it has a fairly high deposition rate. The E-6012 electrode has been chosen for this lesson for variety. It is important for you to be able to handle the popular electrodes currently in use.

You must be careful to set the arc current properly. Always guard against flux from the puddle coming into contact with the end of the electrode. This causes the puddle to become agitated and cloudy. When this happens, the arc should be broken and the crater chipped clean. Make sure there is no lack of fusion or slag inclusion.

MATERIAL AND EQUIPMENT
1. Personal welding equipment
2. 2 pcs. 10 gauge or 12 gauge × 3 in. × 10 in. long carbon steel.
3. $1/8$ in. (3.2 mm) diameter E-6012 electrode

GENERAL INSTRUCTIONS
1. Adjust the welding current—alternating current or either direct current polarity, however, direct current straight polarity is preferred, DCSP. Ampere range: 80 to 140.

2. Tack the weldment in the position shown in Figure 20B-1. Clean away the slag and all intended weld areas.

3. Secure the weldment in the 4-G position as shown in Figure 20B-2.

4. Place the electrode in the holder as in the previous lesson.

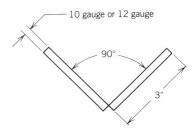

Figure 20B-1 Fit up.

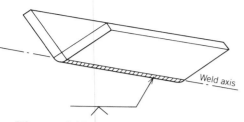

Figure 20B-2 4-G position.

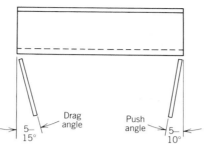

Figure 20B-3 Drag angle, push angle.

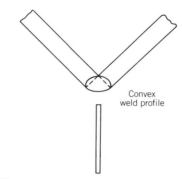

Figure 20B-4 Work angle.

PROCEDURE

1. Start welding at the left side of the joint. Use a drag angle of 5 to 15° as shown in Figure 20B-3. Use a work angle perpendicular to the joint as shown in Figure 20B-4.
2. Make sure the puddle covers the entire joint, overlapping the toss slightly. The metal edges will melt and become part of the joint.
3. Use the drag technique or a slight whipping motion to weld the joint. Make sure you obtain complete fusion. The weld metal should not penetrate through to the opposite side of the joint. Hold a medium arc length.
4. Maintain the electrode drag angle to about the last 1½ in., then you may have to switch to a push angle of 5 to 10°. This is illustrated in Figure 20B-3. This is necessary only if the joint is overheating and there is a possibility of melt through.
5. Clean, cool, and examine the weld for defects.
6. Check for proper bead contour (see Figure 20B-4).
7. Present it to your instructor.

POINTS TO REMEMBER

1. Set the welding current correctly.
2. Maintain the suggested electrode angles.
3. Guard against undercut, melt through, and overwelding.
4. If there is a chance of melt through, change from a drag to a push angle near the end of the joint.

LESSON 20C
WELDING SHEET METAL, WELDING A SIX-SIDED BOX, IN THE 2-G, 3-G, and 4-G POSITIONS, (E-6011 AND E-6013 ELECTRODES)

OBJECTIVE
Upon completion of this lesson you should be able to safely and correctly weld a six-sided box with 12 vertical, horizontal, and overhead type joints, using accepted methods and E-6011 and E-6013 electrodes.

NOTE
This lesson tests your ability to fit and tack as well as your welding skill. Some of the welding problems are new.

To weld a box means you must weld around corners and join welds that meet each other. The tacking procedure, welding sequence, and corner joints must all be done skillfully or the results will be unfavorable.

You should remember to never stop welding directly on a corner or at a junction of two joints. Problems are greater when welding sheet metal rather than plate because the joints are usually made in one pass. In many cases, the joints must be air- or water-tight. Poor welds are unacceptable.

MATERIAL AND EQUIPMENT
1. Personal welding equipment
2. 12 oz. ballpeen hammer
3. 3/32 in. (2.4 mm) × 1/8 in. (3.2 mm) diameter E-6011 and E-6013 electrodes
4. 2 pcs. 10 gauge × 6 in. × 8 in. long hot rolled sheet metal

5. 2 pcs. 10 gauge × 6 in. 10 in. long hot rolled sheet metal
6. 2 pcs. 10 gauge × 8 in. × 10 in. long hot rolled sheet metal

GENERAL INSTRUCTIONS

1. Adjust the welding current as follows:

Electrode Diameter Welding Current (AC. DCSP Electrode Negative)	Ampere Range
E-6011 ³⁄₃₂ in. (2.4 mm)	40 to 80
E-6011 ⅛ in. (3.2 mm)	75 to 125
E-6013 ³⁄₃₂ in. (2.4 mm)	45 to 90
E-6013 ⅛ in. (3.2 mm)	80 to 130

2. Be sure all sheets are joined at a right angle. Use a try square to aid in positioning them.
3. Clean all joints as in previous lessons. Maintain a close fit throughout.
4. Use a ³⁄₃₂ in. (2.4 mm) electrode for the tacking procedure.
5. Use the ballpeen hammer to close the joint openings.

PROCEDURE

1. Tack end pieces 1 and 2 to the bottom piece 3 as in Figure 20C-1.
2. Tack side pieces 5 and 6 to form a hollow box as in Figure 20C-2.
3. Tack top piece 4 to complete the box as in Figure 20C-2. Check to be sure all the joints are fit properly. There should be no openings. Use the smaller diameter electrode to tack weld *all* corners for approximately ¾ in. on each side of the vertical joint. This is shown in Figure 20C-3.
4. Clean all the tacks and the intended weld areas.
5. Place the weldment in position as in Figure 20C-4. It must remain fixed in this position until all welding is completed.
6. Weld the four vertical joints. Use ⅛ in. (3.2 mm) diameter E-6011 electrodes on DCSP and the downhill technique. Start on the tack at the top, slightly to one side of the joint, as in Figure 20C-5. Finish the same way at the bottom of the joint. This will allow you to overlap the connection with the remaining welds. It will also help eliminate poor connections at the junction of the welds. The contour of all welds is shown in Figure 20C-6.
7. After welding all four vertical joints using this procedure clean and weld the bottom joint. Use the ⅛ in. (3.2 mm) diameter E-6013 electrode on DCSP. The work angle of the electrode should be approximately 30 to 40° off the vertical leg as in Figure 20C-7.
8. Complete the box by welding the top joint with the ⅛ in. (3.2 mm) diameter E-6011 electrode on DCSP. The work angle should be approximately 45° off the horizontal leg of the joint. See Figure 20C-7.
9. Clean all welds and allow to cool. Examine for any defects.
10. Present it to your instructor.

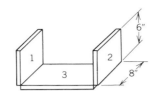

Figure 20C-1 Tacking procedure.

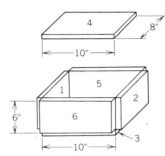

Figure 20C-2 Tacking procedure.

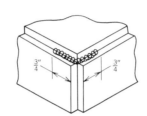

Figure 20C-3 Typical corner tack.

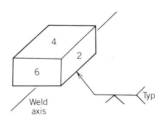

Figure 20C-4 Welding position.

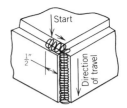

Figure 20C-5 Vertical downhill procedure.

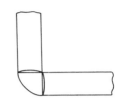

Figure 20C-6 Weld profile.

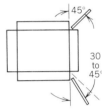

Figure 20C-7 Electrode work angles.

POINTS TO REMEMBER

1. Never stop at a corner. Always weld around them to help reduce chances of defects at that point.
2. Maintain the suggested electrode angles.
3. Conform to the welding procedure at all times.
4. Do not overweld. This is easy to do with 1/8 in. electrode on this metal.
5. Maintain proper weld profiles.

CHAPTER 21
WELDING WITH STAINLESS STEEL ELECTRODES

Stainless steels are in widespread use through industry. Oil refineries, chemical plants, food processors, pharmaceutical manufacturers, and distilleries are a few examples of stainless steel users. Stainless steel is durable, corrosion resistant, and attractive. Its uses are many.

The chrome-nickel steels are known as the "300" series. They are called chrome-nickel because of their high chromium and nickel content. The steels in this series have a chromium content ranging from 15 to 29 percent and a nickel content of from 9 to 35 percent.

These steels do not rust and resist chemical attack by many materials. The chrome-nickel series has an identification number in the three hundreds, such as, E-308-15, E-308-16, E-308L-15, E-308L-16, E-309-15, E-309-16, E-309CB-15, C-309CB-16, E-309 MO-15, and so on.

The "E" indicates the Electrode is for Electric Arc Welding.

The first three digits after the "E" indicate the composition of the core wire. (Refer to Figure 21-1.) The letter(s) following the digits indicate an alloying element has been added or other special qualities of the filler method.

You will remember from Chapter 6 the meaning of the suffixes:

L	Low carbon content
ELC	Extra low carbon content
CH	Columbium added
MO	Molybdenum added

The two digits following the hyphen, either 15 or 16, indicates the type of coating on the electrode and the type of welding current that can be used. The 15 means the electrode coating is titania and the electrode may be used on either alternating current (AC) or direct current reverse polarity (DCRP) electrode positive (+).

You should review Lesson 6D in Chapter 6 before you begin the welding lessons in this chapter.

Stainless steel electrodes should be no problem to you if you learned to use low hydrogen electrodes.

Just as with low hydrogen electrodes, a short arc should be maintained. Electrode oscillation should be kept to a minimum. Never whip or remove it from the puddle, except to finish the weld or change electrodes.

The slag deposit from this electrode is fairly heavy. The wrong electrode motions or too little arc current can cause slag inclusion. Use the stringer technique whenever possible to reduce the possibility of slag inclusion. Use great care with the weave technique. You must insure there is sufficient heat to maintain a fluid puddle. Also, the width of the bead should be kept as narrow as possible.

Stainless steel electrodes of the nickel-chrome variety, such as you will use in this chapter, need less current than mild steel electrodes. The deposited weld metal takes longer to cool than mild steel. Figure 21-2 shows the recommended welding current for various stainless steel electrodes.

Stainless steel electrodes do not throw off as many sparks as mild steel electrodes but they do give off more fumes. You should always take precautions to guard against breathing welding fumes. Position yourself to stay out of the fumes. Use plenty of ventilation whenever you do any type of welding. It is also possible to remove fumes directly at the arc source. There are many products available for this purpose. Figures 21-3 and 21-4 show two typical types that are available. In Figure 21-5 a large shop-type fume exhauster is shown.

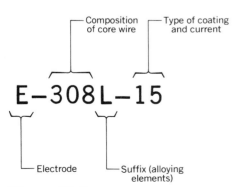

Figure 21-1

Electrode Designation (All Classifications)	Electrode Diameter		Average Arc Current, A	Maximum Arc Voltage, V
	in.	mm		
−15 and −16	1/16	1.6	35 to 45	24
	5/64	2.0	45 to 55	24
	3/32	2.4	65 to 80	24
	1/8	3.2	90 to 110	25
	5/32	4.0	120 to 140	26
	3/16	4.8	160 to 180	27
	1/4	6.4	220 to 240	28

Figure 21-2 Recommended welding current for chromium and chromenickel steel-covered electrodes. (Courtesy American Welding Society)

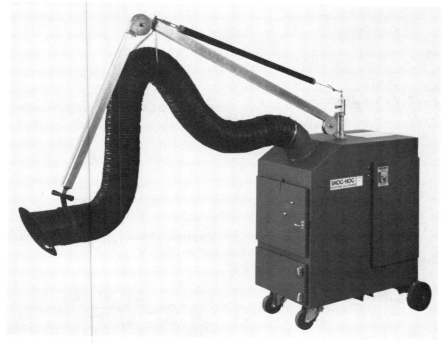

Figure 21-3 Type of fume exhauster that removes fumes at the source. (Courtesy United Air Specialists, Inc.)

Figure 21-4 Typical exhaust system used in welding schools.

Figure 21-5 Typical fume exhauster used in welding shops. (Courtesy United Air Specialists, Inc.)

Before you start the practice exercises in this chapter, there is one more word of caution. The slag covering the weld has a tendency to pop off, or explode, during the cooling process. The slag particles can travel a long distance, and the pieces are very hot and can burn you or your clothing. Always protect your skin and be careful of the type and condition of the clothing you wear. You should also wear safety glasses; keep the sleeves, pockets, and neck buttoned on your shirt. Remove all flammable objects from the shirt pockets. Wear high top shoes, with trousers long enough to cover the tops of the shoes when you are sitting down.

When you weld in a sitting position, use a leather apron or heat-resistant canvas to cover your lap and upper legs.

LESSON 21A
PADDING A PLATE, FLAT OR DOWNHAND, AND HORIZONTAL POSITION, SERIES 300 CHROME-NICKEL ELECTRODES

OBJECTIVE
Upon completion of this lesson you should be able to safely and correctly weld a pad of stringer beads, half in the flat position and half in the horizontal position, using chrome-nickel electrodes.

NOTE
The purpose of this lesson is to familiarize you with the characteristics of stainless steel electrodes. These steels are in widespread use. There is a great demand for qualified welders with skill in welding stainless steel.

As you are welding, notice the arc is quiet, the sound is subdued, and only a small amount of spatter is produced. The puddle is fairly fluid, but surface tension keeps it from flowing too freely.

MATERIAL AND EQUIPMENT

1. Personal welding equipment
2. Hand-held grinder
3. ¼ in. × 3 in. 6 in. long carbon or stainless steel plate
4. ⅛ in. (3.2 mm) diameter stainless steel electrodes (300 series).

GENERAL INSTRUCTIONS

1. Clean the entire top surface by grinding. Remove any mill scale, oxides, or foreign material.
2. Adjust the welding current: direct current reverse polarity, electrode positive (+) DCRP. Ampere range: 90 to 110. This amperage is slightly lower than for a carbon steel electrode of similar diameter. Test on scrap.
3. Place the electrode in the holder at a 90° angle.
4. Secure the weldment in the flat position as shown in Figure 21A-1.

CAUTION: The slag left from stainless welding has a tendency to "pop off." Wear your safety glasses.

PROCEDURE

1. Weld the first stringer bead by starting at the left side, ½ in. from the long edge of the plate farthest from you. (Refer to Figure 21A-2.)
2. Hold a perpendicular work angle as in Figure 21A-3. Use a 5 to 10° drag angle as in Figure 21A-4. Maintain a steady travel rate. Keep the electrode oscillation to a minimum when welding with low hydrogen electrodes. **DO NOT WHIP OUT OF THE PUDDLE.**
3. Maintain a stringer width of approximately 2½ times the diameter of the electrode (⅛ in. electrode = ⅛ + ⅛ + 1/16 = 5/16 in.) (Refer to Figure 21A-2.) Clean away the slag when the bead is finished.

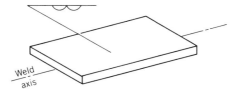

Figure 21A-1 Plate position—first half.

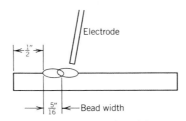

Figure 21A-2 Bead width.

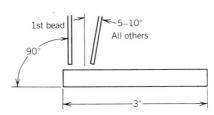

Figure 21A-3 Work angle.

4. Run the next stringer, and all remaining stringers, at a work angle of 5 to 10° into the first stringer. (Refer to Figure 21A-3.) The top edge of the electrode coating should be aimed directly at the toe of the first bead as in Figure 21A-2. The puddle should extend over about one-third to one-half of the first bead. Clean the completed bead.
5. Continue to run stringers until you reach the center of the plate. You should see no more than about one-half of each stringer, except the last one.
6. Reposition the weldment and secure it in the horizontal position as in Figure 21A-5. Place the completed section at the bottom.
7. Complete the pad with horizontal stringer beads. Use the same electrode angles, with respect to the plate, as before. Be careful of undercut at the top of the bead.
8. Cool, clean, and compare with Figure 21A-6. In this instance, the lower section was welded from right to left.
9. Present it to your instructor.

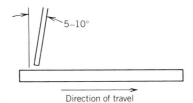

Figure 21A-4 Lead or drag angle.

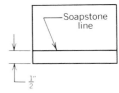

Figure 21A-5 Position of weldment for second half of lesson.

Figure 21A-6 Completed pad.

POINTS TO REMEMBER

1. Set your amperage carefully.
2. Avoid excessive electrode manipulation or movement.
3. Maintain correct electrode angles.
4. Maintain a steady rate of travel.
5. Guard against undercut.
6. The surface should be fairly flat, without hills and valleys.

LESSON 21B
LAP JOINT FILLET, 4-F OVERHEAD POSITION, SERIES 300 CHROME-NICKEL ELECTRODES

OBJECTIVE
Upon completion of this lesson you should be able to safely and correctly weld a lap joint fillet, in the 4-F overhead position, using series 300 stainless steel electrodes.

NOTE
There is little difference between welding a lap joint with stainless steel electrodes and welding with low-hydrogen electrodes. The amperage settings are slightly lower and there is less spatter.

Both have heavy slag. Care must be taken to guard against slag inclusions. In both operations it is advisable to maintain a close arc. Keep electrode oscillation or movement to a minimum.

The same electrode angles and bead sequence can be used as was used for carbon steel, for example, when welding a lap joint in the 4-F or overhead position.

MATERIAL AND EQUIPMENT
1. Personal welding equipment
2. Hand-held grinder
3. 2 pcs. 3/8 in. × 3 in. × 10 in. long hot rolled or stainless steel plate
4. 1/8 in. (3.2 mm) or 5/32 in. (4.0 mm) diameter 300 series electrodes.

GENERAL INSTRUCTIONS
1. Clean all intended weld areas. Remove all mill scale, oxides, and foreign material by grinding.
2. Adjust the welding current: direct current reverse polarity, electrode positive (+) DCRP 1/8" (3.2 mm) diameter electrodes. Ampere range: 90 to 110, 5/32 in. (4.0 mm) diameter electrodes. Ampere range: 120 to 140.
3. Tack the plates as shown in Figure 21B-1.
4. Secure the plates in position as shown in Figure 21B-2.
5. Set the welding current on scrap. It is usually slightly higher than for horizontal welding.
6. Place the electrode in the holder (as in previous lessons on overhead welding) directly in line with the handle.

CAUTION: **WEAR YOUR SAFETY GLASSES**

PROCEDURE
1. Strike an arc at the left side of the joint using either the 1/8 in. (3.2 mm) or the 5/32 in. (4.0 mm) diameter electrode. Use a work angle of 30° as shown in Figure 21B-3, and a drag angle of 5 to 10°.
2. Hold a slightly long arc until the puddle is wide and fluid enough to fuse both legs. Then maintain a close arc for the remainder of the stringer. You can use a slight "C" motion of the electrode to move the puddle. Favor the upper plate. Clean thoroughly when the first stringer is completed.
3. Run the second stringer using the same electrode angles. If you are using a 1/8 in. diameter electrode, all but 1/8 in. of the first stringer should be covered. This is shown in Figure 21B-4.
4. One more stringer bead is needed to complete the second pass. Maintain a 5 to 10° work angle as in Figure 21B-3. Favor the top plate. The third stringer should cover approximately one-third to one-half of the second bead. Both legs must be equal if the weld is to be acceptable. This is shown in Figure 21B-4.
5. If you are using the 5/32 in. (4.0

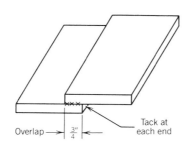

Figure 21B-1 Fit up and tacking.

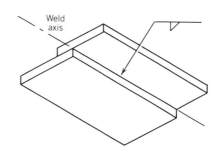

Figure 21B-2 Test position.

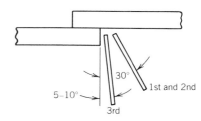

Figure 21B-3 Work angles.

Figure 21B-4 Bead sequence for 1/8 in. (3.2 mm) diameter electrode.

214 WELDING WITH STAINLESS STEEL ELECTRODES

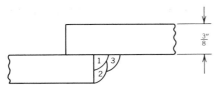

Figure 21B-5 Bead sequence for 5/32 in. (4.0 mm) diameter electrode.

mm) diameter electrode, the two stringers of the second pass will complete the joint. See Figure 21B-5.

6. If you are using the 1/8 in. (3.2 mm) diameter electrode, a third pass or layer will be required to complete the joint. This pass will consist of three stringers. (See Figure 21B-4.) Clean thoroughly between each bead.

7. On the last stringer of each pass it is important to pay special attention to elimination of overlap and undercut. You can move metal to the point you wish by electrode manipulation.

8. Cool, clean, and present to the instructor.

POINTS TO REMEMBER

1. Remove all mill scale or oxides before tacking.
2. Maintain the suggested electrode angles.
3. Favor the top plate. Metal tends to flow downward, so by favoring the top plate you tend to force the metal toward the upper plate.
4. Maintain a close arc. If you hold too long an arc, the molten metal will spill off the end of the electrode and fall.
5. Clean thoroughly between passes.

LESSON 21C
V-GROOVE BUTT JOINT, 3-G VERTICAL POSITION, SERIES 300 CHROME-NICKEL ELECTRODES

OBJECTIVE

Upon completion of this lesson you should be able to safely and correctly weld a V-groove butt joint, in the 3-G vertical position, using series 300 chrome-nickel stainless steel electrodes.

NOTE

The experience you gained in the previous lessons will help you in this one. Welding a V-groove butt joint with stainless steel electrodes is similar to welding the joint with low-hydrogen electrodes as in Lesson 15L.

As with low hydrogen electrodes the first pass is somewhat difficult due to the fluid puddle obtained with these electrodes. To reduce the heat input and puddle size a smaller diameter electrode will be used for the root pass. You can use either the stringer or weave method to finish the weld. Try to limit the bead width to less than three electrode diameters. Smaller welds cool more rapidly, thereby producing the best properties.

MATERIAL AND EQUIPMENT

1. Personal welding equipment
2. Hand-held grinder
3. Two spacers 3/32 in. × 5/8 in. × 5 in. long
4. 2 pcs. 3/8 in. × 4 in. × 10 in. long hot rolled or stainless steel plate
5. 3/32 in. (2.4 mm) and 1/8 in. (3.2 mm) diameter 300 series electrodes.

GENERAL INSTRUCTIONS

1. Prepare the plate bevels as in Figure 21C-1.
2. Remove all scale, oxide, and foreign material from the intended weld areas.
3. Prepare a 3/32 in. (2.4 mm) root face on each bevel by filing or grinding.
4. Adjust the welding current: direct current reverse polarity, electrode positive (+) DCRP 3/32 in. (2.4 mm) diameter electrode. Ampere range 65 to 80 1/8 in. (3.2 mm) diameter electrodes. Ampere: range 90 to 110.
5. Fit and tack the plates as shown in Figure 21C-1.

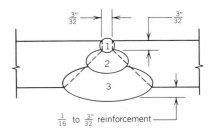

Figure 21C-1 Bead sequence.

Figure 21C-2 Prepositioning.

6. Preposition the plates slightly in the direction of the root and tack as shown in Figure 21C-2.
7. Remember to use the 3/32 in. spacers. Secure the weldment as shown in Figure 21C-3.
8. Place the electrode in the holder as in Lesson 21C or at a 45° angle.

PROCEDURE

1. *First pass:* Start at the bottom of the joint; hold a short arc, use an

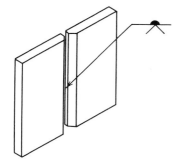

Figure 21C-3 3-G position.

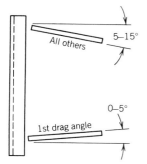

Figure 21C-4 Drag and push angles.

electrode work angle of 90° from the surface of the plate, and weld with a drag angle of 0 to 5° as shown in Figure 21C-4.

2. Drag the electrode up the root. Move at a rate that will allow the crater to fill while a keyhole opens above it. If a keyhole does not open, move more slowly. You may have to stop and increase the amperage. If the keyhole is too large, reduce the current or travel faster.

3. Clean thoroughly. It is important to remove all slag prior to welding the second pass. It is very difficult to burn out slag left from the first bead.

4. Change over to the ⅛ in. (3.2 mm) diameter electrode and use an electrode push of 0 to 10° (Figure 21C-4). *Weld the second pass employing the weave technique.*

5. Pause long enough at the toes of the weld to fuse properly.

6. *Third (final) pass:* This pass is run the same as the second pass. It is more critical in bead width except it is wider. (see Figure 21C-1)

7. Clean, cool, and check for appearance, undercut at the toes, and excess concavity or convexity of the weld reinforcement. The amount of reinforcement at the root and the face of the weld should be consistent for the entire length of the joint.

8. This joint may also be welded using the stringer technique as shown in Figure 21C-5. If stringers are used, don't let the face of the beads become too convex.

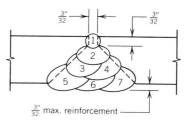

Figure 21C-5 Bead sequence—stringers.

CHAPTER 22
PREPARATIONS FOR WELDING PIPE

There are three important factors in the joining of pipe by welding. First, you must have an understanding of the welder qualification test procedures and of the procedure qualification tests as covered in Lesson 7G. Second, you must know the correct ways to prepare the pipe prior to welding it for testing. Third, you must know the proper alignment, fit up, and tacking procedures to be discussed in this chapter.

In order to be a successful pipe welder you must develop a positive attitude. Pipe welders are considered by many to be the elite of the welding fraternity. In many quarters they are known as prima donnas. In reality, they are perfectionists.

Pipe welders, as a group, insist that every detail of preparation (alignment, fit, and other conditions affecting the quality of the finished product) be exact before starting to weld. They know that once the pipe has been tacked together, they are responsible for the welding soundness of the completed joint.

Every good pipe welder pays attention to details. Proper preparation along with correct alignment, fit up, and tacking, reduces the possibility of failing a test or having a weld rejected.

An old saying among pipe welders is "a joint well prepared is a joint 90 percent completed." Certainly, your welding skill is important, but a properly prepared joint is easier to weld because most of the variables are under control.

Variations in the groove angle, root face, alignment, and so on, can force you to change the recommended amperage, electrode angles, rate of travel, and electrode manipulation. These changes in the variables make it more difficult to produce sound welds.

Even if you are successful in making an X-ray quality weld in a misaligned joint, such as shown in Figure 22-1, it can fail inspection because of misalignment.

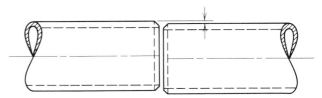

Figure 22-1 Misaligned pipe joint.

LESSON 22A
PREPARATION OF PIPE FOR TEST PURPOSES

OBJECTIVES
Upon completion of this lesson you should be able to:
1. List in proper sequence the steps involved in preparing pipe for tests.
2. List the various methods employed in the preparation of pipe for welding.
3. Safely and correctly flame cut and prepare a pipe joint for welding, using either the hand torch or a mechanical pipe beveler.

NOTE
Improperly prepared pipe is very difficult to weld with any degree of success. Proper preparation is needed to produce quality welds that pass X-ray examination or other stringent tests.

Extra effort in the preparation stage can eliminate possible problems in the welding stage.

MATERIAL AND EQUIPMENT
1. Welding goggles (#5 or 6 lens)
2. Safety glasses
3. Grinding goggles
4. Soapstone
5. Wraparound
6. Hand grinder or sander
7. Chipping hammer
8. Ballpeen hammer
9. Ten-in., half-round, bastard file
10. Wire brush
11. 1 pc. 4 in. or 6 in. schedule 40 pipe
12. Oxy-fuel flame cutting equipment

GENERAL INSTRUCTIONS
1. Hammer the surface of the pipe slightly. Make sure you remove heavy rust deposition on the inside and outside surfaces.
2. Wire brush all loose residue and scale from the surfaces of the pipe.

3. Set up the oxy-fuel flame cutting unit you will use to prepare the pipe.

PROCEDURE

1. Using the wraparound, draw a line with the soapstone, about ¾ in. from the end of the pipe. (Make sure you use a well-sharpened soapstone.)
2. Place another line about 4 in. from the first.
3. If the pipe is long enough, and a pipe beveler is available, make cuts on both lines. If a pipe beveler is not available, use a hand torch.
4. If a hand torch is used, hold the torch at a 30° angle, with the torch tip within ⅛ in. or less (3.2 mm) from the surface of the pipe (see Figure 22A-1).
5. Right-handed students should start cutting at approximately the 10 o'clock position. Then cut clockwise, stopping a little after 12 o'clock as shown in Figure 22A-2. Do not attempt to make cuts longer than from 10 to 12:30. Left-handed students start at 2 and cut counter clockwise toward 12 o'clock.
6. Start the cut ¼ in. from the line, between the line and the scrap end of the pipe. After piercing the metal, make a cut to the line and begin to move along it slowly. Make sure you hold the torch tip at 30°. The torch tip must always point at the center of the pipe.
7. When you reach the limit of travel (12 or 12:30), move the torch tip away from the line toward the end of the pipe, opening a small hole before you release the cutting oxygen lever. This will give you a new starting point when you resume the cut. It also eliminates gouging the bevel when you start the next cut.
8. Rotate the pipe counterclockwise and make the second cut.

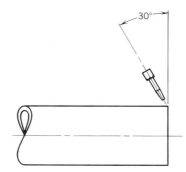

Figure 22A-1 Torch angle.

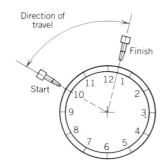

Figure 22A-2 Direction of cut.

Figure 22A-3 Slag removal.

Continue to rotate the pipe after each cut until the full bevel is completed.

9. Tap the feather edge of the bevel lightly with the ballpeen hammer. Try to remove as much of the slag as possible as shown in Figure 22A-3. If the slag cannot be removed, you were moving too slow or your flame was set too high.

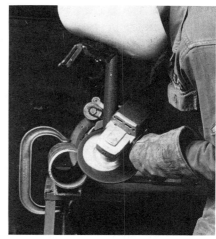

Figure 22A-4 Root preparation.

Figure 22A-5 Filing inner diameters.

10. Use a chipping hammer to remove the remainder of the slag.
11. Place the beveled pipe section in the vise. Then sand or file the remaining slag and scale from the beveled surface.
12. Use the sander or a file to remove the feather edge and form a root face as indicated in Figure 22A-4.
13. Use the file to finish the root face. Do not forget to file all scale from outside of the pipe next to the beveled edge. Then use the rounded side of the file to clean inside the pipe. Refer to Figure 22A-5.
14. After you have completed the bevel, cut off the pipe along the second line.

POINTS TO REMEMBER

1. Maintain the bevel angle around the entire pipe.
2. Clean the beveled edges, the root face, and the areas next to the joint both inside and outside of the pipe.
3. The root face must be the same width throughout.

LESSON 22B
ALIGNMENT, FIT UP, AND TACKING PROCEDURES

OBJECTIVES

Upon completion of this lesson you should be able to:

1. Correctly and safely align, fit up, and tack pipe joints.
2. Explain the relative position of the tacks and the reason for their being placed in such a manner.
3. Explain the importance of maintaining the correct root face and root gap.

NOTE

In Lesson 22A you learned the correct way to prepare pipe for welding. If the pipe is prepared properly and you apply the fundamentals (to correct alignment, fit up and tacking), you will be able to produce welded joints of acceptable quality.

The most important factor is to remember to set up each joint in the same manner. This will consistently produce satisfactory welds. Small differences in the root face, bevel angle, root opening, or fit up may require you to change arc current, rate of travel, and electrode angles. With tight control over alignment, fit up, and tacking, the need to change the other variables is eliminated. Your job will be much easier and the results more consistent and satisfying.

MATERIAL AND EQUIPMENT

1. Two pieces of approximately 4 in. long pipe, prepared as in Lesson 22-A
2. One piece of 3 in. or 4 in. channel iron about a foot long
3. Two spacers, 3/32 in. × 5/8 in. × 6 in. long
4. E-6010 or E-6011 electrodes 1/8 in. (3.2 mm) diameter
5. Safety glasses
6. Welding shield
7. Proper clothing
8. Protective leather and gloves
9. Feather wedge

GENERAL INSTRUCTIONS

1. Check to make sure that the outside surfaces of the pipes do not have any heavy coating or slag on them.
2. Tack or clamp the channel iron to a work table as shown in Figure 22B-1.
3. Adjust the welding current: DCRP, amperes 75 to 125. Set the amperage so that you can drag the electrode without stubbing out. Use scrap metal.

PROCEDURE

1. Set the two pieces of pipe on the toes of the channel iron with the root faces almost touching as in Figure 22B-2.
2. Place the two spacers between the root faces. Place one vertically, to run between 10 and 11 o'clock and 8 and 7 o'clock. Place the second between 1 and 2 o'clock and 4 and 5 o'clock as shown in Figure 22B-3.

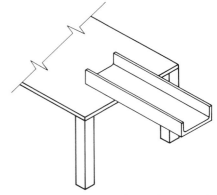

Figure 22B-1 Channel iron is welded or clamped to table top.

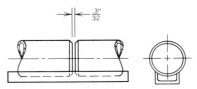

Figure 22B-2 Alignment.

3. The pipes must be pressed firmly together so the spacers cannot move. This will keep the root gap the same throughout the joint.
4. Place the first tack at 12 o'clock as shown in Figure 22B-3. All tacks should be approximately 3/4 in. (19.0 mm) long.
5. Remove the spacers and place them or wedges at 4 and 8 o'clock as shown in Figure 22B-4. If you place them too deep into the root gap, it will be difficult to remove them. Rotate the pipe so you can tack the joint at 6 o'clock.

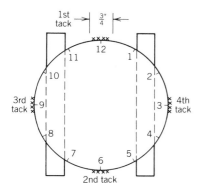

Figure 22B-3 Placement of spacers.

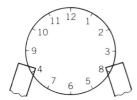

Figure 22B-4

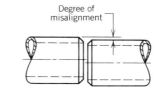

Figure 22B-5 Misaligned joint.

6. Place the second tack at 6 o'clock as shown in Figure 22B-3, then remove the spacers.

7. Check to determine if the spacing at 3 and 9 o'clock are equal. If not, use the feather wedge just below 3 or 9 o'clock to even the spacing. Then place a tack just above the wedge. Remove the wedge. Then place it in the joint, on the opposite side, and tack weld just above the wedge.

8. Examine the joint. If the gap is not evenly spaced all around the pipe, or if the pipe is misaligned as in Figure 22B-5, **CUT THE TACKS OUT WITH A HACKSAW AND START AGAIN FROM THE BEGINNING.**

9. Cool and clean the joint. Present it to your instructor for evaluation.

POINTS TO REMEMBER

1. Equal spacing throughout the joint is very important.
2. After the first tack, DO NOT place the spacers too deeply into the root gap. They will be difficult to remove.
3. Make sure that the pipes set firmly on the channel, otherwise the joint will be misaligned.
4. Make good use of the wedges. They are more easily removed than the spacers when the joint tightens up.

CAUTION: **WEAR GLOVES WHILE TACKING AND HANDLING THE WORKPIECE.**

CHAPTER 23
WELDING PIPE IN THE 1-G POSITION (AXIS OF THE PIPE IN THE HORIZONTAL POSITION)

Pipe welding is different, not difficult. Welding pipe is nothing more than welding a plate that has been rolled into a cylinder. The only real difference is due to the changing contour of the surface. You must constantly change your electrode lead (drag angle) or push angle. The angle should conform to the surface of the pipe.

In Chapter 22 you learned there are factors such as root face, root gap, bevel angle, alignment, and fit up that are extremely important if you are to produce an acceptable joint. This chapter and the remaining pipe welding chapters are concerned with producing quality pipe welds. Apply the knowledge and skills learned in the previous chapter. If you do, your tasks will be easier and acceptable welds will be made almost automatically.

Like plate, you must learn to weld pipe in many positions. Figure 23-1 shows many of the welding positions for pipe and tubing. Each drawing shows the number of degrees off the center line, or the joint axis, a test piece may be tilted. In all but two instances, the maximum is 15° in either direction. These exceptions are the 6-G or 6-GR position.

These joints are at 45° off the vertical. As shown in Figure 23-1 the position of the 6-G (6-GR) joint must be held to within plus or minus 5°.

As in the welding of test plates, do not move the pipe once you have started. The one exception to this is the 1G position. In that position rotate the pipe. All welding

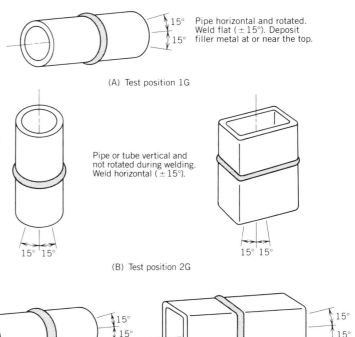

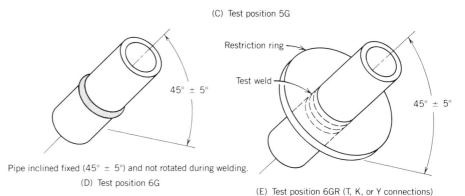

Figure 23-1 Position of test pipe or tubing for groove welds. (Courtesy American Welding Society) (A) Test position 1-G. (B) Test position 2-G. (C) Test position 5-G. (D) Test position 6-G. (E) Test position 6-GR (T, K, or Y connections).

should be done at the top, near 12 o'clock, as when using a positioner. This is illustrated in Figure 23-2.

There is a great need for qualified pipe welders. As you learn to weld pipe keep in mind your job will be easier if the bevel angle, root face, root opening, amperage, rate of travel, electrode angles, and electrode manipulation are selected properly.

Figure 23-2 Pipe being welded using a positioner. (Courtesy Aronson Machine Co.)

LESSON 23A
WELDING PIPE, AXIS OF THE PIPE HORIZONTAL, PIPE ROTATED ONE-QUARTER TURN AT A TIME (E-6010 or E-6011 ELECTRODES)

OBJECTIVE
Upon completion of this lesson you should be able to safely and correctly weld a V-groove pipe joint in the horizontal position, while turning the pipe clockwise a quarter turn at a time, and welding from 4 to 1 o'clock using E-6010 or E-6011 electrodes, in conformance with accepted pipe welding procedures.

NOTE
This lesson gives you practice in changing your electrode angles to conform to the contour of the pipe. In pipe welding it is important that the electrode angle remain fairly constant. It is good practice to maintain an electrode angle that points almost directly at the axis of the pipe as in Figure 23A-1.

If you maintain the proper angle, you will have no trouble in obtaining the proper penetration, fusion, and in general, greater control of the puddle.

Practicing welding one-quarter pipe sections, in a fixed position, will increase your skill. Many pipe joints are fixed in position and may not be moved during the test procedure.

MATERIAL AND EQUIPMENT
1. Personal welding and cutting equipment
2. E-6010 or E-6011 electrodes, ⅛ in. (3.2 mm) diameter
3. 2 pcs. 4 in. schedule 40 pipe, or similar if 4 in. is not available. Pipe nipples or pieces must be of sufficient length to enable the student to obtain a proper fit.

GENERAL INSTRUCTIONS
1. Prepare pipe as in Lesson 22A.
2. Fit and tack as in Lesson 22B.
3. Secure the pipe with its axis in the horizontal position as shown in Figure 23A-2.
4. Adjust the welding current: DCRP. Ampere range: 75 to 125.

PROCEDURE
1. Place the electrode in the holder, as in Figure 23A-3. Use a 90° angle or a 45° angle away from the end of the holder.
2. Position yourself so you are at a 90° angle to the pipe. Be sure

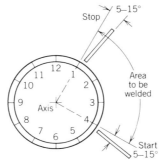

Figure 23A-1 Electrode push angles.

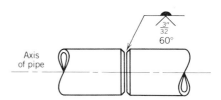

Figure 23A-2 Test position.

WELDING PIPE IN THE 1-G POSITION

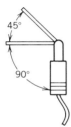

Figure 23A-3

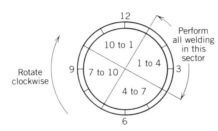

Figure 23A-4

Figure 23A-7 Completed weld.

you are comfortable, as in the previous lesson.

3. Strike the arc, on the bevel, at approximately 3 o'clock. Carry it down to 4 o'clock. Pause long enough for the root faces to melt away and a keyhole form. Then reverse your electrode direction.

4. Utilize the whipping method, as in welding plate in the vertical position, to run the first pass uphill. Use an electrode to push angle 5 to 15° upward as in Figure 23A-1. Whip upward, taking care not to scar the surface of the pipe on either side of the V-groove. Stop when you reach 1 o'clock as shown in Figure 23A-4. Clean thoroughly using the hacksaw blade and the wire brush. Draw the blade across the weld toward you. It will remove the slag from the surface of the weld. Make sure the cutting edge of the teeth are facing toward you, as shown in Figure 23A-5.

5. Turn the pipe toward you one-quarter of a turn. Then proceed in the same manner until the first pass is completed. Be sure to start the next electrode slightly below the crater.

6. The second pass (hot pass) and

Figure 23A-5 Use of the hacksaw blade.

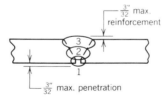

Figure 23A-6 Bead sequence.

third pass (cover pass) can be welded with either the triangle motion or the alternate weave, as in vertical plate welding.

7. Take care to pause at the sides of the joint. Burn out any entrapped slag and fill in any undesirable undercut.

8. Follow the same bead sequence. Adhere to the maximum root and face reinforcement shown in Figure 23A-6.

9. When you make the connection on completing the pass, be sure to overlap slightly. Break the arc by slowly drawing it away from the puddle.

10. Compare your weld with the sample in Figure 23A-7. Present it to your instructor.

POINTS TO REMEMBER

1. Keep the root pass keyhole open at all times.
2. Maintain an electrode angle. It should point directly from the surface of the pipe to the axis of the pipe.
3. Guard against excessive reinforcement and overlap.
4. Be sure to leave a keyhole when you pause to turn the pipe. If you don't, you will probably leave a spot without the proper amount of penetration, or none at all.
5. Be careful on all connections. Hold a long arc on all starts and stops.

CHAPTER 24
WELDING PIPE IN THE 2-G TEST POSITION (AXIS OF THE PIPE IN THE VERTICAL POSITION)

In the 2-G test position the axis of the pipe is in the vertical position. Welding takes place on the horizontal plane. This means you must weld from right to left or left to right, as with plate welded in the horizontal position.

First passes can be either whipped or dragged. As in plate welding use approximately the same electrode angles. It is recommended that all passes be of the stringer bead variety. They can be more easily controlled by the welder.

This does not mean you cannot use the weave method in the 2-G position. However, sag and undercut are more readily controlled by the use of stringers.

Many of the pipe joints you will encounter will be in this position. Therefore, it is important to develop your pipe welding skills.

LESSON 24A
WELDING PIPE, 2-G TEST POSITION (AXIS OF PIPE VERTICAL) FIRST PASS WHIPPED OR DRAGGED (E-6010 OR E-6011 ELECTRODES)

OBJECTIVE
Upon completion of this lesson you should be able to safely and correctly weld a pipe V-groove butt joint, in the 2-G horizontal test position, first pass whipped or dragged using E-6010 or E-6011 electrodes.

NOTE
When you weld pipe in the 2-G test position, its axis is vertical. Because the pipe is in the vertical position, welding is done in the horizontal position, as in Figure 24A-1.

If you successfully pass a pipe test in the 2-G test position, you are qualified to weld complete penetration groove welds on pipe 4 in. in diameter or less, both in that position and the 1-G position according to D1.1 Structural Welding Code-Steel of the AWS. Due to differences in the codes the same does not hold true when using ASME Section IX Boiler and Pressure Vessel Code. Horizontal joints may be welded with either the stringer or the weave method. Usually, the stringer method presents you with fewer problems. There is less chance of undercut and slag inclusion. The weld has a neat appearance and is easier for you to master.

MATERIAL AND EQUIPMENT
1. Personal welding and cutting equipment (as in the previous lessons on pipe welding)
2. E-6010 or E-6011 electrodes, 1/8 in. (3.2 mm) diameter

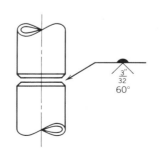

Figure 24A-1 Test position 2-G.

3. 2 pcs. 4 in. schedule 40 pipe, or similar if 4 in. is not available. (Pipe nipples or pieces must be of sufficient length to enable the student to obtain a proper fit.)
4. Two spacers 3/32 in. × 5/8 in. × 6 in. long.

GENERAL INSTRUCTIONS
1. Prepare the pipe as in Lesson 22A.
2. Fit and tack as in Lesson 22B.
3. Secure the pipe so its axis is vertical as in Figure 24A-1.
4. Adjust the welding current: DCRP. Amperage range: 75 to 125.
5. Once the test has started the pipe may not be moved for any reason until the welding is completed. This is standard procedure when taking tests of this type.

PROCEDURE
1. Place the electrode in the holder, at a 90° angle or at a 45° degree angle, as in Lesson 23B.
2. The work angle of the electrode

should be about 5 to 10° below center of the joint as in Figure 24A-2.

3. Use a slight drag angle of 5 to 10° from the axis of the pipe, as shown in Figure 24A-3.
4. Strike an arc and get a keyhole started. If the upper root face burns away excessively, decrease the electrode work angle. You can do this by raising your hand and reducing the angle.
5. As shown in Figure 24A-4 use the whipping motion to carry the arc onto the lower beveled surface. When you return to the keyhole, wash the metal into the root, then pause just long enough to fill the keyhole and reopen it again.
6. Make sure that you hold a long arc when starting a new electrode. This heats the metal where you left off welding so that you get the desired penetration when you enter the keyhole. You must obtain 100 percent penetration at these connections. Otherwise, you will leave voids or unwelded spaces. These will cause a weld to fail in service or during the testing procedure.
7. Clean the root pass thoroughly. Then run a stringer wide enough to cover the root pass. Make sure the weld toes are fused properly to prevent slag inclusions or undercut.
8. The next pass consists of stringers 3 and 4 as shown in Figure 24A-5. These two stringers will be similar to those used in welding plate in the horizontal position. Weld the lower stringer first. Remember: Only two-thirds of the lower stringer should be visible when the pass is completed.
9. The last pass consists of stringers 5, 6, and 7 as shown in Figure 24A-5. If you used heavy beads on the previous passes, this pass may not be necessary.

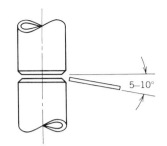

Figure 24A-2 Work angle.

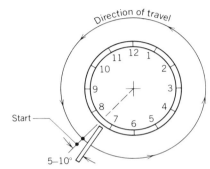

Figure 24A-3 Drag angle.

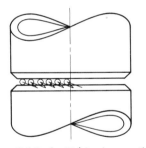

Figure 24A-4 Whipping motion.

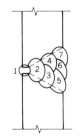

Figure 24A-5 Bead sequence.

10. Apply the three stringers as in horizontal plate welding. Be careful to avoid undercut on the top head. You can use a slight push angle on the last stringer, if undercut is a problem.

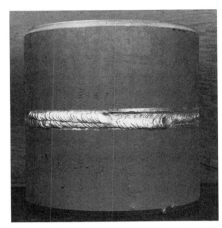

Figure 24A-6 Completed weld, 2-G position.

Figure 24A-7 Welder leaning on a pipe in 2-G position.

11. Cool, clean, and compare with the completed weld in Figure 24A-6 and present to the instructor for evaluation.
12. You should attempt the drag method on your next practice piece, place the electrode in the holder at a 45° angle and assume the position shown in Figure 24A-7.
13. Hold a long arc to start. Slowly move the electrode into the root of the joint as in drag welding first passes on plate. Maintain a 20 to 30° drag angle as shown in Figure 24A-8. Use a work angle of 0 to 10°.
14. Force the electrode gently through the root so the arc blows out the other side of the joint. Listen for a harsh sounding arc.

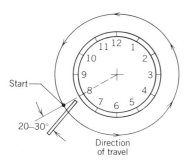

Figure 24A-8 Drag angle root pass.

It tells you that you are achieving penetration.

15. If you do not hear the harsh sound you are not getting enough penetration.
    (a) Slow down the rate of travel.
    (b) Increase the drag angle.
    (c) Stop and increase your amperage.

    If you are blowing through too much you are getting excess penetration.
    (a) Increase the rate of travel.
    (b) Decrease the drag angle.
    (c) Stop and decrease your amperage.

POINTS TO REMEMBER

1. Properly prepare and fit the joint.
2. Use the correct tacking procedure.
3. Remove all slag between passes.
4. The electrode angles, amperage, and rate of travel, all control the depth and width of penetration.
5. Avoid undercut at all times.
6. Apply the finish or cap pass when the filler metal reaches to within 1/16 in. of the pipe surface on the next to last pass.
7. Pay close attention to the sound of the arc when drag welding root passes.

LESSON 24B
WELDING PIPE, 2-G TEST POSITION (AXIS OF PIPE VERTICAL) FIRST PASS DRAGGED (E-6010 OR E-6011 ELECTRODES, SECOND AND OUT E-7018 ELECTRODES)

OBJECTIVE

Upon completion of this lesson you should be able to safely and correctly weld a V-groove pipe joint, in the 2G-position, in conformance with procedures in Section IX of the ASME Boiler and Pressure Vessel Code and D1.1 Structural Welding Code-Steel of the AWS, using E-6010 or E-6011 electrodes for the root pass, and E-7018 electrode for the remaining passes.

NOTE

This lesson gives you more practice in dragging root passes. It introduces you to welding pipe joints, in the 2-G test position, with low hydrogen electrodes.

Low hydrogen electrodes, such as the E-7018 used in this lesson, are used in pipe welding for a number of reasons. However, low hydrogen electrodes should not be used on open roots unless the first pass is welded with another proper covered electrode, or another process. Unless this is done low hydrogen electrodes used for open roots will cause porosity. Joints welded with these electrodes have good low temperature characteristics. Also, they are far less crack-sensitive than those made with E-6010 and E-6011 electrodes, and they can have higher tensile strengths.

A few years ago the E-6010 and E-6011 electrodes were used almost exclusively for pipe welding. Today you need to be skilled in the use of low hydrogen electrodes. This is a necessity if you wish to be a quality pipe welder.

MATERIAL AND EQUIPMENT

1. Personal welding equipment.
2. Two pieces of pipe, prepared as in the previous lesson
3. E-6010 or E-6011 electrodes and E-7018 electrodes, 1/8 in. (3.2 mm) diameter
4. 2 pcs. 4 in. schedule 40 pipe, or similar if 4 in. is not available (Pipe nipples or pieces must be of sufficient length to enable the student to obtain a proper fit.)
5. 2 spacers 1/16 in. × 5/8 in. × 6 in. long

GENERAL INSTRUCTIONS

1. Prepare the pipe as in Lesson 22A.
2. Fit and tack as in Lesson 22B.
3. Adjust the welding current: DCRP. Ampere range: E-6010 or E-6011, 75 to 125; E-7018, 115 to 165. Set the amperage so you can drag the E-6010 or E-6011 electrode along, in contact with the joint, while holding a 20 to 30° drag angle.
4. Fit up and tack two pieces of pipe as in previous lessons.
5. Secure the weldment in the 2-G

position as shown in Figure 24B-1.

PROCEDURE

1. Place the E-6010 or E-6011 electrode in the electrode holder at a 45° angle.
2. Assume a comfortable position. Lean against the pipe or balance yourself as in Lesson 24B. Remember to lean slightly to the left at the start.
3. Strike an arc on the beveled surface and move quickly down into the root gap. Hold the electrode with a 20 to 30° drag angle and a 0 to 10° work angle, upward into the root, as shown in Figures 24B-2 and 24B-3.
4. Obtain penetration and complete the root pass as in Lesson 24A. Hold a long arc on restarts. Be sure to "burn in" when making the connection.
5. Clean the weld area thoroughly. Change over to the E-7018 electrode. Increase the amperage slightly as recommended and try the arc on scrap metal.
6. Strike an arc. Change to a drag angle of 5 to 15°, but use the same work angle as in Figure 24B-2. Hold a short arc and do not whip out of the puddle. Just move the electrode along at a steady rate and pull the puddle with it.
7. The electrode may be oscillated, or moved slightly within the confines of the puddle. Keep the oscillations to a minimum, however.
8. This second pass is a stringer bead. Pay close attention; keep the crater deep enough and hot enough to fuse the sides of the joint and the root pass thoroughly. Welders sometimes call this "Burning out the wagon tracks" with the "hot pass." This is illustrated in Figure 24B-4.
9. Clean the weld thoroughly. Paying close attention to removing any slag that may remain at the toes of the weld. Cover this pass with at least two more stringers as in Figure 24B-5.
10. Clean each pass carefully. Be sure to hold normal arc lengths on restarts. The arc should be struck ½ in. ahead of the crater and slowly moved back to the crater. This will reweld the arc strike area as the weld progresses.
11. The last stringer is critical. Watch the upper left-hand edge of the puddle to be sure that you are filling in all undercut.
12. Clean, cool, and present it to your instructor.

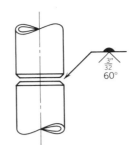

Figure 24B-1 Test position 2-G.

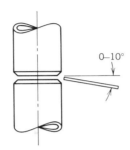

Figure 24B-2 Electrode work angle.

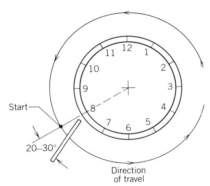

Figure 24B-3 Drag angle—root pass.

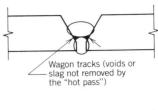

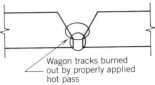

Figure 24B-4

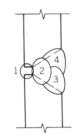

Figure 24B-5 Bead sequence.

POINTS TO REMEMBER

1. Preparation, fit up, and tacking are important.
2. Clean thoroughly.
3. Pay close attention to starts and restarts.
4. Watch out for undercut and slag at the toes of bead.
5. Do not overweld.

CHAPTER 25
WELDING PIPE IN THE 5-G TEST POSITION (AXIS OF PIPE IN THE HORIZONTAL POSITION)

The majority of welded pipeline joints are completed in the 5-G position. Pipe normally rests on sleepers or is supported by hangers. This keeps the pipe axis in the horizontal position, except where the pipe runs uphill or downhill.

The 5-G test position is widely used for testing a welder's ability. Joints in this position can be completed by either the stringer or weave methods. Quite often a combination of both is used.

Pipe joints in the 5-G position are welded by both the uphill and downhill methods, depending upon the pipe code and design service. This chapter provides you with practice welding uphill and downhill, using the stringer bead and weave methods.

LESSON 25A
WELDING PIPE, 5-G TEST POSITION (AXIS OF PIPE HORIZONTAL) FIRST PASS WHIPPED, SECOND PASS AND OUT WEAVED (ALL UPHILL) (E-6010 OR E-6011 ELECTRODES)

OBJECTIVE
Upon completion of this lesson you should be able to safely and correctly weld a pipe open root V-groove butt joint, in the 5-G test position, first pass whipped, second and out weaved using the uphill technique and E-6010 or E-6011 electrodes.

NOTE
More pipe joints are welded in the 5-G position than any other out-of-position joints. Because of this you need to develop your skills to be competitive in the job market.

Welding in this position is slightly more difficult than the 2-G position. You cannot work at one level directly in front of the joint; you must work downhill from the top or uphill from the bottom. As the weld progresses you must move almost constantly. The puddle should always be visible to you. In addition, the curved pipe surface makes it necessary for you to adjust the electrode constantly to maintain the push or drag angles.

MATERIAL AND EQUIPMENT
1. Personal welding equipment
2. E-6010 or E-6011 electrode 1/8 in. (3.2 mm) diameter
3. 2 pcs. 4 in. schedule 40 pipe or similar if 4 in. is not available. (Pipe nipple or pieces must be of sufficient length to enable the student to obtain a proper fit.)
4. 2 spacers 3/32 in. × 5/8 in. × 6 in. long.

GENERAL INSTRUCTIONS
1. Prepare and fit the pipe as in the previous lesson.
2. Secure the pipe in the 5-G (horizontal fixed) position, as in Figure 25A-1. Choose a height that will allow you to comfortably reach the top and bottom of the joint.

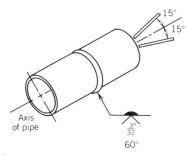

Figure 25A-1 5-G test position.

3. Adjust the welding current: DCRP. Amperage range: 75 to 125. Set the amperage as for whipping the first pass in the 2-G position.
4. Place the electrode in the holder at a 90 or 45° angle, whichever is more comfortable for you.

PROCEDURE
1. Get into as comfortable position as possible. Lean your shoulder or upper arm against the pipe as in Figure 25A-2.

Figure 25A-2 Welder leaning against pipe in 5-G position.

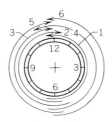

Figure 25A-3 Weld progression.

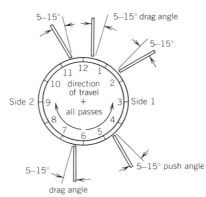

Figure 25A-4 Drag and push angles.

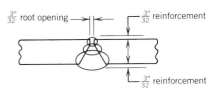

Figure 25A-5 Weld progression.

Figure 25A-6 Completed pipe joint.

2. Strike an arc on the bottom beveled surface at about the 6:30 position. Weld counterclockwise through the 6 o'clock position, up the side one of the pipe. This is shown in Figure 25A-3.
3. Use an electrode work angle of 90° and a drag angle of about 5 to 15° as shown in Figure 25A-4.
4. Maintain a constant drag angle. Hold a close arc, similar to the drag technique, but without the electrode touching the pipe. Keep the arc visible while you whip the bead uphill toward 5 o'clock. Keep the keyhole open.
5. At slightly past 5 o'clock switch from a drag to a 5 to 15° push angle. Continue to weld upward as shown in Figure 25A-4.
6. Use the same whipping technique as in vertical welding of the grooves in plate. Remember to maintain the slight push angle until you reach between the 1 and 2 o'clock position. At this point switch to a 5 to 15° drag angle, continue welding until 12:30 as in Figure 25A-4.
7. Clean and inspect the weld. Move to side two of the pipe. (See the weld progression shown in Figure 25A-3.) Now make the second bead. Remember to hold a long arc when overlapping starts and stops on the first bead. Clean and inspect the second bead.
8. Return to side one. Start the second pass (third bead) by welding uphill as shown in Figure 25A-3. Use the weave method. You can start by moving the electrode from side to side; then continue unless the puddle becomes too fluid. If that is the case, switch to the triangle motion. The triangle motion may become necessary in the area of 4 to 2 o'clock. Maintain the electrode drag and push angles as shown in Figure 25A-4. Clean and inspect the bead.
9. Move to side two. Start bead four and complete the second pass. Hold a long arc on the starts and stops. Clean and inspect the completed pass. See bead sequence shown in Figure 25A-5.
10. Start the final (cap) pass on side one. Begin at approximately 6 o'clock and weld uphill to 12 o'clock using the weave technique. Keep the weld width complete and the reinforcement to a minimum. (Refer to Figure 25A-4.)
11. Weld side two of the cap pass. Make sure to overlap at 6 o'clock and 12 o'clock.
12. Clean and examine your weld for undercut, gas pockets, slag holes on the surface, penetration, width, and reinforcement. Compare with the sample weld in Figure 25A-6. Clean, cool, and present it to your instructor.

POINTS TO REMEMBER

1. Stagger all starts and stops as shown in Figure 25A-3.
2. Keep the keyhole open at all times when making a root pass.
3. Maintain a constant work angle. Keep it at 90° to the surface of the pipe.
4. Maintain the proper drag and push angles for the various sections of the weld.
5. Limit the root and face reinforcement to $3/32$ in. Excessive root penetration can cause internal erosion problems. Excessive face reinforcement creates stress risers that can lead to weld failure.
6. Do not strike the arc outside of the intended weld area. Always strike arcs inside the prepared V-joint.

LESSON 25B
WELDING PIPE, 5-G TEST POSITION (AXIS OF PIPE HORIZONTAL) FIRST PASS DRAGGED (DOWNHILL) (E-6010 OR E-6011 ELECTRODES), SECOND AND OUT (UPHILL) (E-7018 ELECTRODES)

OBJECTIVE
Upon completion of this lesson you should be able to safely and correctly weld a pipe V-groove butt joint, in the 5-G test position, first pass dragged using the downhill technique and E-6010 or E-6011 electrodes, second and out passes using the uphill technique with E-7018 electrodes.

NOTE
Low hydrogen electrodes are becoming increasingly popular for pipe welding because they are less crack sensitive, can withstand low ambient temperatures, and have higher tensile strength.

They have one drawback: They are difficult to use when welding pipe root passes in open root V-groove butts. Many contractors offset this problem by using E-6010 electrodes for the root pass. The fast freezing E-6010 has a digging arc that produces excellent penetration.

This lesson combines the E-6010 or the E-6011, which have similar characteristics, and the E-7018 low hydrogen electrode.

MATERIAL AND EQUIPMENT
1. Personal welding equipment
2. E-6010 or E-6011 and E-7018 electrode ⅛ in. (3.2 mm) diameter 2 pcs. 4 in. schedule 40 pipe or similar if 4 in. is not available. (Pipe nipple or pieces must be of sufficient length to enable the student to obtain a proper fit.)
3. 2 spacers 1/16 in. × 5/8 in. × 6 in. long

GENERAL INSTRUCTIONS
1. Prepare and fit the pipe as in previous lessons.
2. Secure the pipe in the 5-G (horizontal fixed) position as in Figure 25B-1 at a comfortable welding height, to reach the top and bottom of the joint.
3. Adjust the welding current: DCRP. Ampere range: E-6010 and E-6011, 75 to 125; E-7018, 115 to 165.
   Set the amperage as you would for dragging the root pass on a pipe in the 2-G position.
4. Place the electrode in the holder at a 90 or 45° angle, whichever is more comfortable for you.

PROCEDURE
1. Get into as comfortable position as possible, just as in the last lesson. Strike an arc on the beveled surface. Begin at about 11:30 as shown in Figure 25B-2. Weld clockwise, using the downhill technique as shown in Figure 25B-3.
2. Keep the electrode work angle at 90° to the surface of the pipe. The drag angle should be from 10 to 25°, depending on the amount of penetration you obtain. If the drag angle is too large, stop and increase your amperage. This will give you more penetration and reduce the required electrode drag angle. As shown in Figure 25B-2, weld downhill.
3. Stop just past 6 o'clock, as shown in Figure 25B-2. Clean the bead and move to side two. Complete the second downhill root pass. Be sure to properly overlap the first bead.
4. Second pass and out use E-7018 low hydrogen electrodes. Use a drag angle of 10 to 15°. Start the third bead somewhere near 5 to 7 o'clock, taking care to overlap and tie in the connection between the two root beads. Weld

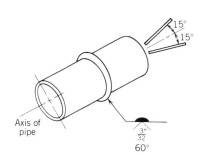

Figure 25B-1 5-G test position.

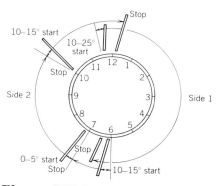

Figure 25B-2 Drag angles.

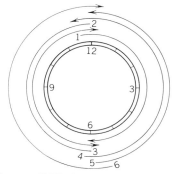

Figure 25B-3 Weld progression.

upward on side two. Use a slight weave. Take care not to leave the puddle or oscillate excessively.

5. As the weld passes 7 o'clock, gradually change from a drag angle of 0 to 5°. Continue until 10 o'clock is reached. Then switch to a drag angle of 10 to 15°. Stop at about 12 o'clock.

6. Clean, and move to side one. Weld the fourth bead uphill, as shown in Figure 25B-3. Hold a long arc on the starts and stops to make good tie in connections.

7. After cleaning use uphill beads to weld the finish pass. Start on side one. Do not weave the pass any wider than 3/32 in. past the edge of the bevels. As shown in Figure 25B-4 keep the reinforcement to approximately 3/32 in. Beads with low hydrogen electrodes should be no wider than three electrode diameters. If the groove is wider than this, it would be better to finish the joint with stringers.

8. Cool and clean the completed weld. Examine for undercut, surface defects, bead shape, width, and appearance of the weld and reinforcement. Present it to your instructor.

POINTS TO REMEMBER

1. Stagger all starts and stops.
2. Make sure you are obtaining adequate penetration.
3. Maintain a constant work angle of 90° to the surface of the pipe. The electrode may be moved slightly to the side to keep the arc from burning one side of the root more than the other.
4. Maintain a root and face reinforcement of approximately 3/32 in.
5. Do not strike the arc outside of the weld area.
6. Maintain a close arc with low hydrogen electrodes.

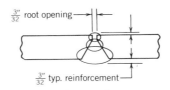

Figure 25B-4 Bead sequence.

LESSON 25C
WELDING PIPE. 5-G TEST POSITION (AXIS OF THE PIPE HORIZONTAL). FIRST PASS DRAGGED, SECOND AND OUT WEAVED (ALL DOWNHILL) (E-6010 OR E-6011 ELECTRODES)

OBJECTIVE
Upon completion of this lesson you should be able to safely and correctly weld a pipe V-groove butt joint, in the 5-G test position, first pass dragged downhill, second and out weaved downhill, with E-6010 or E-6011 electrodes.

NOTE
Most welding on overland oil and gas transmission lines is done downhill. Downhill welding is ideally suited for joints with small root openings and small bevel angles, especially on light wall pipe.

The technique requires fairly high amperages and fast travel. All passes are welded from the top of the pipe to the bottom.

You must take care to maintain correct electrode angles and move rapidly at a constant rate of travel, otherwise the molten pool and slag will sag and may fall. If the slag runs ahead of the puddle, break the arc. Then clean the bead before beginning to weld again. Failure to stop and clean will result in slag inclusions and surface holes.

MATERIAL AND EQUIPMENT
1. Personal welding equipment
2. 2 pcs. 4 in. schedule 40 pipe or similar if 4 in. is not available. (Pipe nipple or pieces must be of sufficient length to enable the student to obtain a proper fit.)
3. 2 spacers 1/16 in. × 5/8 in. × 6 in. long
4. E-6010 or E-6011 electrodes 1/8 in. (3.2 mm) and 5/32 in. (4.0 mm)

GENERAL INSTRUCTIONS
1. Prepare and fit the pipe as in previous lessons.
2. Secure the pipe in the 5-G (horizontal fixed) position, as shown in Figure 25C-1. Use a comfortable height so you can reach both top and bottom of the joint.
3. Adjust the welding current: DCRP. Ampere range: 1/8 in. (3.2 mm) 75 to 125; 5/32 in. (4.0 mm) 110 to 170. Set the amperage as you would for dragging the root pass in previous lessons.

LESSON 25C

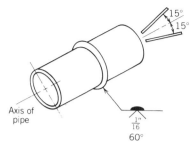

Figure 25C-1 5-G test position.

4. Place the electrode in the holder at a 90 or 45° angle, whichever is more comfortable for you.

PROCEDURE

1. Assume a comfortable welding position just as in Lesson 25C.
2. Drag the first pass downhill, on side one. Weld from 11:30 to approximately 6 o'clock and as shown in Figures 25C-2 and 25C-3. Maintain a 90° work angle and 10 to 25° drag angle.
3. Clean the bead and weld side two of the root pass as shown in Figure 25C-2. Be sure you are obtaining the correct melt through or penetration.
4. Start the second pass at the top of side two. The electrode angles remain the same as on the root pass, except the drag angle can be reduced to zero between 7 o'clock and the bottom on side two, and 5 o'clock and the bottom on side one.
5. Clean weld bead three and complete the second pass by welding bead four on side one as in Figure 25C-2. Be sure to overlap at the start and stop.
6. Clean the second pass thoroughly, then examine it for evenness of deposit. You should have deposited the same amount of weld metal throughout the entire joint. Then finish off with a cap pass using the E-6010 or E-6011

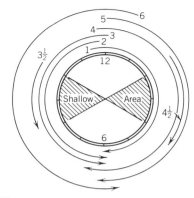

Figure 25C-2 Weld progression.

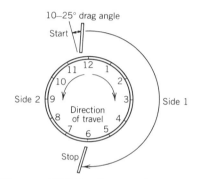

Figure 25C-3 Drag angles.

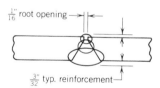

Figure 25C-4 Bead sequence.

5/32 in. (4.0 mm) electrode. (See Figure 25C-4.)

7. Many times the weld is shallow between 2 to 4 and 8 to 10 o'clock. Then, as shown in Figure 25C-2, you must run short beads between these points. These extra beads are shown as 3½ and 4½ in Figure 25C-2. These beads will allow you to weld a uniform finish pass around the entire joint.
8. Cool, clean, and compare your workpiece with the sample in Fig-

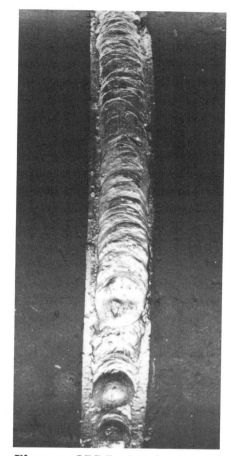

Figure 25C-5 Completed weld downhill pipe.

ure 25C-5. Present it to your instructor.

POINTS TO REMEMBER

1. Set your amperage exactly.
2. Don't leave unwelded or overwelded places in the root pass. Fill in or grind out as needed.
3. Hold a long arc when overlapping on a starting point.
4. Stay within the recommended maximum reinforcement.
5. Clean the joint carefully between passes. Leftover slag can cause gas pockets, slag holes, or voids in the next pass.
6. Do not strike arcs outside of the weld area.

CHAPTER 26
WELDING PIPE IN THE 6-G TEST POSITION

Welding butt joints in the 6-G position presents problems not found on many other types of joints. Because the pipe does not rotate and is tilted, it is easy to obtain root pass undercut on the inside of the top piece of pipe. The root pass undercut is due to heat buildup in that area. The same holds true for finish pass undercut on the outside of the top piece.

To eliminate or at least minimize the problem, the stringer bead method is recommended.

Puddle visibility is better when welding pipe in the 6-G position than when welding in the 5-G position. But it is slightly more difficult to control the flow of the puddle. For this reason the puddle should be small and the heat input kept to a minimum.

A welder who passes a qualification test in the 6G position also qualifies for all other pipe positions according to QW-303.1 Section IX ASME Boiler and Pressure Vessel Code. If the test is taken under 5.23.2.4 AWS D1.1 Structural Welding Code-Steel, the welder is qualified for all position groove and all position fillet welding of pipe, tubing, and plate.

LESSON 26A
WELDING PIPE, 6-G TEST POSITION (AXIS OF PIPE AT A 45° ANGLE), FIRST PASS WHIPPED, SECOND AND OUT STRINGERS (ALL UPHILL) (E-6010 OR E-6011 ELECTRODES)

OBJECTIVE
Upon completion of this lesson you should be able to safely and correctly weld a pipe V-groove butt joint in the 6-G test position, first pass whipped, second pass and out made with stringers using E-6010 or E-6011 electrodes.

NOTE
Table 5.26.1 and paragraph 5.23.2.4 AWS Structural Welding Code-Steel, indicates a welder who passes a 6-G (Inclined fixed) position, 4-in. schedule 40 pipe test is qualified to weld pipe of that wall thickness, from ¾ to 4 in. in diameter, as well as "all position groove and all position fillet welding of pipe, tubing and plate."

The 6-G position requires, that the pipe axis remain tilted at 45° angle plus or −5° throughout the test.

You can readily see that this is a difficult test. It is considered a good test of welder's capabilities.

MATERIAL AND EQUIPMENT
1. Personal welding and cutting equipment
2. E-6010 or E-6011 electrodes, ⅛ in. (3.2 mm) diameter
3. 2 pcs. 4 in. schedule 40 pipe or similar if 4 in. is not available. (Nipples or pieces must be of sufficient length to enable the student to obtain a proper fit.)
4. 2 spacers, ³⁄₃₂ in. × ⅝ in. × 6 in. long
5. 1 large hacksaw blade (one end taped for use as a handle).

GENERAL INSTRUCTIONS
1. Adjust the welding current: DCRP. Ampere range: 75 to 125. (Set the amperage as for whipping the first pass in a butt joint in the 5-G position.)
2. Prepare and fit the pipe as in the previous lessons.
3. Secure the pipe in the 6-G position, as shown in Figure 26A-1. Select a height that is comfortable for you to reach.
4. Place the electrode in the holder at a 90 or a 45° angle, whichever is the most comfortable for you.

PROCEDURE
1. Lean your arm against the pipe or a firm rest of some sort. Strike

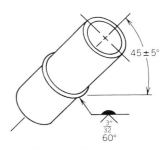

Figure 26A-1 Test position 6-G.

the arc on the beveled surface, between 6 and 7 o'clock, as shown in Figure 26A-2. Point the electrode at a spot slightly off the center of the pipe at all times, as shown in Figure 26A-2.

2. Use a work angle of approximately 0 to 10° toward the upper part of the joint, as shown in Figure 26A-3.

3. This section of the weld is similar to the overhead position. However, you must pay careful attention to the root face of the upper piece of pipe. Proper electrode angles and manipulation must be used or there will be a tendency for the upper root face to melt excessively and burn away. The extra heat will cause the bead to sag and leave undercut (or a low spot) at the top edge of the root.

4. Use the same work angle for the entire joint, but change the push angle as the welding progresses. Note in Figure 26A-2 how the electrode is pointed slightly off center.

5. The "C" motion can be used on the entire root pass. However, you can also try the variation shown in Figure 26A-4. The variation reduces the amount of heat put into the upper root face. Pause slightly at the root gap, then move the electrode quickly down to the left. This allows the puddle to cool before re-entering the root.

6. Stop the first half of the root pass at approximately 12 o'clock. Clean and inspect the finished portion of the root.

7. Move to the other side of the pipe. Use the same electrode angles to start the next root pass by ¾ in. Make sure you obtain complete penetration as you enter the open root. This is an important factor in the welding procedure. Lack of complete penetration at the starts and stops

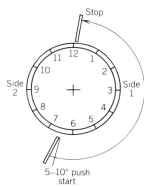

Figure 26A-2 Push angles.

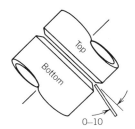

Figure 26A-3 Work angle.

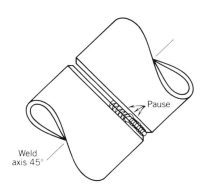

Figure 26A-4

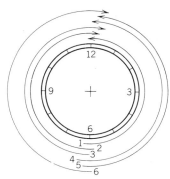

Figure 26A-5 Weld progression.

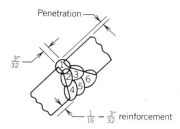

Figure 26A-6 Bead sequence.

of the root pass is the most important reason for weld failures.

8. The position of the arc starts and stops is also important. You should always stagger starts and stops. None of them should occur at the same point as those of a previous bead (see Figure 26A-5). Don't start or stop at tacks, either.

9. The second pass will consist of one or two stringer beads. The lower stringer should be welded first, as shown in Figure 26A-6. Use the same electrode angles and start on the side of the pipe you just welded. Make sure your amperage is set high enough. You want to obtain complete fusion and remove the remaining slag, without causing undercut.

10. The cap or finish pass consists of three stringer beads. Start with the lower one first, as shown in Figure 26A-6. The last stringer is most important. The problem of undercut is always present when puddles are hot. If this is the case, reduce your amperage slightly. Reduced current can correct the undercut problem. When working in the field, you won't encounter this situation too often. The long pipe will usually conduct the heat away from this joint.

11. After each bead clean thoroughly. Check your penetration and your starts and stops, reinforcement, undercut, sag, uni-

formity of beads, and overall appearance before presenting the weld to your instructor.

POINTS TO REMEMBER
1. Keep the root pass keyhole open at all times.
2. The electrode angle should always be slightly off the center of the pipe.
3. Maintain the correct work angle.
4. Do not overheat the top root face.
5. Be sure to stagger the starts and stops.
6. Keep the reinforcement below 3/32 in. Excessive reinforcement can create stress points that can lead to failure in service.
7. Strike the arc in the joint on the beveled surface, not on the pipe surface.

LESSON 26B
WELDING PIPE, 6-G TEST POSITION (AXIS OF PIPE AT A 45° ANGLE) FIRST PASS DRAGGED, SECOND AND OUT STRINGERS (ALL DOWNHILL) (E-6010 OR E-6011 ELECTRODES)

OBJECTIVE
Upon completion of this lesson you should be able to safely and correctly weld a pipe V-groove butt joint, in the 6-G test position, first pass dragged, second and out stringers, using the downhill method and E-6010 or E-6011 electrodes.

NOTE
Under certain conditions, such as in construction of oil or gas transmission lines, downhill welding is not only acceptable, but is often a required procedure.

When welding downhill, the first pass should be dragged. The required penetration will be obtained with a minimum of effort and weld pile up.

Downhill welding is quick, and acceptable welds are made with a minimum of filler metal.

MATERIAL AND EQUIPMENT
1. Personal welding and cutting equipment
2. E-6010 or E-6011 electrodes 1/8 in. (3.2 mm) diameter
3. 2 pcs. 4 in. schedule 40 pipe, or similar, if 4 in. is not available. (Nipples or pieces must be of sufficient length to enable the student to obtain a proper fit.)
4. 2 spacers, 3/32 in. × 5/8 in. × 6 in. long
5. One large hacksaw blade (one end taped for use as a handle)

GENERAL INSTRUCTIONS
1. Prepare and fit the pipe as in previous lessons.
2. Secure the pipe in the 6-G position as shown in Figure 26B-1. Use a height that is comfortable for you to reach.
3. Adjust the welding current: DCRP. Ampere range: 75 to 125. Set the amperage as you normally would for dragging root passes in either plate or pipe.
4. Place the electrode in the holder at a 90° angle.
5. Position yourself as in previous lessons.

PROCEDURE
1. Strike the arc on the beveled surface of the joint, between 11:30 and 12 o'clock. Use a drag angle of 20 to 30° to the surface of the pipe as shown in Figure 26B-2. Maintain this angle for the entire first pass.
2. Maintain a work angle between 0 to 10° off the perpendicular, as shown in Figure 26B-3. The angle you use will be determined by the way the root faces melt. Both root faces may melt evenly, or one may tend to melt faster and burn away. If one face melts faster, angle the electrode toward it. This forces weld metal

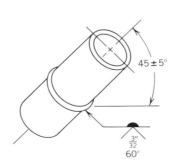

Figure 26B-1 Test position 6-G.

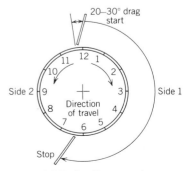

Figure 26B-2 Drag angles.

Figure 26B-3 Work angle.

into the area so the joint may fuse uniformly.

3. As you weld keep a constant pressure on the electrode. Keep lowering your hand and the electrode holder. This maintains the proper drag angle as the electrode burns away while you move along the changing contour of the pipe. Do not burn the electrode any shorter than 2½ to 3 in. It may overheat and cause the electrode to stick to the joint.

4. Before starting a second electrode the completed portion of the weld should be thoroughly cleaned. One electrode should be enough for you to complete the first pass, from 11:30 to 6:30.

5. Start the second half of the root pass on the opposite side of the pipe, side 2. Overlap the first bead by approximately ¾ in. Hold a long arc until the root opening is reached. Then force the electrode gently into the joint until the arc almost disappears. When the electrode is inside the joint, a harsh sound is heard. This sound is the best indication that you are obtaining penetration.

6. When you reach the end of the root opening at the bottom of the pipe, pull out to a normal arc length. Continue to weld until you overlap the first bead at least ¾ in.

7. Clean the pass thoroughly before you start the second, or "hot pass." (Use your chipping hammer, the saw blade, and wire brush.) Begin on side one. Start welding somewhere between 12 and 1 o'clock. Be sure to stagger the starts and stops.

8. For the hot pass, and all other passes, the electrode drag angle should be between 10 to 20°. Also, you should always aim the electrode at a point slightly off the center of the pipe as in Figure 26B-2. This amount of electrode angle tends to push the molten pool in the direction you wish. It also gives you better control of the weld deposit.

9. The hot pass not only deposits filler metal, it "burns out" any trapped slag in the toes of the root pass. Make sure you set the current high enough so the arc melts the crown of the root pass. When the puddle is concaved and fluid, you probably are obtaining a good bead. Use a slight weaving motion as shown in Figure 26B-4.

10. Do not pause as long at the joint sides as in uphill welding. Pausing will cause the center of the weld to sag; however, you should pause long enough to properly fuse the toes of the joint.

11. Repeat this on side two. Remember to overlap at least ¾ in. at the start and end of the preceding pass.

12. The next pass will be comprised of two stringer beads as shown in Figure 26B-5. Run the first one on the lower part of the joint. Use the same electrode manipulation as you used for the hot pass.

13. Do not start both stringers at the same point. Stringers must start and stop on a staggered basis. Watch out for undercut on the upper stringer.

14. Clean the pass completely. Finish the joint with a three stringer cap pass as shown in Figure 26B-5.

NOTE: The number of passes required to complete the pipe joint depends on the wall thickness and size of the stringers. Sometimes you can complete the joint by depositing a root pass, a hot pass, and a cap pass of two stringers.

POINTS TO REMEMBER

1. Exert just enough pressure on the electrode to force the arc through the joint when dragging first passes.
2. Listen for the harsh arc sound which indicates you are obtaining penetration.
3. Maintain the suggested electrode angles.
4. Be sure to stagger all starts and stops.
5. Strike the arc in the joint. Do not start on the pipe surface.
6. Do not burn the electrode down to less than 2½ to 3 in. when dragging first passes. A short electrode tends to overheat and may stub out.

Figure 26B-4 Downhill weave.

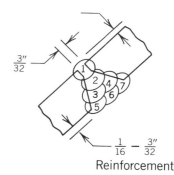

Figure 26B-5 Bead sequence.

LESSON 26C
WELDING PIPE, 6-G TEST POSITION (AXIS OF PIPE AT A 45° ANGLE) FIRST PASS DRAGGED (DOWNHILL) (E-6010 OR E-6011 ELECTRODES), SECOND AND OUT STRINGERS (UPHILL) (E-7018 ELECTRODES)

OBJECTIVE
Upon completion of this lesson you should be able to safely and correctly weld a pipe V-groove butt joint, in the 6-G test position, first pass dragged (downhill) using E-6010 or E-6011 electrodes, second and out stringers (uphill) using E-7018 electrodes.

NOTE
This lesson provides practice with the low hydrogen electrodes and increases your ability to produce acceptable root passes.

The welding industry uses these electrodes more than ever. Your ability to handle them in all positions is important if you wish to stand out.

MATERIAL AND EQUIPMENT
1. Personal welding and cutting equipment
2. E-6010 or E-6011 electrodes ⅛ in. (3.2 mm) diameter and E-7018 electrodes ⅛ in. (3.2 mm) diameter
3. 2 pcs. 4 in. schedule 40 pipe or similar if 4 in. is not available. (Nipples or pieces must be of sufficient length to enable the student to obtain a proper fit.)
4. Two spacers, 1/16 in. × 5/8 in. × 6 in. long
5. One large hacksaw blade (one end taped for use as a handle)

GENERAL INSTRUCTIONS
1. Prepare and fit the pipe as in previous lessons.
2. Secure the pipe in position as in previous lessons and as shown in Figure 26C-1.
3. Adjust the welding current: DCRP Ampere range: E-6010, 75 to 125; E-7018, 115 to 165.
Set the amperage as for dragging root passes in plate or pipe as in previous lessons.

PROCEDURE
1. Drag the first pass with E-6010 or E-6011 electrodes. Use the same procedure as in Lesson 26B. Make sure to use the electrode angles indicated in Figures 26C-2 and 26C-3.
2. Clean the first pass thoroughly, then switch to E-7018 electrodes. Adjust your amperage on scrap plate.
3. The iron powder in the coating of the E-7018 low hydrogen electrode puts more filler metal in the deposit than E-6010 or E-6011 electrodes. This means it will take fewer passes to fill the joint.
4. Start the second pass near the bottom of the pipe, between 5 and 6 o'clock. Run a stringer bead toward the top of side two. Use a slight weaving motion as shown in Figure 26C-4. Do not oscillate the electrode excessively. Do not whip or remove the electrode from the puddle at any time. Use an electrode drag angle of 0 to 5°.

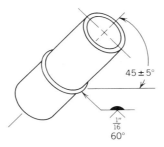

Figure 26C-1 Test position 6-G.

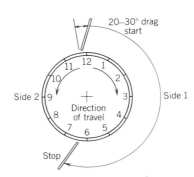

Figure 26C-2 Drag angles.

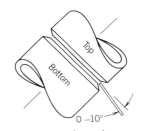

Figure C-3 Work angle.

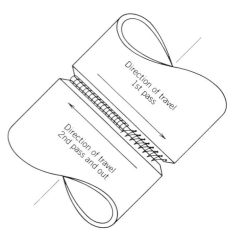

Figure 26C-4 Downhill weave.

5. Clean the bead thoroughly, using the slag hammer and wire brush. Remember that the slag formed by low hydrogen electrodes is difficult to remove. It is almost impossible to burn out with the next pass.
6. Weld the other side of the pipe. Be careful of the overlap at both the start and stop. Follow the bead sequence shown in Figure 26C-5.
7. The finish pass is run with stringer beads three and four. The lower stringer should be welded first. Use the same electrode angles and electrode motion as before. Be sure to stagger the starts and stops. Figure 26C-6 shows a sample of the type of finish that is desired.

POINTS TO REMEMBER
1. You can feel, hear, and see how the root pass is progressing.
2. Maintain the suggested electrode angles.
3. Stagger all starts and stops.
4. Do not strike the arc outside the prepared weld joint. Arc strikes on the surface of the pipe create hard spots that can weaken its structure.
5. Do not overweld the joint by adding too much reinforcement or making the finish pass too wide.

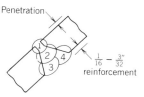

Figure 26C-5 Bead sequence.

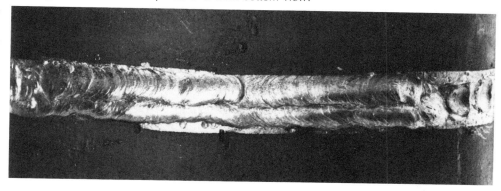

Figure 26C-6 Completed 6-G from bottom view.

CHAPTER 27
WELDING BRANCH CONNECTIONS ON PIPE

Branch connections on pipes are just like branches on trees. Pipeline connections can come together at various angles from different directions. This chapter will acquaint you with some typical branch connections and how to join them properly. There are two major joint preparations used for branch connections: One is the saddle type, the other is the insert type.

In the saddle type a hole is put into the main pipe. Then the branch pipe is cut to fit around the hole, against the shape of the main pipe. Then the branch is welded to the main pipe. Sometimes the end of the branch is beveled before welding; other times it is not. It all depends on the intended service for the joint.

In the insert type the branch fits into the hole cut into the main pipe. The hole in the main pipe may be cut at a right angle, or it may be beveled to fit the angle of the branch. Figures 27-1, 27-2, and 27-3 are examples of some of the branch connections.

Regardless of the type of preparation, the joint should be well designed and readily weldable.

The three lessons in this chapter will provide you with welding practice on both types of connections. If you learn to weld these joints, you should be able to weld other combinations you might encounter. Note regarding personal welding equipment: In this chapter, "Personal welding equipment" means you should have all of the following:

1. Proper protective clothing and equipment, such as safety glasses, welding helmet, and so on.
2. Two feather wedges
3. A few electrode stubs (for use as wedges)
4. Ballpeen hammer
5. Try square
6. Soapstone
7. Centerpunch
8. Pipe templates for the size pipe used
9. Grinder
10. File

Remember to refer back to this note if you are in doubt of the equipment required.

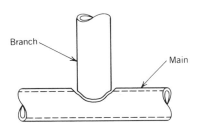

Figure 27-1 Main—beveled to receive the branch to form a single bevel joint allowing for a root gap between. The branch is cut at 90°, but conforms with the contour of the main.

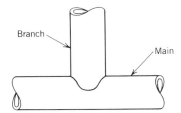

Figure 27-2 Saddle—main and branch cut at 90° with the diameter of the orifice in the main equal to the I.D. of the branch.

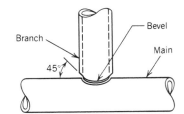

Figure 27-3 Saddle—main cut at 90° to conform to the I.D. of the branch. The branch is contoured to fit and beveled at a 45° angle.

LESSON 27A
PREPARATION, FITTING, AND TACKING OF BRANCH CONNECTIONS

OBJECTIVE
Upon completion of this lesson you should be able to safely and correctly prepare, fit, and tack branch connections in accordance with accepted procedures in the industry.

NOTE
Preparation, fitting, and tacking a joint are just as important as welding the

LESSON 27A

joint. If you prepare the job properly, you will have an easier time welding the joint. In addition, proper preparation greatly reduces the time and effort required to produce acceptable welds.

Branch connections are sometimes called: connections, Y-connections, K-connections, crosses, and so on.

MATERIAL AND EQUIPMENT
1. Personal welding equipment
2. Oxy-fuel cutting outfit
3. One piece 4, 5, or 6 in. schedule 40 pipe, 6 to 12 in. long
4. One piece schedule 40 pipe smaller in diameter than the other piece.
5. E-6010, or E-6011 electrodes ⅛ in. (3.2 mm) diameter.

GENERAL INSTRUCTIONS
1. Use a template to layout the hole to be cut in the main (larger diameter) pipe. The diameter of the opening must be equal to the inside diameter of the branch.
2. Use a template to layout the cut at the end of the branch.
3. Flame cut the main pipe at a 90° angle.
4. Bevel cut the branch at approximately 45°.
5. Grind and file both sections to fit well when placed together. Clean away all slag.

PROCEDURE
1. Secure the main with the cut opening facing up, and the axis pipe in the horizontal position, as shown in Figure 27A-1.
2. Place the branch over the hole in the main. The axis of the branch should be vertical as shown in Figure 27A-1. Hammer four welding stubs flat for use as wedges. Place them at points A, B, C, and D to obtain the root opening. Refer to Figures 27A-1 and 27A-2.
3. Maintain a consistent root opening between 3/32 and 1/8 in. Check with the try square to see if the branch is at 90° to the main, as shown in Figure 27A-3.
4. If it is reasonably square, tack the joint at point B or D, whichever has the best root opening. Refer to Figure 27A-4.
5. Remove the stubs. Then square up the pipe with a feather wedge placed in the joint opposite and tack. Remove the wedge and use the hammer to move the branch toward the side with the larger opening. The root opening at points A and C are as equal as possible. Tack the side opposite the first tack.
6. Place a tack at points A and C. If one side has a larger opening than the other tack that side first. Refer to Figure 27A-4.
7. Save this weldment for use in the next lesson.

POINTS TO REMEMBER
1. Take care. A good fit will make your job easier.
2. Keep the root opening as consistent as possible.
3. Tack according to the procedure; otherwise, you may find the pipe cannot be moved after the first two tacks.
4. Use the try square.

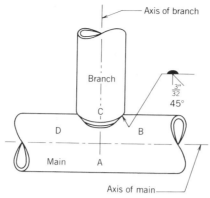

Figure 27A-1 Side view.

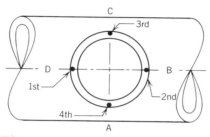

Figure 27A-2 Top view.

Figure 27A-3 Branch and square.

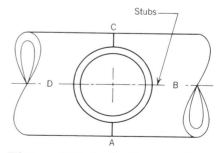

Figure 27A-4 Tacking procedure.

LESSON 27B
SADDLE-TYPE BRANCH CONNECTION, AXIS OF THE MAIN HORIZONTAL. AXIS OF THE BRANCH VERTICAL, AT A 90° ANGLE, ABOVE THE MAIN (E-6010, E-6011, OR E-7018 ELECTRODES)

OBJECTIVE
Upon completion of this lesson you should be able to safely and correctly weld a saddle-type branch connection to a horizontal main, with the branch axis vertical, above the main. The weld conforms to industry standards, using E-6010, E-6011, or E-7018 electrodes.

NOTE
The main pipe is marked and cut to fit the inside diameter of the branch. The branch is shaped to rest on the surface of the main as in the previous lesson.

Be sure you take time to properly prepare and fit up the joint. If the fit is poor, the joint will be very difficult to weld.

Fit and tack the joint carefully. Too large a root opening or not enough opening can lead to an unacceptable weld.

You should have attained fairly good welding skills by now. This lesson will acquaint you with the procedures for attacking problems of this type. It will also provide you with practice using low hydrogen electrodes on this type of joint configuration.

MATERIAL AND EQUIPMENT
1. Personal welding and cutting equipment
2. Oxy-fuel cutting outfit
3. One piece, 4, 5, or 6 in. schedule 40 pipe 6 to 12 in. long.
4. One piece schedule 40 pipe smaller in diameter than the other piece
5. E-6010, E-6011, and E-7018 electrodes ⅛ in. (3.2 mm) and ⁵⁄₃₂ in. (4.0 mm) diameter

GENERAL INSTRUCTIONS
1. Mark off and flame cut a hole in the main. Its diameter should be comparable to the inside diameter of the branch.
2. Use a template to mark off the branch. Flame cut the branch to fit the contour of the main. Make the cut at a 90° angle, then grind or file to obtain a good fit.
3. Secure the main with its axis horizontal and the prepared hole facing up. The branch will be placed vertically, as shown in Figure 27B-1.
4. Place the branch over the hole and tack in place. Use the try square to insure the branch is at a right angle as in the previous lesson.
5. Tack the branch in four places, as in Lesson 27A.
6. Adjust the welding current: DCRP. Amperage range: E-6010 and E-6011 ⅛ in. (3.2 mm) 75 to 125; E-6010 ⁵⁄₃₂ in. (4.0 mm) 110 to 170; E-7018 ⅛ in. (3.2 mm) 115 to 165; E-7018 ⁵⁄₃₂ in. (4.0 mm) 150 to 220.
7. Set the amperage as you would for horizontal welding using the "C" motion.

PROCEDURE
1. Use the ⅛ in. (3.2 mm) E-6010 or E-6011 electrodes to weld the root pass. Start a little to the left of point (A) in Figure 27B-2 and travel toward point (B). Use the "C" or whipping motion. Make sure you have adequate amperage to obtain good penetration.
2. Clean the bead thoroughly. Weld the second bead from (A) to (D). Start a little to the right of (A) so the first bead is overlapped approximately ¾ in. Stop just on the other side of (D).
3. Follow the contour of the joint, but make sure the electrode always points into the weld area at a slight drag angle. (Refer to Figure 27B-3.) Be careful not to burn away the beveled edge.
4. Move to point (C) and weld toward (B), then (D), in the same manner as on the first side. As you weld the joint remember to stag-

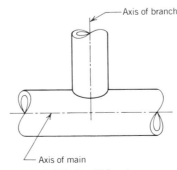

Figure 27B-1 Side view.

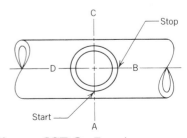

Figure 27B-2 Top view.

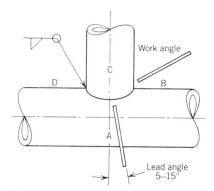

Figure 27B-3 Test Position.

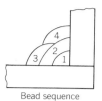

Figure 27B-4 Bead sequence.

ger your starts and stops. Complete the root pass and clean it thoroughly.

5. Change to the 5/32 in. (4.0 mm) E-6010 or E-6011 electrodes for the second pass, as shown in Figure 27B-4. The increased diameter of the electrode will create a wider bead. It will easily cover the root pass. Run a stringer with a "C" motion. Be on guard for sag and undercut.

6. Clean thoroughly and change to the E-7018 electrodes. This is not a normal procedure, but it will provide you with practice in the use of low hydrogen electrodes in out-of-position work.

7. The cover pass is comprised of beads three and four. Remember that you cannot use a whipping motion or excessively oscillate low hydrogen electrodes. Take care in setting your amperage. Bead three is run as you would normally run a stringer. However, you must pay attention to the changing contour of the joint. Stagger the starts and stops. Finish the cover pass with bead four. The completed weld should appear as shown in Figure 27B-5.

8. If the weld does not have the proper contour at this point, finish with beads 5, 6, and 7.

9. Sometimes, due to a poor fit, the beads at points (A) and (C) cannot be completely covered. In these cases you can run a short stringer to finish the weld. The stringer should cover the area while traveling in one direction. This means that a small portion of the stringer will probably be run in the downhill mode. Clean the weld when it is completed. Compare your results with Figure 27B-5.

POINTS TO REMEMBER

1. Stagger all starts and stops.
2. A poor fit will make the welding difficult.
3. Constantly adjust the electrode angles to match the changing contour of the joint.
4. At points (B) and (D) the face of the weld should be at a 45° angle to the pipe.
5. Don't overweld by putting in too many passes. You should be able to estimate how much or how little filler metal is required to obtain the desired joint.

Figure 27B-5 Completed weld.

LESSON 27C
SADDLE-TYPE BRANCH CONNECTION, BRANCH BEVELED, AXIS OF THE MAIN VERTICAL, AXIS OF THE BRANCH HORIZONTAL AT A 90° ANGLE TO THE SIDE OF THE MAIN (E-6010 OR E-6011 ELECTRODES)

OBJECTIVE
Upon completion of this lesson you should be able to safely and correctly weld a saddle-type branch connection, with the branch beveled, to a vertical main, with the branch axis horizontal, at a 90° angle to the side of the main using E-6010 or E-6011 electrodes.

NOTE
In this lesson the branch end is beveled at 45° to form a single bevel joint. You must obtain excellent penetration on the first pass with this type of joint. It is typical of those used for high pressure service. The weld may be completed using either the stringer or weave methods. If you use the weave method, take care not to overheat the joint. Overheating can cause undercut or sag.

After the first pass is completed it is good practice to reverse the direction of travel for each pass.

MATERIAL AND EQUIPMENT
1. Personal welding and cutting equipment
2. Oxy-fuel cutting outfit
3. One piece 4, 5, or 6 in. schedule 40 pipe 6 to 12 in. long
4. One piece schedule 40 pipe smaller in diameter than the other piece
5. E-6010, E-6011 electrodes ⅛ in. (3.2 mm) and 5/32 in. (4.0 mm) diameter

GENERAL INSTRUCTIONS
Use the weldment you prepared in Lesson 27A or:

1. Mark off and flame cut a hole in the main. Its diameter should be comparable to the inside diameter of the branch.
2. Mark off and flame cut the branch to conform to the contour of the main. Bevel the end of the branch (angle of bevel, 45° as shown in Figure 27C-1).
3. Prepare as in the previous lessons.
4. Adjust the welding current: DCRP. Ampere range: ⅛ in. (3.2 mm) 75 to 125; DCRP. Ampere range: 5/32 in. (4.0 mm) 110 to 170.

PROCEDURE
1. Flatten four electrode stubs to use as spacers (approximately 3/32 in. (2.4 mm) in thickness.)
2. Secure the main in the horizontal position with the opening facing up. Refer to Figure 27C-2.
3. Place four of the flattened welding stubs at points one, two, and three.
4. Set the contoured edge of the branch connection over the opening. Place the leading edge of the bevel even with the edges of the opening, as shown in Figure 27C-1. It is important that the opening in the main is equal to the inside diameter of the branch. If the opening is too large, it will be very difficult to obtain an acceptable root pass. If it is too small, it will cause a restriction in the line. A restriction can slow the flow of material through the line. It can also cause the pipe to erode faster at point of restriction.
5. If the root opening is acceptable, place a tack between points two and three, as shown in Figure 27C-2. After tacking remove all three spacers. Then use either the spacers or a feather wedge to obtain the correct root opening at point one. Use the try square to set the branch at a 90° angle to the main. Then place a tack at point one.

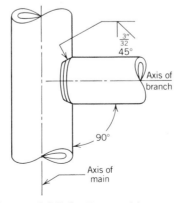

Figure 27C-1 Test position.

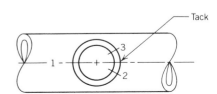

Figure 27C-2 Top view.

LESSON 27C

6. Equalize the root openings on the two sides. Use a hammer and a wedge if necessary, then tack both sides. Clean the tacks thoroughly.

7. Remove the weldment and secure it with the main in the vertical position. The branch will stick out at a 90° angle, as shown in Figure 27C-3, and Figure 27C-1.

8. Starting at slightly to the left of 6 o'clock, weld counter clockwise toward 12. Use either the whip or "C" motion. Favor the branch from six to three and the main from three to twelve, as shown in Figure 27C-3. If the root opening is tight, or closes as the welding progresses, it is possible to drag that portion if you are careful. See bead sequence 11, Figure 27C-4. Clean and repeat on side two.

9. Use the 5/32 in. (4.0 mm) diameter electrodes for the second pass and out. Utilize the "C" motion and remember to keep the puddle on a horizontal plane at all times. It you fail to do so, the puddle will run and sag. Weld the second pass clockwise. Remember to stagger the start and stop.

10. When using this method, remember to "play" or favor the branch on the lower half. This means to start the puddle at the point to be favored and pause slightly before carrying the puddle across. Follow the bead sequence shown in Figure 27C-4.

11. Cover the second pass with two stringers. Use the same method as for pass number two.

12. The cap pass can be completed with two or three stringers at the welder's discretion. Watch out for undercut on the main during the last stringer.

13. Notice the smooth transition of the finish pass. This is done by slightly overlapping the ends of the weld. (See Figure 27C-5.)

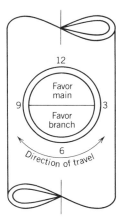

Figure 27C-3 Weldment position.

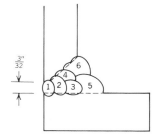

Figure 27C-4 Bead sequence.

POINTS TO REMEMBER

1. Stagger all starts and stops.
2. Pay close attention to the electrode angles.
3. Be sure to fill the entire joint. Do not fail to extend the weld past the edge of the joint.
4. The bottom and top portion of the joint should be finished at a 45° angle. The angle on the remainder of the joint will depend on the joint at each point.
5. Do not overweld.
6. Be careful to guard against excessive melt through or lack of penetration on the first pass.

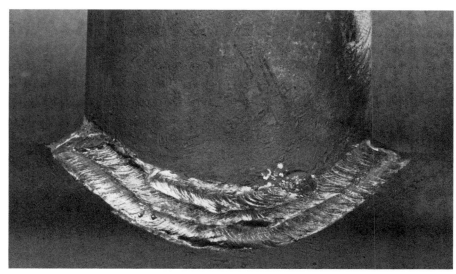

Figure 27C-5 Overlapping stringer at bottom of branch.

GLOSSARY

AAC. American Welding Society abbreviation for the Air Carbon Arc Cutting process.

Acceptable weld. A weld that meets all the requirements and the acceptance criteria prescribed by the welding specifications.

Actual throat. See Throat of a fillet weld.

Air carbon arc cutting (AAC). An arc cutting process in which metals to be cut are melted by the heat of a carbon arc and the molten metal is removed by a blast of air.

All-weld-metal test specimen. A test specimen with the reduced section composed wholly of weld metal.

Alloying elements. Material added to weld metal through electrode coatings, and the electrode core wire, to change the composition of the metal.

Alternate weave. Similar to the weave, but the electrode should form a slightly convex path as it moves across the face of the weld (Figure 1)

Alternating current. Electric current that changes direction at a regular frequency. When it reverses 120 times a second, it is referred to as 60 hertz. Each complete cycle (two reversals) is one hertz.

Ambient temperature. The temperature in the immediate area. The temperature of the atmosphere in the vicinity.

Amperage range. The minimum and maximum amperes allowed for a given electrode.

Annealing. A controlled heat treatment. In addition to relieving stress, it also increases weld ductility.

Angle of bevel. See preferred term Bevel angle.

ANSI Z49.1. The American National Standard Safety in Welding and Cutting published by the American Welding Society.

Arc. A sustained electrical discharge, where current flows through the gap between two electrodes.

Arc blow. The deflection of an electric arc from its normal path because of magnetic forces.

Arc cutting (AC). A group of cutting processes that melts the metals to be cut with the heat of an arc between an electrode and the base metal.

Arc gap. The distance between the tip of two electrodes, normally between an electrode and the workpiece. Also known as arc length.

Arc gouging. An arc cutting process variation used to form a bevel or groove.

Arc stream. The visible portion of an arc through which the particles of molten metal are transferred.

Arc voltage. The voltage across the welding arc.

Arc welding (AW). A group of welding processes that produces coalescence of metals by heating them with an arc, with or without the application of pressure and with or without the use of filler metal.

Arc welding electrode. A component of the welding circuit through which current is conducted between the electrode holder and the arc. See Arc welding.

As welded. The weld remains untouched by any finishing method after welding.

AWS. American Welding Society.

Axis of a weld. A line through the length of a weld, perpendicular to and at the geometric center of its cross section.

Back bead. See preferred term Backing weld.

Back blow. Arc blow opposite to the direction of travel.

Backfire. The momentary recession of the flame into the welding tip or cutting tip followed by immediate reappearance or complete extinction of the flame.

Back gouging. The removal of weld metal and base metal from the other side of a partially welded joint to assure complete penetration upon subsequent welding from that side.

Backhand welding. A welding technique in which the welding torch or gun is directed opposite the progress of welding. Sometimes referred to as the "pull gun technique" in GMAW and FCAW. See Travel angle, Work angle, and Drag angle. (Figure 2)

Backing. A material (base metal, weld metal, carbon, or granular material) placed at the root of a weld joint for the purpose of supporting molten weld metal.

Backing bead. See preferred term Backing weld.

Backing filler metal. See Consumable insert.

Backing pass. A pass made to deposit a backing weld.

Backing ring. Backing in the form of a ring, generally used in the welding of piping.

Backing strap. See preferred term Backing strip.

Backing strip. Backing in the form of a strip.

Backing weld. Backing in the form of a weld. (Figure 3)

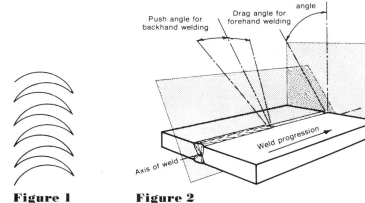

Figure 1   **Figure 2**

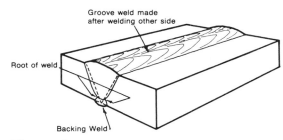

Figure 3

Backstep sequence. A longitudinal sequence in which the weld bead increments are deposited in the direction opposite the progress of welding the joint. See Block sequence, Cascade sequence, Continuous sequence, Joint buildup sequence, and Longitudinal sequence. (Figure 4)

Back weld. A weld deposited at the back of a single groove weld. (Figure 5)

Bare electrode. A filler metal electrode consisting of a single metal or alloy that has been produced into a wire, strip, or bar form and that has had no coating or covering applied to it other than that which was incidental to its manufacture or preservation.

Base metal. The metal to be welded, brazed, soldered, or cut.

Base metal test specimen. A test specimen composed wholly of base metal.

Bead. See preferred term Weld bead.

Bead sequence. The method used to deposit the beads.

Bead weld. See preferred term Surfacing weld.

Bend test. A test to determine the ductility of weld metal.

Bevel. An angular type of edge preparation. (Figure 6)

Bevel angle. The angle formed between the prepared edge of a member and a plane perpendicular to the surface of the member.

Bevel groove. See Groove weld.

Block sequence. A combined longitudinal and buildup sequence for a continuous multiple pass weld in which separated lengths are completely or partially built up in cross section before intervening lengths are deposited. See also Backstep sequence, Longitudinal sequence, and so on. (Figure 7)

Blowhole. See preferred term Porosity.

Bottle. See preferred term Cylinder.

Boxing. The continuation of a fillet weld around a corner of a member as an extension of the principal weld. (Figure 8)

Brittleness. The inability to deform under load without cracking or breaking.

Buildup sequence. See Joint buildup sequence.

Burner. See preferred term Oxygen cutter.

Burning. See preferred term Oxygen cutting.

Burnoff rate. See preferred term Melting rate.

Burn-thru. A term erroneously used to denote excessive melt-thru or a hole. See Melt-thru.

Butt joint. A joint between two members aligned approximately in the same plane. (Figure 9)

Buttering. A surfacing variation in which one or more layers of weld metal are deposited on the groove face of one

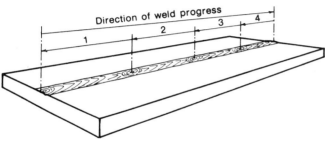
Figure 4

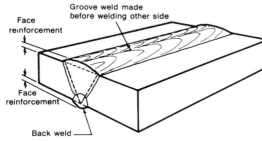
Figure 5

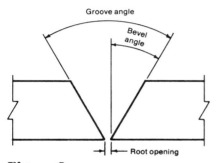
Figure 6

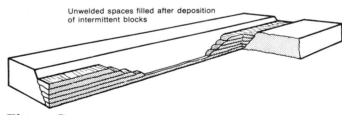
Figure 7

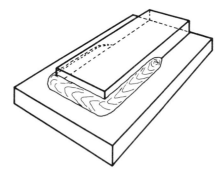
Figure 8

Figure 9

Applicable welds
Square-Groove   J-Groove
V-Groove   Flare-V-Groove
Bevel-Groove   Flare-Bevel-Groove
U-Groove   Edge-Flange

member (for example, a high alloy weld deposit on steel base metal that is to be welded to a dissimilar base metal). The buttering provides a suitable transition weld deposit for subsequent completion of the butt joint.

Carbon electrode. A non filler material electrode used in arc welding or cutting, consisting of a carbon or graphite rod, which may be coated with copper or other coatings.

Carburizing flame. See preferred term Reducing flame.

Cascade sequence. A combined longitudinal and buildup sequence during which weld beads are deposited in overlapping layers. See also Backstep sequence, Block sequence, Buildup sequence, Longitudinal sequence. (Figure 10)

Caulk weld. See preferred term Seal weld.

Ceramic rod flame spray gun. A flame spraying device using heat provided by an oxy-fuel gas flame. The material to be sprayed is in ceramic rod form.

Chain intermittent fillet welding. Two lines of intermittent fillet welds on a joint in which the fillet weld increments on one side are approximately opposite those on the other side of the joint. (Figure 11)

Chamfer. See preferred term Bevel.

Chemical flux cutting (FOC). An oxygen cutting process in which metals are severed using a chemical flux to facilitate cutting.

Chill ring. See preferred term Backing ring.

Cladding. A relatively thick layer (> 1 mm., 0.04 in.) of material applied by surfacing for the purpose of improved corrosion resistance or other properties.

Closed circuit. The completed welding circuit when welding is taking place and current is flowing.

Coalescence. The growing together or growth into one body of the materials being welded.

Coated electrode. See preferred term Covered electrode.

Code welder. Someone proven capable by test. Able to do high-quality welding on specific applications.

Complete fusion. Fusion that has occurred over the entire base material surfaces intended for welding, and between all layers and passes.

Complete joint penetration. Joint penetration in which the weld metal completely fills the groove and is fused to the base metal throughout its total thickness. (Figure 12)

Complete penetration. See preferred term Complete joint penetration.

Concave fillet weld. A fillet weld having a concave face. (Figure 13)

Concave root surface. A root surface that is concave. (Figure 14)

Concavity. The maximum distance from the face of a concave fillet weld perpendicular to a line joining the toes. (Figure 13)

Configuration. The shape of a weld joint.

Connector. A device used to join lengths of welding cable or to connect the cable to the welder.

Consumable insert. Preplaced filler metal that is completely fused into the root of the joint and becomes part of the weld.

Consumables. Items that are used up or consumed in welding. For example, in shielded metal arc welding electrodes are consumable.

Continuous sequence. A longitudinal sequence in which each pass is made continuously from one end of the joint to the other. See Backstep sequence, Longitudinal sequence, and so on.

Continuous weld. A weld that extends continuously from one end of a joint to the other. Where the joint is essentially circular it extends completely around the joint.

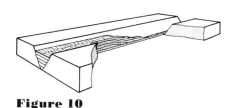

Figure 10

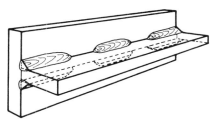

Figure 11

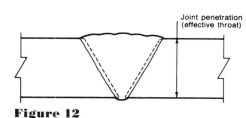

Figure 12

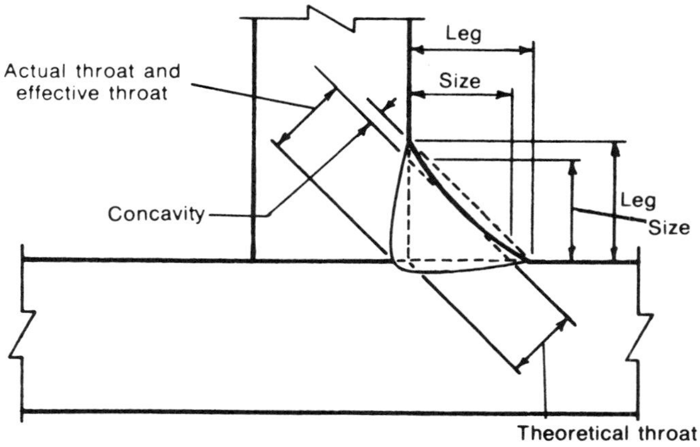

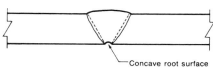

Figure 13

Figure 14

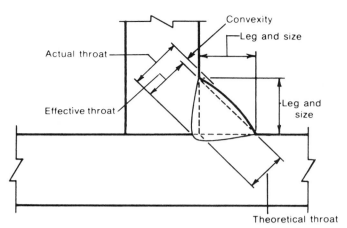

Figure 15

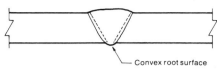

Figure 15A

Contraction. The shrinkage of hot metal while it is cooling.

Convex. Curved with the center higher than the edges.

Convex fillet weld. A fillet weld having a convex face. (Figure 15)

Convex root surface. A root surface that is convex. (Figure 15A)

Convexity. The maximum distance from the face of a convex fillet weld perpendicular to a line joining the toes. (Figure 15)

Core wire. The wire in the center of a shielded metal arc electrode.

Corrective lens (eye protection). A lens ground to the wearer's individual corrective prescription.

Covered electrode. A composite filler metal electrode consisting of a core of a bare electrode or metal cored electrode to which a covering sufficient to provide a slag layer on the weld metal has been applied. The covering may contain materials providing such functions as shielding from the atmosphere, deoxidation, and arc stabilization, and can serve as a source of metallic additions to the weld.

Cover plate (eye protection). A removable pane of colorless glass, plastic-coated glass, or plastic, that covers the filter plate and protects it from weld spatter, pitting, or scratching when used in a helmet, hood, or goggle.

Crack sensitive. Cracks easily under certain conditions. (A metal that is not crack sensitive will withstand those conditions which crack sensitive metals cannot.)

Crater. In arc welding, a depression at the termination of a weld bead or in the molten weld pool.

Crater crack. A crack in the crater of a weld bead.

Cutting head. The part of a cutting machine or automatic cutting equipment in which a cutting torch or tip is incorporated.

Cutting nozzle. See preferred term Cutting tip.

Cutting process. A process that brings about the severing or removal of metals. See Arc cutting and Oxygen cutting.

Cutting tip. That part of an oxygen cutting torch from which the gases issue.

Cutting torch. A device used in oxygen cutting for controlling and directing the gases used for preheating and the oxygen used for cutting the metal.

Cylinder. A portable container used for transportation and storage of a compressed gas.

DCRP. Direct Current Reverse Polarity. The electrode is positive and the workpiece is negative. The current flows from electrode to the work, and returns to the welder through the workload cable.

DCSP. Direct Current-Straight Polarity. The electrode is negative and the workpiece is positive. The current flows from the workpiece through the arc to the electrode, and returns to the welder through the torch cable.

Defect. A discontinuity or discontinuities which by nature or accumulated effect (i.e., total crack length) render a part or product unable to meet minimum applicable acceptance standards or specifications. This term designates rejectability. See Discontinuity and Flaw.

Defective weld. A weld containing one or more defects.

Deposited metal. Filler metal that has been added during a welding operation.

Deposition efficiency (arc welding). The ratio of the weight of deposited metal to the net weight of filler metal consumed, exclusive of stubs.

Deposition rate. The weight of material deposited in a unit of time. It is usually expressed as kilograms per hour (kg/h) (pounds per hour, lb/h).

Deposition sequence. The order in which the increments of weld metal are deposited. See Longitudinal sequence, Buildup sequence, and Pass sequence.

Deposit sequence. See preferred term Deposition sequence.

Depth of fusion. The distance that fusion extends.

Destructive testing. Testing of welds or metal during which the item being tested suffers physical damage.

Dilution. The change in chemical composition of a welding filler material caused by the admixture of the base material or previously deposited weld material in the deposited weld bead. It is normally measured by the percentage of base material or previously deposited weld material in the weld bead. (Figure 16)

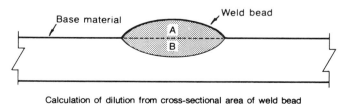

Calculation of dilution from cross-sectional area of weld bead
$$\text{Dilution} = \frac{B}{A+B}$$

Figure 16

GLOSSARY

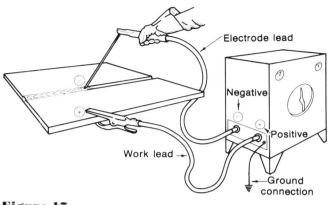

Figure 17

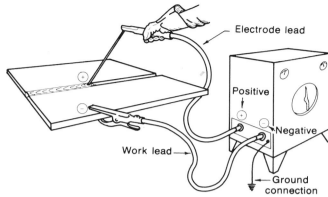

Figure 18

Dimensional discrepancies. Measurement differences due to the failure of a weld to meet specified dimensions. Variations in the size and shape of welds when compared with the required dimensions.

Direct current electrode negative. The arrangement of direct current arc welding leads in which the work is the positive pole and the electrode is the negative pole of the welding arc. See also Straight polarity. (Figure 17)

Direct current electrode positive. The arrangement of direct current arc welding leads in which the work is the negative pole and the electrode is the positive pole of the welding arc. See also Reverse polarity. (Figure 18)

Direct current reverse polarity (DCRP). See Reverse polarity and direct current electrode positive.

Direct current straight polarity (DCSP). See Straight polarity and Direct current electrode negative.

Discontinuity. An interruption of the typical structure of a weldment, such as a lack of homogeneity in the mechanical, metallurgical, or physical characteristics of the material or weldment. A discontinuity is not necessarily a defect. See Defect, and flaw.

Distorted. Out of line, mishaped, or warped.

Double-bevel-groove weld. A type of groove weld. (Figure 19)

Double-flare-bevel-groove weld. A type of groove weld. (Figure 20)

Double-flare-V-groove weld. A type of groove weld. (Figure 21)

Double-J-groove weld. A type of groove weld. (Figure 22)

Double-square-groove weld. A type of groove weld. (Figure 23)

Figure 19

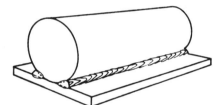

Figure 20

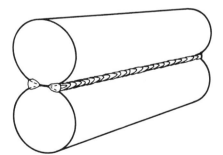

Figure 21

Figure 22

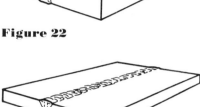

Figure 23

Figure 24

Figure 25

Double-U-groove weld. A type of groove weld. (Figure 24)

Double-V-groove weld. A type of groove weld. (Figure 25)

Downhand welding. Welding that is accomplished with the weldment in the flat position.

Downhill welding. Welding where the welding starts at the top of the joint and proceeds toward the bottom when the weldment is in the vertical position.

Drag angle. The travel angle when the electrode is pointing backward. See also Backhand welding. (Figure 26)

Dross. Waste matter attached to the bot-

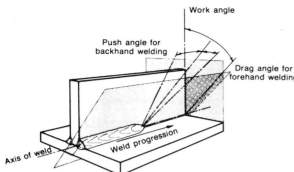

Figure 26

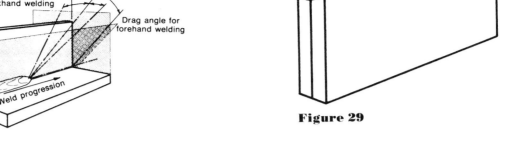

Figure 29

Figure 30

Figure 27

Figure 28

Applicable welds
Plug
Slot
Square-Groove
Bevel-Groove
V-Groove
U-Groove
J-Groove
Edge-Flange
Corner-Flange
Spot
Projection
Seam
Edge

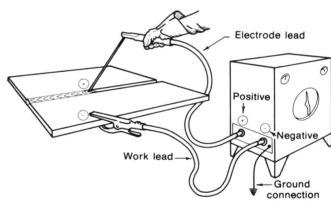

Figure 31

tom edges and surfaces of metal cut by oxy-fuel flames or arcs.

Ductile. A metal is ductile if it can be bent easily and permanently to a new shape without cracking.

Ductility. The ability of metal to stretch, bend, or twist without breaking or cracking.

Duty cycle. The percentage of time during an arbitrary test period, usually 10 minutes, during which a power supply can be operated at its rated output without overloading.

Edge-flange weld. A flange weld with two members flanged at the location of welding. (Figure 27)

Edge joint. A joint between the edges of two or more parallel or nearly parallel members. (Figure 28)

Edge preparation. The surface prepared on the edge of a member for welding.

Edge weld. A weld in an edge joint. (Figure 29)

Effective length of weld. The length of weld throughout which the correctly proportioned cross section exists. In a curved weld, it shall be measured along the axis of the weld.

Effective throat. The minimum distance from the root of a weld to its face less any reinforcement. See also Joint penetration. (Figure 30)

Elasticity. The ability of a metal to return to its original shape after the load is released.

Elastic limit. The point after which the metal will not return to its original shape.

Electrode. See Covered electrode the preferred term.

Electrode classification. A method used to enable one to determine the tensile strength, welding position, welding current, type of coating, type of arc, and type of penetration through the use of a numbering system.

Electrode coating. The material that covers the metal core wire of the welding electrode.

Electrode holder. A device used for mechanically holding the electrode while conducting current to it.

Electrode lead. The electrical conductor between the source of arc welding current and the electrode holder. (Figure 31)

Elongation. The distance or amount the metal stretches during the pulling process.

End return. See preferred term Boxing.

Engine driven welders. Welding ma-

GLOSSARY

chines that are powered by internal combustion engines such as gasoline and diesel engines.

Eye size (eye protection). The nominal size of the lens-holding section of an eye frame expressed in millimeters.

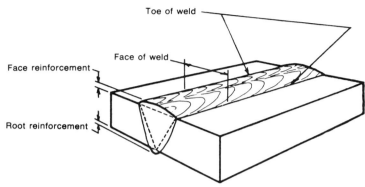

Figure 32

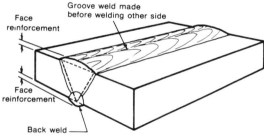

Figure 33

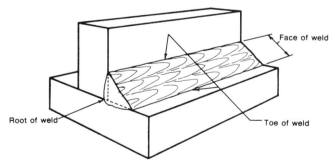

Figure 34

Figure 35

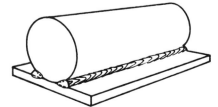

Figure 36

Face bend. A test where the face or finished pass of the weld must be bent over a greater arc than the root side of the joint.

Face of weld. The exposed surface of a weld on the side from which welding was done. (Figure 32)

Face reinforcement. Reinforcement of weld at the side of the joint from which welding was done. See also Root reinforcement. (Figure 33)

Face shield (eye protection). A device positioned in front of the eyes and a portion of, or all of, the face, whose predominant function is protection of the eyes and face. See also Hand shield and Helmet.

Faying surface. That mating surface of a member that is in contact or in close proximity with another member to which it is to be joined.

Ferrous metal. Metal that contains iron.

Filler metal (material). The metal (material) to be added in making a welded, brazed, or soldered joint.

Fillet weld. A weld of approximately triangular cross section joining two surfaces approximately at right angles to each other in a lap joint, T-joint, or corner joint. (Figure 34)

Fillet weld size. See preferred term Size of weld.

Filter glass. See preferred term Filter plate.

Filter lens (eye protection). A round filter plate.

Filter plate (eye protection). An optical material that protects the eyes against excessive ultraviolet, infrared, and visible radiation.

Fissure. A small, cracklike discontinuity with only slight separation (opening displacement) of the fracture surfaces. The prefixes *macro* or *micro* indicate relative size.

Fixture. A device designed to hold parts to be joined in proper relation to each other.

Flame cutting. See preferred term Oxygen cutting.

Flange weld. A weld made on the edges of two or more members to be joined, at least one of which is flanged. (Figure 35)

Flare-bevel-groove weld. See also Single-flare-bevel-groove weld and double-flare-bevel-groove weld. (Figure 36)

Flare-V-groove weld. See also Single-flare-V-groove weld and Double-flare-V-groove weld. (Figure 37)

GLOSSARY

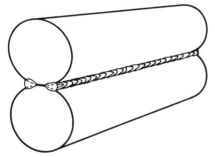

Figure 37

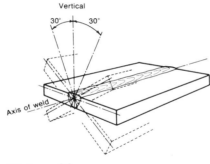

Figure 38

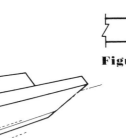

Figure 39

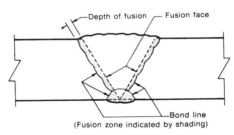

Figure 40

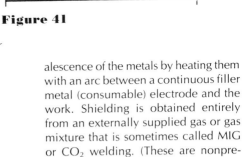

Figure 41

Flash or arc eye. A condition caused by exposure to an open arc that causes a burn on the exterior surface of the eye.

Flashback. A recession of the flame into or back of the mixing chamber of the torch.

Flashback arrester. A device to limit damage from a flashback by preventing propagation of the flame front beyond the point at which the arrester is installed.

Flash goggles. The same as safety glasses, but with a tinted lens.

Flat position. The welding position used to weld from the upper side of the joint; the face of the weld is approximately horizontal. (Figure 38, Figure 39)

Flaw. A near synonym for discontinuity, but with an undesirable connotation. See Defect and Discontinuity.

Flux lines. Invisible lines of magnetic current flow.

Forehand welding. A welding technique in which the welding torch or gun is directed toward the progress of welding. See also Work angle and Push angle.

Foreward blow. A condition where the arc is blown in the direction of travel.

Fracture. That place at which a weld breaks or separates.

Freezing point. See preferred terms Liquidus and Solidus.

Fuel gases. Gases usually used with oxygen for heating, such as acetylene, natural gas, hydrogen, propane, methylacetylene propadiene stabilized, and other synthetic fuels and hydrocarbons.

Full fillet weld. A fillet weld whose size is equal to the thickness of the thinner member joined.

Fused zone. See preferred term Fusion zone.

Fusing (thermal spraying). See preferred term Fusion.

Fusion. The melting together of filler metal and base metal (substrate), or of base metal only, which results in coalescence. See Depth of fusion.

Fusion welding. Any welding process or method that uses fusion to complete the weld.

Fusion zone. The area of base metal melted as determined on the cross section of a weld. (Figure 40)

Gas cutter. See preferred term Oxygen cutter.

Gas cutting. See preferred term Oxygen cutting.

Gas gouging. See preferred term Oxygen gouging.

Gas metal arc welding (GMAW). An arc welding process that produces coalescence of the metals by heating them with an arc between a continuous filler metal (consumable) electrode and the work. Shielding is obtained entirely from an externally supplied gas or gas mixture that is sometimes called MIG or CO_2 welding. (These are nonpreferred terms.)

Gas pocket. See preferred term Porosity.

Gas regulator. See preferred term Regulator.

Gas shielded arc welding. A general term used to describe gas metal arc welding, gas tungsten arc welding, and flux cored arc welding when gas shielding is employed.

Gas test. Taking a sample of the atmosphere in the job area for the purpose of determining the existence of any explosive gases.

Gas torch. See preferred terms Welding torch and Cutting torch.

Gas tungsten arc welding (GTAW). An arc welding process that produces coalescence of metals by heating them with an arc between a tungsten (nonconsumable) electrode and the work. Shielding is obtained from a gas or gas mixture. Pressure may or may not be used and filler metal may or may not be used. (This process has sometimes been called TIG welding, a nonpreferred term.)

Globular transfer (arc welding). A type of metal transfer in which molten filler metal is transferred across the arc in large droplets. (Figure 41)

Gouging. The forming of a bevel or groove by material removal. See also Back gouging and Arc gouging.

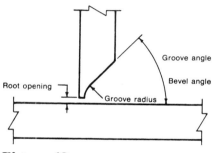

Figure 42

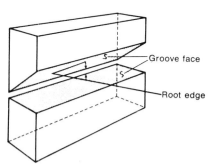

Figure 43

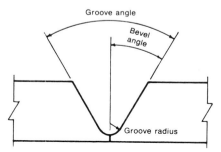

Figure 44

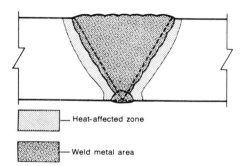

Figure 45

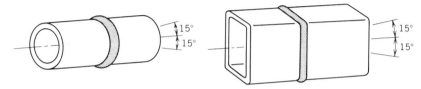

Pipe or tube horizontal fixed (±15°) and not rotated during welding. Weld flat, vertical, overhead.

(C) Test position 5G

Figure 46

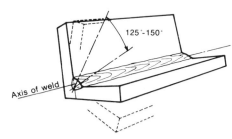

Figure 47

Groove. An opening or channel in the surface of a part or between two components which provides space to contain a weld.

Groove angle. The total included angle of the groove between parts to be joined by a groove weld. (Figure 42)

Groove face. That surface of a member included in the groove. (Figure 43)

Groove radius. The radius used to form the shape of a J- or U-groove weld joint. (Figure 44)

Groove type. The geometric configuration of a groove.

Groove weld. A weld made in the groove between two members to be joined.

Ground connection. A electrical connection of the welding machine frame to the earth for safety. See also Work connection and Work lead.

Ground lead. See preferred term Work lead.

Guided bend test. A test in which the specimen is forced to bend in a predetermined and controlled manner.

Hand shield. A protective device used in arc welding for shielding the eyes, face, and neck. A hand shield is equipped with a suitable filter plate and is designed to be held by hand.

Hard facing. A particular form of surfacing in which a coating or cladding is applied to a surface for the main purpose of reducing wear or loss of material by abrasion, impact, erosion, galling, or cavitation.

Hardness. The ability to resist penetration by a harder material.

Hardness test. A test that determines the ability of a metal to resist penetration by a harder metal or diamond.

Heat-affected zone (HAZ). That portion of the base metal which has not been melted, but whose mechanical properties or microstructure have been altered by the heat of welding, brazing, soldering, or cutting. (Figure 45)

Heat treating. A group of heat applications applied to impart certain qualities to the weldment.

Helmet (eye protection). A protection device, used in arc welding, for shielding the eyes, face, and neck. A helmet is equipped with a suitable filter plate and is designed to be worn on the head.

High alloy steels. Steels that contain large amounts of alloy ingredients.

Horizontal fixed position (pipe welding). In pipe welding the position of a pipe joint in which the axis of the pipe is approximately horizontal and the pipe is not rotated during welding. (Figure 46)

Horizontal position.

(a) **Fillet weld.** The position in which welding is performed on the upper side of an approximately horizontal surface and against an approximately vertical surface. (Figure 47)

(b) **Groove weld.** The position of welding in which the axis of the weld lies in an approximately horizontal plane and the face of the weld lies in an approximately vertical plane. (Figure 48)

(c) **Horizontal rolled position (pipe**

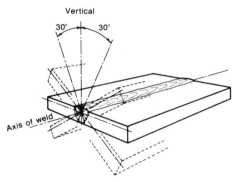

Figure 48

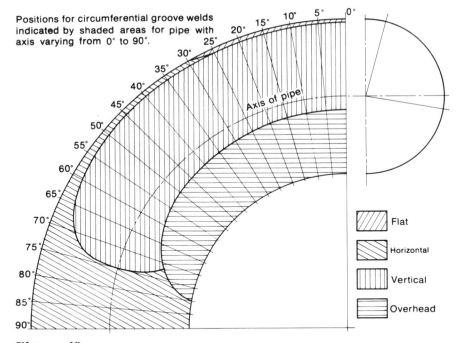

Figure 49

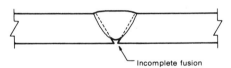

Figure 50

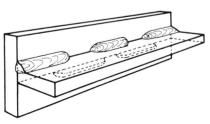

Figure 51

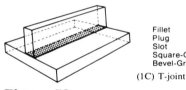

Figure 52

welding). The position of a pipe joint in which the axis of the pipe is approximately horizontal, and welding is performed in the flat position by rotating the pipe. (Figure 49)

Horseplay. The kidding or fooling around that causes many accidents.

Hot tap. A method of welding and cutting into vessels and piping while the liquid is flowing through them.

"Hot work." Any work where heat or sparks are generated.

Housekeeping. The day-to-day cleanup in your immediate work area.

Hydrostatic test. A test where air and water are used to increase the pressure within a vessel or pipeline for the purpose of determining its ability to maintain that pressure, thus ensuring that it is leakproof at that pressure.

Impact test. A test that determines the impact strength of metal. A sudden application of energy is used to break a specimen. The specimen normally has a groove cut into it.

Impurities. Unwanted material.

Inadequate joint penetration. Joint penetration which is less than that specified.

Included angle. See preferred term Groove angle.

Incomplete fusion. Fusion that is less than complete. (Figure 50)

Inclusions. Small spaces in a weld that are filled with slag or other impurities.

Induction. The electrical property that can cause current to flow in one circuit without contact with a second circuit.

Inert gas. A gas that does not normally combine chemically with the base metal or filler metal. See also Protective atmosphere.

Inert-gas metal arc welding. See preferred term Gas metal arc welding.

Inert-gas tungsten arc welding. See preferred term Gas tungsten arc welding.

Input current. The line current used to operate a welder supplied by the power company. (NEMA) National Electrical Manufacturers Association.

Intermittent weld. A weld in which the continuity is broken by recurring unwelded spaces. (Figure 51)

Interpass temperature. In a multiple-pass weld, the temperature (minimum or maximum as specified) of the deposited weld metal before the next pass is started.

Iron powder. Fine particles of iron that are sometimes added to the coating of the electrode. It is used to increase the amount of filler metal added to the joint.

Jaws. The copper pieces of an electrode holder that grip the electrode.

Joint. The junction of members or the edges of members that are to be joined or have been joined. (Figure 52)

Joint buildup sequence. The order in which the weld beads of a multiple-pass weld are deposited with respect

GLOSSARY

to the cross section of the joint. See also block sequence and longitudinal sequence. (Figure 53)

Joint clearance. The distance between the faying surfaces of a joint. In brazing this distance is referred to as that which is present either before brazing, at the brazing temperature, or after brazing is completed.

Joint design. The joint geometry together with the required dimensions of the welded joint.

Joint efficiency. The ratio of the strength of a joint to the strength of the base metal (expressed in percent).

Joint geometry. The shape and dimensions of a joint in cross section prior to welding.

Joint penetration. The minimum depth a groove or flange weld extends from its face into a joint, exclusive of reinforcement. Joint penetration may include root penetration. See also Complete joint penetration, Root penetration, and Effective throat. (Figure 54)

Joint welding procedure. The materials, detailed methods, and practices employed in the welding of a particular joint.

Joint welding sequence. See preferred term Joint buildup sequence.

Kerf. The width of the cut produced during a cutting process. (Figure 55)

Keyhole. A technique of welding in which a concentrated heat source penetrates completely through a workpiece, forming a hole at the leading edge of the molten weld metal. As the heat source progresses, the molten metal fills in behind the hole to form the weld bead.

Lack of fusion. See preferred term Incomplete fusion.

Lack of joint penetration. See preferred term Inadequate joint penetration.

Lap joint. A joint between two overlapping members. (Figure 56)

Land. See preferred term Root face.

Layer. A stratum of weld metal or surfacing material. The layer may consist of one or more weld beads laid side by side. (Figure 57)

Lead angle. See preferred term Travel angle.

Lead burning. An erroneous term used to denote the welding of lead.

Leg of fillet weld. The distance from the root of the joint to the toe of the fillet weld. (Figure 58)

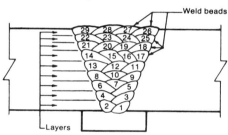

Figure 53

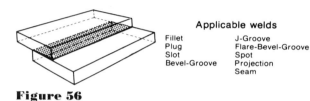

Figure 56

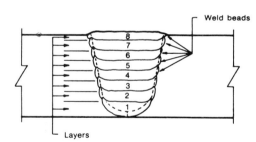

Figure 57

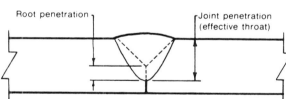

Figure 54

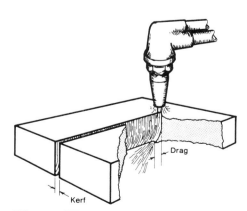

Figure 55

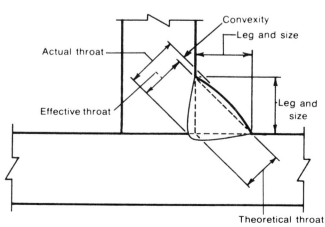

Figure 58

Lens. See preferred term Filter lens.
Liquidus. The lowest temperature at which a metal or an alloy is completely liquid.
Local preheating. Preheating a specific portion of a structure.
Local stress relief heat treatment. Stress relief heat treatment of a specific portion of a structure.
Locked-up stress. See preferred term Residual stress.
Longitudinal. Running the length of a weld.
Longitudinal sequence. The order in which the increments of a continuous weld are deposited with respect to its length. See Backstep sequence, Block sequence, and so on.
Low alloy steels. Steels that contain small amounts of alloying ingredients.
Low hydrogen. Electrodes that have a low hydrogen content.

Malleability. The ability of metal to deform permanently when hammered, compressed, or rolled.
Manual oxygen cutting. A cutting operation performed and controlled completely by hand.
Manual welding. A welding operation performed and controlled completely by hand.
Maximum amperage. The highest current a manufacturer recommends for a particular diameter electrode.
Mechanical properties. The strength, hardness, ductility, etc., of a metal.

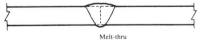
Melt-thru

Figure 59

Melting point. The temperature at which the metal melts.
Melting range. The temperature range between solidus and liquidus.
Melting rate. The weight or length of electrode melted in a unit of time.
Melt-thru. Complete joint penetration for a joint welded from one side. Visible root reinforcement is produced. (Figure 59)
Metal arc cutting (MAC). Any of a group of arc cutting processes that severs metals by melting them with the heat of an arc between a metal electrode and the base metal. See Shielded metal arc cutting and Gas metal arc cutting.
Metal arc welding. See Shielded metal arc welding, Flux cored arc welding, Gas metal arc welding and Gas tungsten arc welding.
Metal cored electrode. See Covered electrode.
Metal electrode. See Covered electrode.
Metal powder cutting (POC). An oxygen cutting process which severs metals through the use of powder, such as iron, to facilitate cutting.
Method. An orderly arrangement or set form of procedure to be used in the application of welding or allied processes.
Microscopic testing. Similar to macroscopic, but the magnifications are much higher, ranging from 50 to 5000 times.
Mil specs. Manufacturing specifications set up by the Armed Forces to govern the manufacture of items for the military.
Mismatch. The amount of offset of the centerline of two members to be joined.
Mixing chamber. That part of a welding or cutting torch in which a fuel gas and oxygen are mixed.
Molten weld pool. The liquid state of a weld prior to solidification as weld metal.
Motor driven welders. Welding generators that are driven by electric motors.
Multipass. More than one pass.

Neutral flame. An oxy-fuel gas flame in which the portion used is neither oxidizing nor reducing. (Figure 60)
Nondestructive testing. Testing of welds or metal when there is no damage done to the item being tested.
Nonmetallic. A material other than metal.
Notch. A section of the weld that has a sharp break or cut.
Normalizing. Heat treatment used to relieve stresses. It makes the metal stronger and more ductile, but not as ductile as with annealing.
Nondestructive testing. A group of tests that cause no damage to the weld or metal.
Notch toughness. The ability to withstand sudden forces in an area that has been weakened by a notch.
Nozzle. A device that directs shielding media.

Open-circuit voltage. The voltage between the output terminals of the welding machine when no current is flowing in the welding circuit.
OSHA. Occupational Safety and Health Act. A law that assures safe work and a safe place to work for all workers.
Out of position. Welding that is accomplished in positions other than flat or downhand such as vertical, horizontal, and overhead.
Overhead position. The position in which welding is performed from the underside of the joint. (Figure 61)

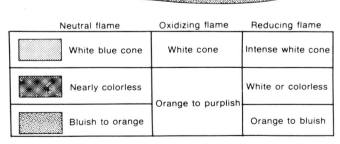
cone, neutral flame, oxidizing flame, reducing flame

Figure 60

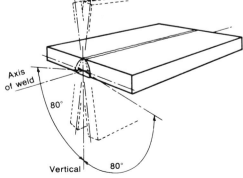

Figure 61

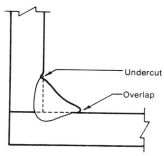

Figure 62

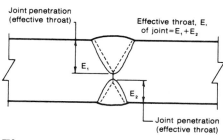

Figure 63

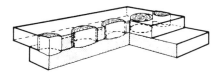

Figure 64

Overlap. The protrusion of weld metal beyond the toe, face, or root of the weld; in resistance seam welding the area in the preceding weld remelted by the succeeding weld. (Figure 62)

Overwelding. Depositing more filler metal than required.

Oxidizing flame. An oxy-fuel gas flame having an oxidizing effect (excess oxygen).

Oxyacetylene cutting (OFC-A). An oxyfuel gas cutting process used to sever metals by means of the chemical reaction of oxygen with the base metal at elevated temperatures. The necessary temperature is maintained by gas flames resulting from the combustion of acetylene with oxygen.

Oxy-fuel gas cutting (OFC). A group of cutting processes used to sever metals by means of the chemical reaction of oxygen with the base metal at elevated temperatures. The necessary temperature is maintained by means of gas flames obtained from the combustion of a specified fuel gas and oxygen. See Oxygen cutting and Oxyacetylene cutting.

Oxygen arc cutting (AOC). An oxygen-cutting process used to sever metals by means of the chemical reaction of oxygen with the base metal at elevated temperatures. The necessary temperature is maintained by an arc between a consumable tubular electrode and the base metal.

Oxygen cutter. One who performs a manual oxygen cutting operation.

Oxygen cutting (OC). A group of cutting processes used to sever or remove metals by means of the chemical reaction of oxygen with the base metal at elevated temperatures. In the case of oxidation-resistant metals the reaction is facilitated by the use of a chemical flux or metal powder. See Oxygen arc cutting, oxy-fuel gas cutting, Oxygen lance cutting, Chemical flux cutting, and Metal powder cutting.

Oxygen cutting operator. One who operates machine or automatic oxygen cutting equipment.

Oxygen gouging. An application of oxygen cutting in which a bevel or groove is formed.

Oxygen grooving. See preferred term Oxygen gouging.

Oxygen lance cutting (LOC). An oxygen-cutting process used to sever metals with oxygen supplied through a consumable lance or hollow tube; the preheat to start the cutting is obtained from other means.

Parent metal. See preferred term Base metal.

Partial joint penetration. Joint penetration that is less than complete. (Figure 63) See also Complete joint penetration.

Pass. A single progression of a welding or surfacing operation along a joint, weld deposit, or substrate. The result of a pass is a weld bead, layer, or spray deposit.

Pass sequence. See Deposition sequence.

Paste brazing filler metal. A mixture of finely divided brazing filler metal with an organic or inorganic flux or neutral vehicle or carrier.

Peel test. A destructive method of inspection that mechanically separates a lap joint by peeling.

Peening. The mechanical working of metals using impact blows.

Penetration. See preferred terms Joint penetration and Root penetration.

Plano lens (eye protection). A lens that does not incorporate correction.

Plug weld. A circular weld made through a hole in one member of a lap or T-joint fusing that member to the other. The walls of the hole may or may not be parallel and the hole may be partially or completely filled with weld metal. (A fillet welded hole or a spot weld should not be construed as conforming to this definition.) (Figure 64)

Polarity. See Direct current electrode negative, Direct current electrode positive, Straight polarity, and Reverse polarity.

Porosity. Cavity-type discontinuities formed by gas entrapment during solidification.

Positioned weld. A weld made in a joint that has been so placed as to facilitate making the weld.

Position of welding. See Flat, Horizontal, Vertical, and Overhead positions, and Horizontal rolled, Horizontal fixed, and Vertical pipe welding positions. (Figure 65)

Postheating. The application of heat to an assembly after a welding, brazing, soldering, thermal spraying, or cutting operation. See Postweld heat treatment.

Postweld heat treatment. Any heat treatment subsequent to welding.

Power source. The welder. The welding machine that provides the welding current.

Preheat. See preferred term Preheat temperature.

Preheating. The application of heat to the base metal immediately before welding, brazing, soldering, thermal spraying, or cutting.

Preheat temperature. A specified temperature that the base metal must attain in the welding, brazing, soldering, thermal spraying, or cutting area immediately before these operations are performed.

Primary circuit. Where the input current induces a current in the secondary circuit by induction and without physical contact.

Primary coil. The transformer coil into which the input current enters.

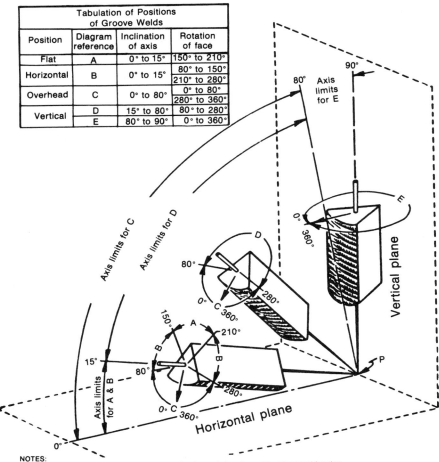

Tabulation of Positions of Groove Welds			
Position	Diagram reference	Inclination of axis	Rotation of face
Flat	A	0° to 15°	150° to 210°
Horizontal	B	0° to 15°	80° to 150° 210° to 280°
Overhead	C	0° to 80°	0° to 80° 280° to 360°
Vertical	D	15° to 80°	80° to 280°
	E	80° to 90°	0° to 360°

NOTES:
1. The horizontal reference plane is taken to lie always below the weld under consideration.
2. Inclination of axis is measured from the horizontal reference plane toward the vertical.
3. Angle of rotation of face is determined by a line perpendicular to the theoretical face of the weld and which passes through the axis of the weld. The reference position (0°) of rotation of the face invariably points in the direction opposite to that in which the axis angle increases. The angle of rotation of the face of weld is measured in a clockwise direction from this reference position (0°) when looking at point P.

Figure 65

Procedure. The detailed elements (with prescribed values or ranges of values) of a process or method used to produce a specific result.

Procedure qualification. The demonstration that welds made by a specific procedure can meet prescribed standards.

Progressive block sequence. A block sequence during which successive blocks are completed progressively along the joint, either from one end to the other or from the center of the joint toward either end.

Protective atmosphere. A gas envelope surrounding the part to be brazed, welded, or thermal sprayed, with the gas composition controlled with respect to chemical composition, dew point, pressure, flow rate, and so on. Examples are inert gases, combusted fuel gases, hydrogen, and vacuum.

Puddle. See preferred term Molten weld pool.

Push angle. The travel angle when the electrode is pointing forward. See also Forehand welding. (Figure 66)

Qualification. See preferred terms Welder performance qualification and Procedure qualification.

Quick quench. Term given to the act of rapidly cooling metal by methods such as immersion in water.

Random sequence. See preferred term Wandering sequence.

Rate of deposition. See Deposition rate.

Reactor (arc welding). A device used in arc welding circuits for the purpose of minimizing irregularities in the flow of welding current.

Rectifier. Straightens out the stepdown current of a transformer and causes it to flow in one direction providing current to weld with.

Reducing flame. A gas flame having a reducing effect (excess fuel gas). (Figure 67)

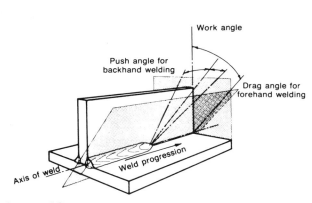

Figure 66

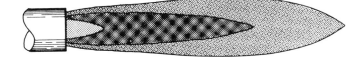

	Neutral flame	Oxidizing flame	Reducing flame
	White blue cone	White cone	Intense white cone
	Nearly colorless	Orange to purplish	White or colorless
	Bluish to orange		Orange to bluish

cone, neutral flame, oxidizing flame, reducing flame

Figure 67

GLOSSARY

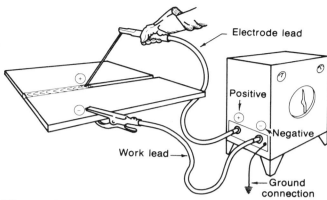

Figure 68

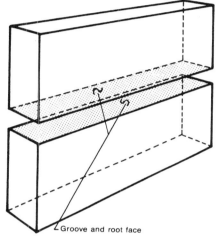

Figure 69

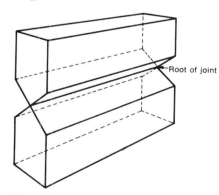

Figure 70

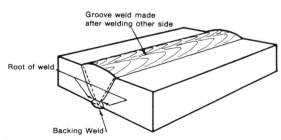

Figure 71

Regulator. A device for controlling the delivery of gas at some substantially constant pressure.

Reinforcement of weld. Weld metal in excess of the quantity required to fill a joint. See Face reinforcement and Root reinforcement.

Residual stress. Stress remaining in a structure or member as a result of thermal or mechanical treatment or both. Stress arises in fusion welding primarily because the melted material contracts on cooling from the solidus to room temperature.

Responsibilities. The items that you as the welder are responsible for. A good craftsman is never concerned with the weight of responsibility because that person knows that he or she does *only* quality work, and so does not worry whether it is done properly or not because good workmanship is an everyday occurrence to him or her.

Reverse polarity. The arrangement of direct current arc welding leads with the work as the negative pole and the electrode as the positive pole of the welding arc. A synonym for direct current electrode positive. (Figure 68)

Rewelding. Welding over a weld usually for the purpose of making repairs.

Root. See preferred terms Root of joint and Root of weld.

Root bend. A test where the root of the weld must be bent over a greater radius than the face of the weld.

Root crack. A crack in the weld or heat-affected zone occurring at the root of a weld.

Root edge. A root face of zero width. See Root face.

Root face. That portion of the groove face adjacent to the root of the joint. (Figure 69)

Root gap. See preferred term Root opening.

Root of joint. That portion of a joint to be welded where the members approach closest to each other. In cross section the root of the joint may be either a point, a line, or an area. (Figure 70)

Root of weld. The points, as shown in cross section, at which the back of the weld intersects the base metal surfaces. (Figure 71)

Root opening. The separation between the members to be joined at the root of the joint. (Figure 72)

Root penetration. The depth that a weld extends into the root of a joint measured on the centerline of the root cross section. (Figure 73)

Root radius. See preferred term Groove radius.

Root reinforcement. Reinforcement of

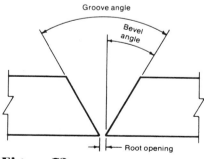

Figure 72

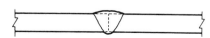

Figure 73

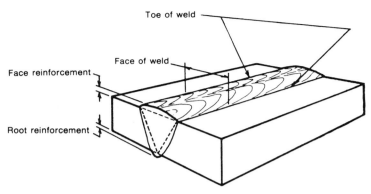

Figure 74

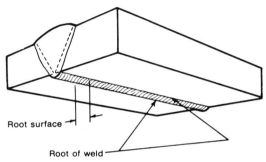

Figure 75

Figure 76

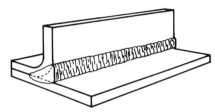

Figure 77

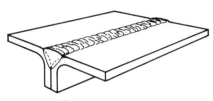

Figure 78

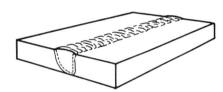

Figure 79

weld at the side other than that from which welding was done. (Figure 74)

Root surface. The exposed surface of a weld on the side other than that from which welding was done. (Figure 75)

Safety glasses. Specially designed glasses with lenses that will not shatter upon impact and have a protective side shield to protect the eye from flying objects or flashes that might otherwise enter the eye from the side.

Safety person. Your link to the outside when you are working inside of a vessel or other confined area.

Seal weld. Any weld designed primarily to provide tightness against leakage.

Secondary circuit. That portion of a welding machine which conducts the secondary current between the secondary terminals of the welding transformer and the electrodes, or electrode and work.

Secondary coil. The coil that has a current induced in it without coming into physical contact with the primary coil.

Selective block sequence. A block sequence in which successive blocks are completed in a certain order selected to create a predetermined stress pattern.

Sever. To cut or to part.

Shielded metal arc cutting (SMAC). A metal arc cutting process in which metals are severed by melting them with the heat of an arc between a covered metal electrode and the base metal.

Shielded metal arc welding (SMAW). An arc welding process that produces coalescence of metals by heating them with an arc between a covered metal electrode and the work. Shielding is obtained from decomposition of the electrode covering. Pressure is not used and filler metal is obtained from the electrode.

Shielding atmosphere. The area surrounding the arc where gases displace the atmospheric gases and protect the molten metal from them.

Shielding gas. Protective gas used to prevent atmospheric contamination.

Shoulder. See preferred term Root face.

Shrinkage stress. See preferred term Residual stress.

Shrinkage void. A cavity-type discontinuity normally formed by shrinkage during solidification.

Side bend. A test where the specimen is bent neither on the face or the root side but to either side to the center line of the weld so that one side of the weld is compressed.

Single-bevel-groove-weld. A type of groove weld. (Figure 76)

Single-flare-bevel-groove weld. A type of groove weld. (Figure 77)

Single-flare-V-groove weld. A type of groove weld. (Figure 78)

Single-J-groove weld. A type of groove weld. (Figure 79)

Single-square-groove weld. A type of groove weld. (Figure 80)

Single-U-groove weld. A type of groove weld. (Figure 81)

GLOSSARY

Figure 80

Figure 81

Figure 82

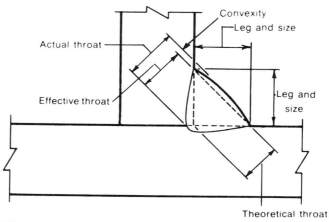

Figure 83

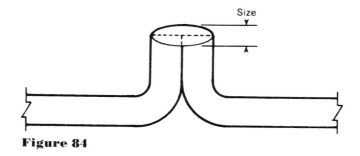

Figure 84

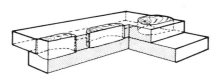

Figure 85

Single-V-groove weld. A type of groove weld. (Figure 82)

Single-welded joint. In arc and gas welding, any joint welded from one side only.

Size of weld.

(a) Groove weld. The joint penetration (depth of bevel plus the root penetration when specified). The size of a groove weld and its effective throat are one and the same.

(b) Fillet weld. For equal leg fillet welds, the leg lengths of the largest isosceles right triangle that can be inscribed within the fillet weld cross section. (Figure 83)

(c) Flange weld. The weld metal thickness measured at the root of the weld. (Figure 84)

Skip sequence. See preferred term Wandering sequence.

Skull. The unmelted residue from a liquated filler metal.

Slag inclusion. Nonmetallic solid material entrapped in weld metal or between weld metal and base metal.

Slot weld. A weld made in an elongated hole in one member of a lap or T-joint joining that member to that portion of the surface of the other member which is exposed through the hole. The hole may be open at one end and may be partially or completely filled with weld metal. (A fillet welded slot should not be construed as conforming to this definition.) (Figure 85)

Slugging. The act of adding a separate piece or pieces of material in a joint before or during welding that results in a welded joint not complying with design, drawing, or specification requirements.

Solidus. The highest temperature at which a metal or alloy is completely solid.

Spacer strip. A metal strip or bar prepared for a groove weld, and inserted in the root of a joint to serve as a backing and to maintain root opening during welding. It can also bridge an exceptionally wide gap due to poor fitup. (Figure 86)

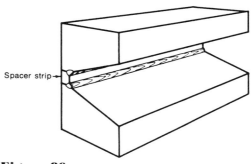

Figure 86

Spatter. In arc and gas welding, the metal particles expelled during welding that do not form a part of the weld.

Spatter loss. Metal lost due to spatter.

Specimen. Or coupon, the piece or part of the weld that is going to be tested.

Square groove butt. A joint comprised of plates whose edges are at 90° to the surface of the plate.

Square-groove weld. A type of groove weld. (Figure 87)

Staggered intermittent fillet welding. Two lines of intermittent fillet welding on a joint in which the fillet weld increments in one line are staggered with respect to those in the other line. (Figure 88)

Stepback sequence. See preferred term Backstep sequence.

Stepdown transformer. Decreases the voltage and increases the amperage of line current.

Stick electrode. See Electrode, Covered electrode.

Stick electrode welding. See preferred term Shielded metal arc welding.

Straight polarity. The arrangement of direct current arc welding leads in which the work is the positive pole and the electrode is the negative pole of the welding arc. A synonym for direct current electrode negative. (Figure 89)

Strain. The distortion caused by the application of a load or force.

Stranded electrode. See Electrode.

Strength. The ability of the metal to resist being pulled apart.

Stress. The force acting against a metal.

Stress raisers. Points, usually found in welded structures, which cause excessive stress to be placed in a small area.

Stress relief cracking. Intergranular cracking in the heat-affected zone or weld metal that occurs during the exposure of weldments to elevated temperatures during postweld heat treatment or high temperature service.

Stress relief heat treatment. Uniform heating of a structure or a portion thereof to a sufficient temperature to relieve the major portion of the residual stresses, followed by uniform cooling.

Stress relieving. The most widely used heat treatment for the purpose of relieving stresses caused by the welding operation.

Stringer bead. A type of weld bead made without appreciable weaving motion. See also Weave bead. (Figure 90)

Structural discontinuity. A pause or break in the structure—a section not completely filled with solid weld metal.

Suck-back. See preferred term Concave root surface.

Surface preparation. The operations necessary to produce a desired or specified surface condition.

Surface roughening. A group of procedures for producing irregularities on a surface.

Surface tension. A force that attracts or holds liquids to a surface.

Surfacing. The deposition of filler metal (material) on a base metal (substrate) to obtain desired properties or dimensions. See also Buttering, Cladding, Coating, and Hard facing.

Surfacing weld. A type of weld composed of one or more stringer or weave beads deposited on an unbroken surface to obtain desired properties or dimensions. (Figure 91)

Tack weld. A weld made to hold parts of a weldment in proper alignment until the final welds are made.

Taps. Connections to a transformer winding that are used to vary the transformer turns ratio, thereby controlling welding voltage and current.

Temperature indicator. A device or product used for the purpose of accurately determining the temperature of the weldment or metal.

Temporary weld. A weld made to attach a piece or pieces to a weldment for temporary use in handling, shipping, or working on the weldment.

Tensile strength. The number of pounds

Figure 87

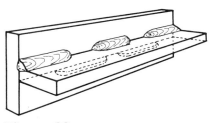

Figure 88

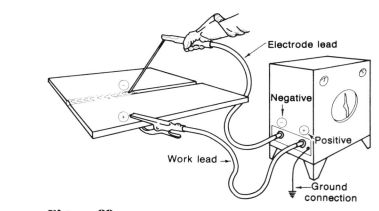

Figure 89

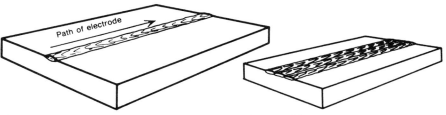

Figure 90   Figure 91

GLOSSARY

per square inch required to pull a specimen apart.

Theoretical throat. See Throat of a fillet weld.

Thermal cutting. A group of cutting processes that melts the metal (material) to be cut, such as Arc cutting, Oxygen cutting, Electron beam cutting, and Laser beam cutting.

Thermal expansion. Substances expand or become larger when elevated in temperature.

Thermal stresses. Stresses in metal resulting from nonuniform temperature distributions.

Throat depth (resistance welding). The distance from the center line of the electrodes or platens to the nearest point of interference for flat sheets in a resistance welding machine. If a resistance seam welding machine has a universal head, the throat depth is measured with the machine arranged for transverse welding.

Throat height (resistance welding). The unobstructed dimension between arms throughout the throat depth in a resistance welding machine.

Throat of a fillet weld. See below.
(a) **Theoretical throat.** The distance from the beginning of the root of the joint perpendicular to the hypotenuse of the largest right triangle that can be inscribed within the fillet weld cross section. (Figure 92)
(b) **Actual throat.** The shortest distance from the root of a fillet weld to its face. (Figure 92)
(c) **Effective throat.** The minimum distance minus any reinforcement from the root of a weld to its face. (Figure 92)

Throat of a groove weld. See preferred term Size of weld.

TIG welding. See preferred term Gas tungsten arc welding.

T-joint. A joint between two members located approximately at right angles to each other in the form of a T. (Figure 93)

Toe crack. A crack in the base metal occurring at the toe of a weld. (Figure 94)

Toe of weld. The junction between the face of a weld and the base metal. (Figure 95)

Torch. See preferred terms Welding torch and Cutting torch.

Torch tip. See preferred terms Welding tip or Cutting tip.

Travel angle. The angle that the electrode makes with a reference line perpendicular to the axis of the weld in the plane of the weld axis. See also Drag angle and Push angle. (Figure 96)
Note: This angle can be used to define the position of welding guns, welding torches, high energy beams, welding rods, thermal cutting and thermal spraying torches, and thermal spraying guns.

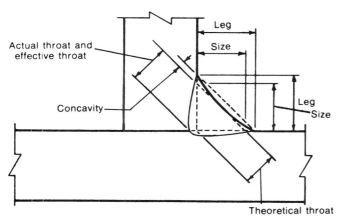

Figure 92

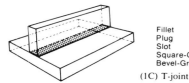

Figure 93

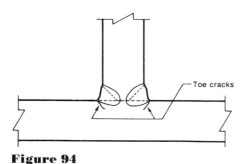

Figure 94

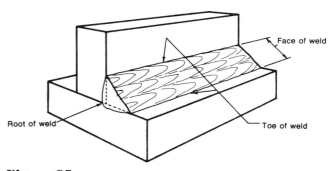

Figure 95

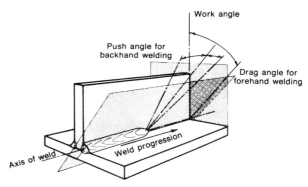

Figure 96

Travel angle (pipe). The angle that the electrode makes with a reference line extending from the center of the pipe through the puddle in the plane of the weld axis. (Figure 96)
Note: This angle can be used to define the position of welding guns, welding torches, high energy beams, welding rods, thermal cutting and thermal spraying torches, and thermal spraying guns.

Travel speed or rate of travel. The rate of movement of the electrode along the weld path measured in inches per minute.

Tungsten electrode. A nonfiller metal electrode used in welding and cutting, made principally of tungsten. (It is virtually nonconsumable.

Underbead crack. A crack in the heat-affected zone generally not extending to the surface of the base metal. (Figure 97)

Undercut. A groove melted into the base metal adjacent to the toe or root of a weld and left unfilled by weld metal. (Figure 98)

Underfill. A depression on the face of the weld or root surface extending below the surface of the adjacent base metal. (Figure 99)

Underwelding. Depositing less filler metal than required.

Upsetting. A condition where a piece of metal held in restraint will expand in another direction when heated.

Variable. Those items that can have different sizes or shapes, for instance, roots and root faces.

Vertical position. The position of welding in which the axis of the weld is approximately vertical. (Figure 100)

Vertical position (pipe welding). The position of a pipe joint in which welding is performed in the horizontal position and the pipe may or may not be rotated. (Figure 101)

Voltage regulator. An automatic electrical control device for maintaining a constant voltage supply to the primary of a welding transformer.

Wandering sequence. A longitudinal sequence in which the weld bead increments are deposited at random.

Weave bead. A type of weld bead made with transverse oscillation. (Figure 102)

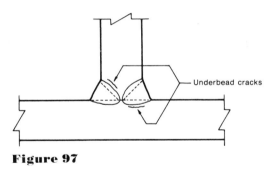

Figure 97

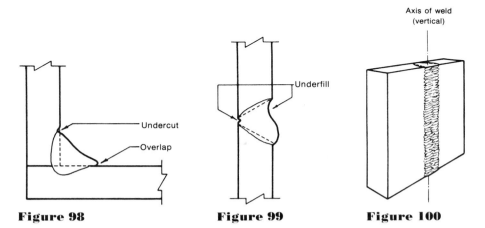

Figure 98 **Figure 99** **Figure 100**

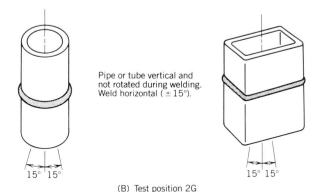

Figure 101

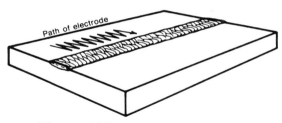

Figure 102

GLOSSARY

Weld. A localized coalescence of metals or nonmetals produced either by heating the materials to suitable temperatures, with or without the application of pressure, or by the application of pressure alone, and with or without the use of filler material.

Weldability. The capacity of a material to be welded under the fabrication conditions imposed into a specific, suitably designed structure, and to perform satisfactorily in the intended service.

Weld bead. A weld deposit resulting from a pass. See Stringer bead and Weave bead. (Figure 103)

Weld crack. A crack in weld metal.

Welder or weldor. One who performs a manual or semiautomatic welding operation. (Sometimes erroneously used to denote a welding machine.)

Welder or weldor certification. Certification in writing that a welder has produced welds meeting prescribed standards.

Welder or weldor performance qualification. The demonstration of a welder's ability to produce welds meeting prescribed standards.

Welder or weldor registration. The act of registering a welder certification or a photostatic copy thereof.

Weld gauge. A device designed for checking the shape and size of welds.

Welding. A materials joining process used in making welds.

Welding current. The current in the welding circuit during the making of a weld.

Welding electrode. See preferred term Electrode.

Welding generator. A generator used for supplying current for welding.

Welding goggles. Protective safety gear to protect the eyes while welding or cutting.

Welding ground. See preferred term Work connection.

Welding head. The part of a welding machine or automatic welding equipment in which a welding gun or torch is incorporated.

Welding leads. The work lead and electrode lead of an arc welding circuit.

Welding machine. Equipment used to perform the welding operation. For example, spot welding machine, arc welding machine, seam welding machine, and so on.

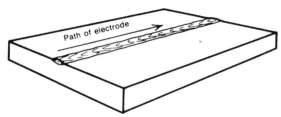

Figure 103

Welding operator. One who operates machine or automatic welding equipment.

Welding procedure. The detailed methods and practices including all joint welding procedures involved in the production of a weldment. See Joint welding procedure.

Welding rectifier. A device in a welding machine for converting alternating current to direct current.

Welding rod. A form of filler metal used for welding or brazing that does not conduct the electrical current.

Welding sequence. The order of making the welds in a weldment.

Welding technique. The details of a welding procedure that are controlled by the welder or welding operator.

Welding tip. A welding torch tip designed for welding.

Welding torch. A device used in oxy-fuel gas welding or torch brazing for mixing and controlling the flow of gases.

Welding transformer. A transformer used for supplying current for welding. See also Reactor (arc welding).

Welding wire. See preferred terms Electrode and Welding rod.

Weld interface. The point at which the weld metal and the base metal meet.

Weld length. See preferred term Effective length of weld.

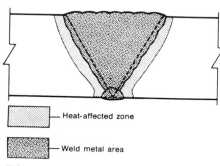

Figure 104

Weldment. An assembly whose component parts are joined by welding.

Weld metal. That portion of a weld which has been melted during welding.

Weld metal area. The area of the weld metal as measured on the cross section of a weld. (Figure 104)

Weldor. See preferred term Welder.

Weld penetration. See preferred terms Joint penetration and Root penetration.

Weld size. See preferred term Size of weld.

Work angle. The angle that the electrode makes with the referenced plane or surface of the base metal in a plane perpendicular to the axis of the weld. See also Drag angle and Push angle. (Figure 105)

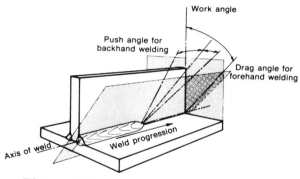

Figure 105

Work angle (pipe). The angle that the electrode makes with the referenced plane or surface of the pipe in a plane extending from the center of the pipe through the puddle. (Figure 106)
Note: This angle can be used to define the position of welding guns, welding torches, high energy beams, welding rods, thermal cutting and thermal spraying torches, and thermal spraying guns.

Work connection. The connection of the work lead to the work. (Figure 107)

Worklead. See preferred term workpiece lead.

Workpiece. The weldment you are working on. Also known as the parent metal, weldment, work or testpiece.

Workpiece lead. The electric conductor between the source of arc welding current and the work.

Yield point. The point at which the metal begins to "neck down" or narrow. Elastic limit.

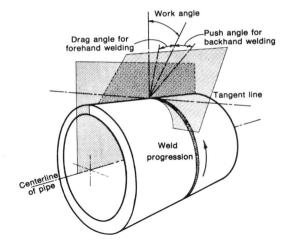

Figure 106

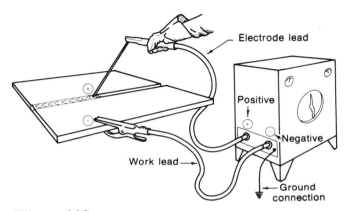

Figure 107

INDEX

Accidents, 7
Air carbon arc process, 126–135
    equipment, 126–127
        assembling, 131
        maintenance, 129–130
    safety, 128–129, 131
    troubleshooting, 129
AISI, see American Iron and Steel Institute
Alternating current welding circuit, 28
American Iron and Steel Institute:
    stainless steel classification, 65
    standards, 36–37
American Petroleum Institute standards, 72
American Society of Mechanical Engineers, 72
    standards for test specimens, 103
American Welding Society (AWS):
    electrode classification, 63–64
        stainless steel, 65–66
    electrode standards, 25–26, 61–62
    structural welding code-steel, 72
    test positions, 70–71
    test specimen standards, 103
Amperage range, 137
Annealing, 51
Arc:
    change occuring in, 3–4
    gap, 2
    length, 142
        and voltage change, 19–20
    stream, 3
    striking of, 2–3, 138–140
Arc blow, 29–31
    causes, 29
    reduction of, 30–31
Arc burn, 8
    treatment of, 8–9
Arc strike, 55
Arrow, broken, 85, 87
Arrow side, 84–86
AWS, see American Welding Society

Backfire, 115
Backing, symbol for, 92–93
Backing ring, 76–77
Backing strip, 76
Back-step welding, 47–48, 66
Backup bar, 76, 150–151
Bead sequence, 68–69
Bend Test, 98–99
Bevel:
    angle, 72
    single, butt joint, 163–164
Beveling:
    pipe, 123–125
    steel plate, 121–123
Box, welding of, 207–209
Branch connections, 238–243
    preparations for, 238–239
    saddle type, 240–243
    tacking, 239
    types, 238
Bridging, 55
Brittleness, 34
Butt joint:
    single bevel with backup bar, 163–164
    square groove, 150–151
    V groove, 165–166

C motion, 144, 146, 155
    and iron powder electrode, 147
Cable, 23–24
    connectors, 24
    copper, 23
Capacitor, 22
Cap, 12
Carbon content of steel, 38
Cascade welding, 47–48, 66
Cast iron, spark test, 42
Certification, 78–83
Charpy V-notch, 63, 100, 102
    tester, 34, 35, 101
Chipping hammer, 26, 27, 73
Chip test, 40
Chrome nickel steel, 210
Chromium in stainless steel electrode, 65
Classification number:
    for electrodes, 63–64
    for steel, 36–38
Closed circuit voltage, 19
Coalescence, 2
Cold cracking, 57
Confined area, 16
Constant current machine, 19
Consumable insert ring, 77–78
Cooling, rapid, 50
Corner joint, 152–153
    open:
        downhill, 177–178
        uphill, 175–176
Crack, 57
    crater, 148–149
Crater, 148–149

INDEX

Crayon, temperature stick, 54
Current, 142
    recommended for air carbon arc, 127–128
Cutting:
    air carbon arc process, 126–135
    pipe, 217
    steel plate, 116–120
        examples, 120
        heavy, 134–135
        light, 133–134
Cylinder:
    oxygen, *see* Oxygen cylinder
    safety procedures, 16

Destructive testing, 98–102
Dimensional discrepancy, 57
Direct current:
    reverse polarity, 29
    straight polarity, 28–29
    welding circuit, 28–29
Discontinuities, 54–57
Distortion, 45–48
    prevention, 46–48
Downhand welding, 136–138
Downhill welding, 167
    open corner joint, 177–178
    pipe welding, 230–231, 234–237
    on sheet metal, 200–204
    V groove butt joint, 181–182
    weave, 235
Ductility, 33
Duty cycle, 21–22

Ear protection, 12
Eddy current testing, 110
Elasticity, 34
Electrode, 2
    for air carbon arc, 127
    ampere ranges for, 137
    angle, 142
    care of, 58–60
    classification of, 63–64
    coating, 3, 58
    correct position of, 3
    group characteristics, 61–62
    iron powder, 147–148, 150–151
    length of, 26
    low hydrogen, *see* Low hydrogen electrode
    operating characteristics of, 64
    oven, 58, 60
    reconditioning of, 58–59
    selection of, 6, 60–62, 142
    stainless steel, 65–66, 210–215
    standard diameters, 25
    stub disposal, 12
Electrode holder, 14, 25
    for air carbon arc, 126–127
Electromagnetic testing, 110
Elongation, 100
Engine driven generator, 6, 23
Equipment:
    maintenance, 14
    personal, 238
    required, 23–27
Eye safety, 8–10
    in air carbon arc cutting, 128

F-1 group electrode, 61
F-2 group electrode, 61
F-3 group electrode, 62
F-4 group electrode, 62
Face bend, 98–99, 102, 105
Fast fill electrode, 61, 136
Fast follow electrode, 61
Fast freeze electrode, 62
Field weld, 87–88
File test, 40
Fillet gauge, 96–97
Fillet weld, 69
    dimensions of, 89
    visual inspection of, 96–97
Finishing symbols, 88
Flashback, 115
Flash burn, 8
Flash goggles, 8, 9
Flat position, 136–138
Flux, molten, 3
Fracture test, 40
Fume exhauster, 210–211
Fusion, incomplete, 55

Gamma rays, 106
Gas pocket, 54
Gas tungsten arc process, 55, 109
    and consumable rings, 77–78
    symbol, 86
Generator:
    engine-driven, 23
    motor, 5, 22
Gloves, 9
Gouging:
    air carbon arc, 126–135
    plate, 132–133
    Ground effect, 29–30
    Groove, shape of, 47
    Groove weld symbol, 90–91

INDEX

Hardness test, 39–40, 100–101
Hardness tester, 34
HAZ (heat affected zone), 49–50
Hazardous materials container welding, 16–17
Heat affected zone (HAZ), 49–50
Heat sink, 48
Horizontal welding, 154–166
Hot cracking, 57
Hot taps, 17
Hot work:
    permit, 15, 16
    safety in, 15
Hydrostatic test, 110

Inductor, 22
Inspection, 94–97
    visual, 95–97
Inspector, 106
Interpass temperature, 51
Iron powder electrode, 4, 147–148
    on T joint, 162–163
Izod procedure, 100, 102

Joint:
    inadequate penetration, 109–110
    parts of, 69
    types, 67–68
    welded from both sides, 68

Kerf, 118, 119
Keyhole, 152–153

Lap joint fillet, 74, 157–160
    overhead, 188–189
    on sheet metal, 198–199, 202–206
    with stainless steel electrode, 213–214
    vertical, 171–173
Leak test, 111
Life line, 16
Low hydrogen electrode, 62, 156
    on butt joint, 173–175
    on lap joint, 172–173
    on T joint, 169–170
    for pipe welding, 229–230, 236–237

Magnetic particle inspection, 106
Magnetic test, 39
Malleability, 34
Melt through symbol, 87–88

Metal:
    identification of, 38–42
    mechanical properties of, 32–34
    preheating temperatures for, 52–53
    upsetting, 45
Mismatch, 57
Motor generator, 5, 22

Nameplate, 21
National Electrical Code, 13
National Electrical Manufacturers Association (NEMA), 13, 21
Navy weld gauge, 96
NEMA (National Electrical Manufacturers Association), 13, 21
Neutral flame, 115–116
Nickel, spark test for, 42
Nondestructive testing, 105–111
    electromagnetic, 110
    liquid penetrant, 106
    magnetic particle, 106
    proof test, 110
    radiographic, 106–110
    ultrasonic, 109–110
Normalizing, 51
Notch toughness, 34, 35
Nub, 76

Occupational Safety and Health Act, (OSHA) 8
Open circuit voltage, 19
Open corner joint, 175–178
    overhead, 190–191
    in sheet metal, 195–196, 200–202, 206–207
OSHA (Occupational Safety and Health Act), 8
Out of position welding, 136
    electrodes for, 61–62
    with stainless steel electrode, 65–66
    see also Overhead welding
Overhead welding, 185–194
    lap joint, 188–189, 213–214
    open corner joint, 190–191
    T joint, 185–188
    V groove butt joint, 191–194
Overlap, 56
Overwelding, 57
Oxy-fuel flame cutting, 112–116
    cutting tip, 115
    equipment, 113
        depressurizing, 114
    on steel plate, 116–120
Oxygen cylinder, 112–113
    leak testing, 113–114
    valve, 114

Peening, 48
Penetration, inadequate, 55
Performance Qualification test, 2, 78–79, 95
Pipe:
    beveling, 123–125
    full of flammable liquid, 17
    thermal expansion in, 35
    working inside, 16
Pipe welding:
    aligning before, 218–219
    at angle, 232–237
    backing rings and, 76–77
    branch connections See Branch connections
    in horizontal position, 220–222, 227–231
    preparation for, 216–219
    tacking, 219
    in vertical position, 223–226
Plate preparation, 71–73
    for tests, 72–73
Pneumatic test, 110
Porosity, 54–55, 73, 107
    worm hole, 107–108
Postheating, 51
Power source, 5–6
    engine driven, 6
Preheating, 49–51
    purpose of, 49
    temperatures for metals, 52–53
    weldments needing, 50–51
Preparation depth, 90–91
Primary circuit, 20
Primary input, 13
Procedure qualification, 78–79
    record, 81
Proof test, 110
Protective clothing, 9–12
Protective curtain, 11
Puddle, molten, 3
    of low hydrogen electrode, 156
Pull test, 99–100
Push angle, 132

Qualification test, 2, 78–79
    records, 81–82
Quality control, 94–97
Quick quench, 49
  test piece, 151

Radiographic inspection, 106–110
Rectifier, 22
Reference line, 86–87
    multiple, 88–89

Reverse polarity, 21, 22
Rockwell hardness tester, 34, 35, 100–101
Root bend test, 98–99, 104–105
Root face, 72, 75

SAE (Society of Automotive Engineers), 36, 37
Safety, 7–18
    air carbon arc process and, 128
    in confined area, 16
    electrical equipment, 13–14
    eye, 8–10
    personal, 7–12
    in work area, 12–13
Scratch start, 139
Secondary circuit, 20
Secondary current, 14
Sewer cover, 13
Shade number, 10
Sheet metal welding:
    in flat position, 195–196
    in horizontal position, 197–199
    in overhead position, 205–209
    problems with, 195
    in vertical position, 200–204
Shrinkage, 45–46
    prevention, 46–48
Side bend test, 104–105
Skip welding, see Back-step welding
Slag:
    inclusion, 55, 107–109
    removal, 73
    from stainless steel welding, 211
Society of Automotive Engineers (SAE), 36, 37
Spark test, 41–42
Specimen preparation, 102–104
Spectograph, 38
Square groove butt joint, 150–151, 180–181
Stainless steel classification, 65
Stainless steel electrode, 65–66,
    slag from, 211
    V groove butt joint, 214–215
    weld pad, 211–212
Standards organizations, 72
Steel:
    classification of, 36–38
    identification of, 38
    purchasing, 43
    spark test for, 41–42
    structural shapes, 43–44
Steel plate:
    beveling, 121–123
    cutting, 116–120, 133–135
    gouging, 132–133

INDEX

Stepdown transformer, 19–20
Stick welding, 1
Straight polarity, 21, 22
Strain, 32
Stress, 32
    relieving with postheating, 51
Stringer bead, 140–141
    pad of, 144–145, 154–157
    in T joint, 145–148
Stubbing out, 139–140
Surfacing weld, 92–93
Symbols, 84–93
    for extent of weld, 89–90
    for intermittent weld, 89–90
    for location of weld, 84–85
    purpose of, 84
    references with, 86–87
        multiple, 88–89
    supplementary, 86–89
    weld *vs.* welding, 84

T joint, 74
    overhead, 185–188
    with sheet metal, 197–198
    vertical, uphill, 167–170
    horizontal weld, 160–163
    with iron powder electrode, 162–163
    downhand weld, 145–148
Tacking, 74–75
    branch connectors, 239
    pipes, 219
Temperature:
    differential, 49
    meter, 54
    to recondition electrode, 59
Tensile test, 33, 99–100
Test:
    bend, 33–34, 98–99
    positions, 70–71
    side bend, 104–105
    specimen prepartion for, 102–104
    U bend, 33–34
Testing:
    destructive, 98–105
    nondestructive, 105–111
Thermal expansion, 35
Throat, 69
    and preparation depth, 90–91
Transformer, 5
    movable coil, 20–21
    stepdown, 19–22
    welder, 19–22
Tungsten inclusion, 55, 109, 111

U bend test, 33–34
Ultrasonic inspection, 109–110
Undercut, 55–56, 109
Underwelding, 57
Underwriter's Laboratories, 13
Uphill welding, 167
    on lap joint, 171–173
    open corner joint, 175–176
    pipe welding, 227–229, 232–233, 236–237
    square groove butt joint, 180–181
    on T joint, 167–170
    V groove butt joint, 178–179, 181–182

V groove butt joint, 75, 165–166
    overhead, 191–194
    on pipe, 223–237
    with stainless steel electrode, 214–215
    vertical weld, 178–179, 181, 184
Ventilation, 15–16
Vertical welding, 167–184
    V groove butt joint, 214–215
Visual inspection, 95–97
Voltage change, 19–20

Wandering, 48
Weave, types, 170
Weave bead, 148
    on outside corner joint, 152
    on T joint, 148
Weld all-around symbol, 87
Weld pad, 144–145, 154–157
    with stainless steel electrode, 211–212
Weld profile, 56
Welder:
    job opportunities for, 1–2
    pipe, 216
    qualification test for, 2, 78–79, 81–82
Welding shield, 9
Weldment, 45–46
Wire brush, 26, 27
Work angle, 132
Workpiece:
    clamp, 14, 24
    connector, 24
Wrought iron, spark test for, 41

X-ray 106–110

Yield point, 33

Zones in weld joint, 49–50